Pseudomonas

BIOTECHNOLOGY HANDBOOKS

Series Editors: Tony Atkinson and Roger F. Sherwood

Duramed Europe Ltd.
Oxford, England

Volume 1   *PENICILLIUM* AND *ACREMONIUM*
            Edited by John F. Peberdy

Volume 2   *BACILLUS*
            Edited by Colin R. Harwood

Volume 3   CLOSTRIDIA
            Edited by Nigel P. Minton and David J. Clarke

Volume 4   *SACCHAROMYCES*
            Edited by Michael F. Tuite and Stephen G. Oliver

Volume 5   METHANE AND METHANOL UTILIZERS
            Edited by J. Colin Murrell and Howard Dalton

Volume 6   PHOTOSYNTHETIC PROKARYOTES
            Edited by Nicolas H. Mann and Noel G. Carr

Volume 7   *ASPERGILLUS*
            Edited by J. E. Smith

Volume 8   SULFATE-REDUCING BACTERIA
            Edited by Larry L. Barton

Volume 9   *THERMUS* SPECIES
            Edited by Richard Sharp and Ralph Williams

Volume 10   *PSEUDOMONAS*
            Edited by Thomas C. Montie

Pseudomonas

Edited by

Thomas C. Montie
University of Tennessee
Knoxville, Tennessee

Plenum Press • New York and London

Library of Congress Cataloging in Publication Data

Pseudomonas / edited by Thomas C. Montie.
    p.   cm—(Biotechnology handbooks; v. 10).
    Includes bibliographical references and index.
    ISBN 0-306-45849-7
    1. *Pseudomonas*. 2. *Pseudomonas*—Biotechnology. 3. Molecular microbiology. I. Montie,
Thomas C. II. Series.
QR82.P78P77   1998   98-41288
579.3′32—dc21   CIP

Cover: Electromicrograph of *Pseudomonas aeruginosa* showing polar pili and flagellum as surface structures. (Courtesy of Thomas C. Montie)

ISBN 0-306-45849-7

© 1998 Plenum Press, New York
A Division of Plenum Publishing Corporation
233 Spring Street, New York, N.Y. 10013

http://www.plenum.com

10 9 8 7 6 5 4 3 2 1

Contributors

Robert E. W. Hancock • Department of Microbiology and Immunology, University of British Columbia, Vancouver V6T 1Z4, British Columbia

Toshimitsu Hoshino • Mitsubishi Kasei Institute of Life Sciences, Machida-Shi, Tokyo 194, Japan

Sachiye Inouye • Department of Biochemistry, Yamaguchi University School of Medicine, Ube, Yamaguchi 755-8505, Japan

Estelle J. McGroarty • Department of Biochemistry, Michigan State University, East Lansing, Michigan 48824

Jean-Marie Meyer • Laboratoire de Microbiologie et de Génétique, Unité de Recherche Associée au Centre National de la Recherche Scientifique No. 1481, Université Louis-Pasteur, 67000 Strasbourg, France

Thomas C. Montie • Department of Microbiology, The University of Tennessee, Knoxville, Tennessee 37996-0845

Paul V. Phibbs, Jr. • Department of Microbiology and Immunology, East Carolina University School of Medicine, Greenville, North Carolina 27858

Holly C. Pinkart • U.S. Environmental Protection Agency, Microbial Ecology Branch, Gulf Breeze, Florida 32514

Andrew E. Sage • World Wide Microbiology Group, Millipore Corp., Bedford, Massachusetts 01730

Herbert P. Schweizer • Department of Microbiology, Colorado State University College of Veterinary Medicine, Fort Collins, Colorado 80523

Alain Stintzi • Laboratoire de Microbiologie et de Génétique, Unité de Recherche Associée au Centre National de la Recherche Scientifique No. 1481, Université Louis-Pasteur, 67000 Strasbourg, France

Louise M. Temple • Department of Biology, Drew University, Madison, New Jersey 07940

David C. White   •   Department of Microbiology and Center for Environmental Biotechnology, University of Tennessee, Knoxville, Tennessee 37932

Bernard Witholt   •   Institute of Biotechnology, ETH Hönggerberg, HPT, CH-8093 Zürich, Switzerland

Elizabeth A. Worobec   •   Department of Microbiology, University of Manitoba, Winnipeg R3T 2N2, Manitoba

Marcel G. Wubbolts   •   DSM Research, 6160 MD Geleen, The Netherlands

Preface

The genus *Pseudomonas* represents a large group of medically and environmentally important bacteria. Interest in these bacteria is reflected in the extensive number of publications devoted to original research, reviews, and books on this subject. In this volume selected areas of *Pseudomonas* research are presented in depth by persons who have been active in their fields over many years. The extensive reviews presented are an effort to provide a balanced perspective in a number of areas not readily available in the current literature. In the style of the previous *Biotechnology Handbooks* most of these topics have not been reviewed at all, and several are also presented from a new direction. For example, in addition to structural and compositional aspects, the chapter on lipids provides shifts in lipid parameters that result from environmental changes. This information will be invaluable to a cross section of *Pseudomonas* researchers in pathogenesis and bioremediation.

The chapters presented include basic aspects of plasmid biology and carbohydrate metabolism and regulation. A major emphasis is placed on the *Pseudomonas aeruginosa* cell surface. Chapters cover lipopolysaccharide, capsular polysaccharide and alginate, the outer membrane, transport systems, and the flagellum. Uptake of iron is also necessarily an important portion of the chapter on iron metabolism. Although the emphasis in this text is on *Pseudomonas aeruginosa,* many chapters include detailed and comparative discussions of other important species, such as *Burkholderia cepacia* (formerly *Pseudomonas cepacia*) and *Pseudomonas putida.* A final chapter on selected industrial biotransformations reveals the tremendous catalytic properties and potential of this group of organisms.

This volume should be of great interest to researchers in molecular biology and biotechnology in industry, academia and various research institutes, and as a resource for graduate students and advanced undergraduates.

Contents

Chapter 1

Plasmids 1

Sachiye Inouye

1. Introduction 1
2. General Properties of Plasmid DNA from *Pseudomonas* 1
   2.1. Replication and Conjugation 1
   2.2. Isolation of Plasmid DNA 2
3. Antibiotic-Resistant Plasmids (R-Plasmids) 2
   3.1. RP1, RP4, RK2 3
   3.2. pMG1 4
   3.3. RSF1010 and R1b679 4
   3.4. pMS350 4
   3.5. pSa 6
4. Resistance to Metal Ions, Pesticides, and UV Light 6
   4.1. pMR1, pMERPH 6
   4.2. pPT23D, pPSR1 7
   4.3. pCMS1 8
   4.4 pMG2 8
5. Plasmids for Biosynthesis 9
   5.1. Indole-3-Acetic Acid (IAA) Production 9
   5.2. Ethylene-Forming Enzyme 9
   5.3. Coronatine Production 9
   5.4. Syringolide Production 10
   5.5. Malonate Assimilation 10
6. Degradative Plasmids 10
   6.1. CAM 11
   6.2. OCT 11
   6.3. TOL(pWW0) 13
   6.4. NAH(NAH7) and SAL(SAL1) 16
   6.5. pHMT112 17
7. Cloning Vectors 17

7.1. Incompatibility Group P4/Q 18
7.2. Incompatibility Group P1 19
7.3. Incompatibility Group W 19
8. Conclusions and Perspectives 20
References 21

Chapter 2

Carbohydrate Catabolism in *Pseudomonas aeruginosa* 35

Louise M. Temple, Andrew E. Sage, Herbert P. Schweizer,
and Paul V. Phibbs, Jr.

1. Introduction 35
2. Glucose Utilization via Phosphorylation 41
   2.1. Transport 41
   2.2. Phosphorylation 44
3. Oxidative Pathway of Glucose Utilization 45
4. Fructose Utilization 46
5. Mannitol Utilization 47
6. Glycerol and Glycerol 3-Phosphate Utilization 48
   6.1. Glycerol and Glycerol-P Uptake 49
   6.2. Glycerol Transport 49
   6.3. Uptake of Glycerol-P 52
   6.4. Glycerol-P Dehydrogenase 52
   6.5. The GlpM Protein 53
   6.6. Glycerol Kinase 54
   6.7. Regulatory Loci 55
   6.8. Chromosomal Mapping of Glycerol Metabolism Genes ... 56
7. Central Cycle 57
   7.1. Fructose 1,6-Bisphosphate Aldolase 57
   7.2. Fructose 1,6-Bisphosphatase 57
   7.3. Phosphoglucoisomerase 58
   7.4. Glucose 6-Phosphate Dehydrogenase 58
   7.5. Entner–Doudoroff Dehydratase 58
   7.6. Entner–Doudoroff Aldolase 59
8. Lower Embden–Meyerhoff–Parnas (EMP) Pathway 60
   8.1. Glyceraldehyde 3-Phosphate Dehydrogenase 60
   8.2. Phosphoglycerate Kinase 61
   8.3. Phosphoglycerate Mutase (Pgm), Enolase (Eno),
         and Pyruvate Kinase(Pyc) 61
   8.4. Pyruvate Carboxylase 61

9. Recycling of Glyceraldehyde 3-Phosphate 61
10. Regulation of Central and Lower EMP Pathways 62
11. Catabolite Repression Control (CRC) 64
12. Clustering of Genes for Glycoltyic Enzymes 65
References 66

Chapter 3

Polysaccharides: Lipopolysaccharide and Capsular
Polysaccharide 73

Estelle J. McGroarty

1. Introduction 73
2. Lipopolysaccharide 74
   2.1. Structure 74
   2.2. Lipopolysaccharide Heterogeneity 78
   2.3. Biosynthesis and Genetic Regulation of LPS 80
   2.4. Lipopolysaccharide Physical Properties 83
   2.5. Biological Activity of Lipopolysaccharide 84
   2.6. Immunology of LPS 84
3. Capsular Polysaccharides and Slime 85
   3.1. Introduction 85
   3.2. O-Capsular Polysaccharide—Structure 86
   3.3. O-Capsule Function 87
   3.4. O-Capsule as an Antigen 87
   3.5. Alginate—Introduction 88
   3.6. Alginate—Structure and Physical Properties 88
   3.7. Alginate—Biosynthesis 90
   3.8. Regulation of Alginate Synthesis 92
   3.9. Immunology of Alginate 95
   3.10. Alginate Function 96
References 97

Chapter 4

Lipids of *Pseudomonas* 111

Holly C. Pinkart and David C. White

1. Introduction 111

2. Lipids of the Genus Pseudomonas 111
   2.1. Membrane Lipids 111
   2.2. Fatty Acids 115
   2.3. Storage Lipids 126
   2.4. Exolipids 128
3. Alteration of Lipids in Response to Environmental
   Conditions 130
   3.1. Growth Temperature 130
   3.2. Oxygen Tension 131
   3.3. Desiccation 131
   3.4. Nutrient Deprivation 132
   3.5. Solvent Tolerance 133
   3.6. Antibiotic Resistance 133
4. Summary 134
   References 134

Chapter 5

Outer Membrane Proteins 139

Robert E. W. Hancock and Elizabeth A. Worobec

 1. Introduction 139
 2. Role in Antibiotic Susceptibility 139
 3. IROMPs: FpvA, FptA, PfeA 140
 4. OprC 145
 5. OprJ 146
 6. AlgE 146
 7. OprN 146
 8. OprK 147
 9. OprM 147
10. OprP 148
11. OprO 149
12. OprB 153
13. OprD 154
14. OprE 155
15. E2 156
16. OprF 156
17. OprG 159
18. OprH 159
19. OprL 160

20. OprI 160
    References 161

Chapter 6

Transport Systems in *Pseudomonas* 169

Toshimitsu Hoshino

1. Introduction 169
2. Sugars 171
    2.1. Glucose and Gluconate 171
    2.2. Fructose and Mannitol 173
3. Amino Acids 174
    3.1. Branched-Chain Amino Acids 174
    3.2. Basic Amino Acids (Arginine, Lysine, Histidine) 182
    3.3. Proline 183
    3.4. Aromatic Amino Acids 184
    3.5. Methionine 184
4. Inorganic Ions 185
    4.1. Cations 185
    4.2. Anions 185
5. Other Compounds 187
    5.1. Compounds Catabolized by the β-Ketoadipate
          Pathway 187
    5.2. *Myo*-inositol 190
    5.3. Steroids 191
    References 192

Chapter 7

**Iron Metabolism and Siderophores in *Pseudomonas*
and Related Species** 201

Jean-Marie Meyer and Alain Stintzi

1. Introduction 201
2. Siderophores at the Bench 202
    2.1. Optimization of Siderophore Production 202

2.2.   Detection of Siderophores 203
2.3.   Purification of Siderophores 204
3. Siderophore-Mediated Iron Uptake Systems
   in Fluorescent *Pseudomonas* 209
   3.1.   Fluorescent *Pseudomonas* Siderophores 209
   3.2.   Internalization of Iron 214
   3.3.   Iron Uptake from Other Iron Sources 215
   3.4.   Biosynthesis, Genetic Organization,
          and Regulation of Siderophores 217
4. Siderophore-Mediated Iron Uptake Systems in
   Nonfluorescent *Pseudomonas* and Related Strains 223
   4.1.   *Pseudomonas stutzeri* 223
   4.2.   *Burkholderia* (Formerly *Pseudomonas*) *cepacia* and Related
          Strains 224
   4.3.   Other Strains 226
5. Iron, Siderophores, and Biotechnology 227
   5.1.   Iron, Siderophores, and Plant-Related Biocontrol 227
   5.2.   Iron, Siderophores, and Human Pathogenicity 227
   5.3.   Siderotyping and Searching for New Siderophones:
          Back to the Bench 229
   References 230

Chapter 8

The Flagellum 245

Thomas C. Montie

1. Introduction 245
2. Biochemistry 245
   2.1.   General Morphology 245
   2.2.   Purification 246
   2.3.   Molecular Weights of Flagellins 246
   2.4.   Amino Acid Composition 248
3. Posttranslational Modifications of Flagellins 250
   3.1.   Phosphorylation 250
   3.2.   Glycosylation 252
4. Comparative Analysis of *fliC* Genes 254
   4.1.   Flagellin b-Type Genes 254
   4.2.   Flagellin a-Type Genes 257

5. Flagella and Chemotaxis Genes and Regulation 260
   5.1. Regulation of Flagella Biosynthesis and Assembly 260
   5.2. Chemotaxis Genes 262
6. Antigenicity and Immunogenicity 262
   6.1. Antigenicity 262
   6.2. Immunogenicity and Passive Protection 263
7. Virulence 264
8. Other Pseudomonads 265
   References 266

Chapter 9

Selected Industrial Biotransformations 271

Marcel G. Wubbolts and Bernard Witholt

1. Introduction 271
   1.1. Metabolites of *Pseudomonas* in Biotechnology 272
   1.2. Biocatalysis versus Chemical Catalysis 273
2. Biocatalysts and Their Practical Application 279
   2.1. Oxidoreductases 279
   2.2. Transferases 291
   2.3. Hydrolases 294
   2.4. Lyases 304
   2.5. Isomerases and Racemases 308
   2.6. Ligases 311
3. The Impact of Molecular Biology on Biocatalyst
   Development 311
   3.1. Constructing "Tailor-Made" Biocatalysts 311
   References 313

Index 331

Plasmids

1

SACHIYE INOUYE

1. INTRODUCTION

On the basis of rRNA homologies and the biology of pseudomonads, most species of RNA group I among the five RNA groups, such as *Pseudomonas aeruginosa*, *P. putida*, *P. fluorescens*, and *P. syringae*, belong to this organism (Palleroni *et al.*, 1973). These *Pseudomonas* species play key roles in the environment, including the biodegradation of natural and man-made toxic chemicals and the plant–bacteria interaction. An additional property of pseudomonads, particularly *Pseudomonas aeruginosa*, is their resistance to many antibiotics. Plasmids control some of those various important features in bacterial cells and promote the transfer of genetic information between different taxonomic groups of bacteria.

This chapter summarizes the nature and properties of plasmids isolated from *Pseudomonas* species, including resistance to antibiotics and heavy metal ions, biosynthesis of plant hormones, degradation of toxic pollutants, and cloning vectors for genetic manipulation.

2. GENERAL PROPERTIES OF PLASMID DNA FROM *PSEUDOMONAS*

2.1. Replication and Conjugation

Many plasmids isolated from *Pseudomonas* species have a broad host range and are able to transfer (Tra$^+$) to gram-negative bacteria, whereas some are transfer-deficient (Tra$^-$). For DNA transfer during conjugation, the origin of transfer (*oriT*) and the primase gene (*pri*) are involved (Lanka *et al.*,, 1984). Replication of plasmids requires an origin of repli-

SACHIYE INOUYE • Department of Biochemistry, Yamaguchi University School of Medicine, Ube, Yamaguchi 755-8505, Japan.

Pseudomonas, edited by Montie. Plenum Press, New York, 1998.

cation (*oriV*) and a replication initiator gene (*trfA*) essential for the function of *oriV* (Ayres *et al.*, 1993: Kornacki *et al.*, 1984; Reimmann and Haas, 1986; Shingler and Thomas, 1984). The *oriV*, *trfA*, *oriT*, and *pri* genes are all involved in broad-host-range behavior (Krishnapillai *et. al.*, 1984). The plasmid DNA of Tra⁻ plasmid can be transferred by conjugation promoted by helper-plasmid, when a mobility gene (*mob*) is present on plasmids (Jacoby, 1984; Jacoby *et al.*, 1983).

2.2. Isolation of Plasmid DNA

The method usually carried out in our laboratory for large-scale preparation of plasmid DNA from *P. putida* or *P. aeruginosa* is as follows (Nakazawa *et al.*, 1980): Grow the bacterial cells to late log phase in standard LB medium. Harvest the cells from a 500 ml. culture by centrifugation at 8000 rpm for 10 min at 4 ;°C. Suspend the bacterial pellet in 12.5 ml of ice-cold 0.05 M Tris·HCl buffer (pH8.0) containing 1 mM EDTA and 20% (w/v) sucrose. Add 2.5 ml of a freshly prepared lysozyme solution (0.5 mg/ml), and incubate it for 5 min on ice. Add 5 ml of 25 mM EDTA (pH 8.0) and incubate it for 5 min on ice. Add 20 ml of 50 mM Tris·HCl buffer (pH8.0) containing 1% (w/v) Brij 58, 0.4%(w/v) Na·deoxycholate and 0.0625 M EDTA and incubate it for 5 min on ice. Centrifuge the lysate at 17,000 rpm for 20 min at 4 °C. Precipitate the plasmid DNA in the supernatant with 10% (w/v) polyethylene glycol in the presence of 0.5 M NaCl. Collect the precipitate by centrifugation at 3000 rpm for 1 min at 4 °C. Dissolve the precipitate in 4 ml of 50 mM Tris·HCl (pH8.0) containing 10 mM EDTA and 10 mM NaCl. Purify the plasmid DNA by equilibrium centrifugation in ethidium bromide-CsCl gradients. In the preparation of large plasmids, lysis of cells by alkali or Sarkosyl solution gives good recovery of plasmid DNA (Kao *et al.*, 1982; Rosenberg *et al.*, 1982).

3. ANTIBIOTIC-RESISTANT PLASMIDS (R-PLASMIDS)

Because of their pathogenic nature, many studies to characterize antibiotic resistance have been done with *P. aeruginosa*. Many drug-resistant plasmids (R-plasmids) mediating resistance to β-lactam antibiotics, aminoglycosides, chloramphenicol, streptomycin, sulfonamides, and tetracycline have been found in *P. aeruginosa*. According to a review (Jacoby, 1986), the R-plasmids of *P. aeruginosa* are classified into 13 incompatibility groups. Table I summarizes the properties of some typical R-plasmids. Inactivating enzymes specify the resistance to these antibiotics. Resistance to β-lactam antibiotics, including carbenicillin, is caused

Table I. R-Plasmids in *Pseudomonas*

Plasmid	Incompatibility group	Properties[a]	Size (kb)
RP1, RP4, RK2	P1	Cb, Km, Tc, Tra$^+$	56.6
pMG1	P2	Sm, Su, Gm, Hg, Mob$^+$	~500
RSF1010, Rlb679	P4	Sm, Su, Mob$^+$	8.7
pMS350	P9	Cb, Gm, Su, Tra$^+$	49
pSa	Unclassified (W)[b]	Sp, Sm, Su, Km, Cm, Gm, Tra$^+$	39

[a] Abbreviations for resistances: Cb, carbenicillin; Cm, chloramphenicol; Gm, gentamicin; Hg, mercuric ions; Km, kanamycin; Sm, streptomycin; Sp, spectinomycin; Su, sulfonamides; Tc, tetracycline. Tra$^+$ represents the ability of self-transmissible. Mob$^+$ represents the ability to be mobilized by a conjugative plasmid.
[b] Incompatibility group of enterobacteria.

by the production of many varieties of plasmid-determined β-lactamases. TEM-1 and TEM-2 β-lactamases are the most common types found in *Pseudomonas* (Matthew, 1979). *P. aeruginosa* strains resistant to aminoglycosides, including gentamicin, kanamycin, streptomycin, and amikacin, determine a variety of inactivating enzymes of these drugs. The aminoglycosides are inactivated by modification of the amino and hydroxy groups of antibiotics by physphorylation, acetylation, or adenylation (Bryan, 1984). Chloramphenicol is also inactivated by plasmid-determined chloramphenical acetyltransferase (Matsuhashi *et al.*, 1975). Resistance to tetracycline is caused by decreased permeability and active export of the drug (Hedstrom *et al.*, 1982; Nikaido, 1992; Quinn, 1992). The active efflux of various agents and impermeability of the outer membrane of *P. aeruginosa* are involved in the intrinsic resistance of this organism to a wide variety of antibiotics, not only tetracycline but also chloramphenicol, fluoroquinolones, and β-lactams (Li *et al.*, 1994a, b).

3.1. RP1, RP4, RK2

A general feature of incompatibility group P1 plasmids is the capability of conjugal self-transfer to a wide variety of gram-negative bacteria. Although RP1, RP4, and RK2 have been isolated from different bacteria, they are presumably the same plasmid (Burkardt *et al.*, 1979; Datta *et al.*, 1971; Ingram *et al.*, 1973). Most P1-group plasmids specify resistance to carbenicillin, kanamycin, and tetracycline, and some plasmids confer resistance to gentamicin, chloramphenicol, sulfonamides, and streptomycin (Jacoby, 1986; Chakrabarty, 1976). The molecular size of these plasmids is around 50 to 80 kb. Recombination or transposition of P1-group plasmids with other plasmids or chromosomes is important in the evolution of various antibiotic-resistant genes in microorganisms.

3.2. pMG1

R-plasmids of incompatibility group P2 are readily transmissible between *Pseudomonas* strains but not to *E. coli* or other enterobacteria (Jacoby *et al.*, 1976). Plasmid pMG1 isolated from *P. aeruginosa* encodes resistance to gentamicin, streptomycin, sulfonamides, and mercuric ions, and mobilizes transfer-deficient plasmids (Hansen and Olsen, 1978; McCombie *et al.*, 1983). Some degradative plasmids, such as CAM and OCT, are also included in this group (see Table II).

3.3. RSF1010 and Rlb679

The broad-host-range plasmids RSF1010 and Rlb679 are noncon-jugative, multicopy replicons conferring resistance to streptomycin and sulfonamides (Harring *et al.*,1985; Scherzinger *et al.*, 1991). The complete nucleotide sequence and gene organization of the plasmid RSF1010 hasbeen described (Scholz *et al.*, 1989). A molecule of RSF1010 DNA consists of 8684 bp, and analysis of the nucleotide sequence has revealed the existence of more than 40 open reading frames. The data provide a detailed structure of the plasmid used as a replicon for one of the most versatile cloning systems and for understanding the mechanism of RSF1010 replication. This and other studies identified the following genes: *oriV*, the origin of vegetative DNA replication; *repA*, *repB*, and *repC*, the genes for positive-acting replication proteins; *oriT*, the site of the relaxation complex and the origin of conjugational DNA transfer; *mobA*, *mobB*, and *mobC*, genes encoding *trans*-active proteins involved in plasmid mobilization; and *sul* and *str*, the genes conferring resistance to sulfonamide and streptomycin, respectively. RepA catalyzes unwinding of the extensive stretches of dsDNA in a reaction requiring ATP hydro-lysis. RepC binds the site within the *oriV* region. Three genes, *mobA*, *mobB*, and *mobC* were required, besides *oriT*, for mobilization of RSF1010 in the presence of conjugative plasmids. The *sul* gene product is a dihydropteroate synthase, and the *str* gene products exhibit strep-tomycin phophotransferase activity (Higashi *et al.*, 1994; Honda *et al.*, 1991; Miao *et al.*, 1995; Scherzinger *et al.*, 1991).

3.4. pMS350

Plasmid pMS350 was isolated from an imipenem-resistant strain of *P. aeruginosa* GN17203 (Watanabe *et al.*, 1991) and belongs to P9 group.

Table II. Degradative Plasmids in *Pseudomonas*

Plasmid	Substrate	Size (kb)	Source	Incompatibility group	Reference
CAM	Camphor	~500	*P. putida* PpG1	P2	Rheinwald *et al.*, 1973 Koga *et al.*, 1986
OCT	*n*-Alkanes (C$_6$–C$_{10}$)	~500	*P. putida* PpG6	P2	Chakrabarty *et al.*, 1973 Shapiro *et al.*, 1981
TOL(pWW0)	Xylene, toluene	117	*P. putida* mt-2	P9	Worsey and Williams, 1975 Duggleby *et al.*, 1977
NAH(NAH7)	Naphthalene	83	*P. putida* PpG7	P9	Yen and Gunsalus, 1982 Yen *et al.*, 1983

The plasmid was transferable to *P. aeruginosa* but not to *E. coli* and specifies resistance to β-lactams, gentamicin, and sulfonamides. The β-lactamase gene of pMS350 was cloned and expressed in *E. coli* (Iyobe, *et al.*, 1994). The enzyme hydrolyzes carbapenem and other extended-spectrum β-lactam antibiotics. The incompatibility group P9 includes also degradative plasmid, such as TOL and NAH (see Table III).

3.5. pSa

Plasmid pSa (39 kb) was originally isolated in Japan from a clinical isolate of *Shigella* (Watanabe *et al.*, 1968). The plasmid is conjugative and has a broad host range among gram-negative bacteria, belongs to incompatibility group W, and has a low number of copies per cell (about 3 to 5) (Tait *et al.*, 1982). Plasmid pSa confers resistance to chloramphenicol, kanamycin/gentamycin, streptomycin/spectinomycin, and sulfonamides (Loper and Kado, 1979). A detailed genetic map of the plasmid was constructed (Ireland, 1983; Valentine and Kado, 1989), and nucleotide sequences of genes conferring streptomycin/spectinomycin and sulfonamide resistance were determined (Tait *et al.*, 1985; Valentine *et al.*, 1994). Introduction of pSa into *Agrobacterium tumefaciens* inhibits the cell's ability to induce crown-gall tumors on a host plant (Close and Kado, 1991; Loper and Kado, 1979).

4. RESISTANCE TO METAL IONS, PESTICIDES, AND UV LIGHT

In addition to antibiotic resistance, many *Pseudomonas* plasmids determine resistance to toxic metal ions, pesticides, and ultraviolet (UV) light (Table III).

4.1. pMR1, pMERPH

Plasmids confer resistance to a variety of metal ions, such as cadmium, cobalt, lead, and mercury and to oxyanions of arsenic, boron, chromium, and tellurium. The genetic determinants that specify mercury resistance are usually associated with transposon-containing plasmids. Two such transposable elements, Tn*21* and Tn*501*, have been characterized in detail (Brown *et al.*, 1985 and 1989; Chakrabarty, 1976; Jacoby, 1986). An inducible mercuric reductase converts toxic Hg^{2+} to the relatively nontoxic Hg^0 which can be volatized from the cells (Summers and Lewis, 1973). Organomercurial lyase converts organomercurials to

Table III. Plasmid Encoding Resistance to Metal Ions, Pesticides, and UV Light

Plasmid	Phenotype[a]	Source	Size (kb)
pMR1	Hg, Pm, As, Cd	*P.* sp.strain MR1	146
pMERPH	Hg, IncJ	*P. putrefaciens*	ND[b]
pPT23D	Cu	*P. syringae*	39
pPSR1	Cu, Sm	*P. syringae* pv.*syringae*	68
pCMS1	Op	*P. diminuta* MG	70
pMG2	Gm, Sm, Su, Hg, Pm, UV, IncP2	*P.* sp.	400

[a] Abbreviations for resistances: Cb, carbenicillin; Gm, gentamicin; Km, kanamycin; Sm, streptomycin; Su, sulfonamides; Hg, mercuric chloride; Pm, phenylmercury acetate; Cd, cadmium compounds; As, arsenic compounds; Cu, copper compounds; Op, organophosphorus compounds; UV, ultraviolet.
[b] ND, not determined.

Hg^{2+} and corresponding hydrocarbons, for example, phenylmercury acetate to Hg^{2+} and benzene. The Hg^{2+} is subsequently reduced to Hg^0 by the reductase (Clark *et al.*, 1977). The mercuric reductase is specified by *merA* gene. A regulatory gene *merR* controls the transcription of the *merTCA* operon positively in the presence of Hg^{2+} and negatively in the absence of Hg^{2+} (Brown *et al.*, 1985 and 1989). The *merR* gene negatively regulates its own expression (Foster and Ginnity, 1985).

Pseudomonas sp. strain MR1, isolated from the coastal waters of the Bay of Bengal, contains a single 146-kb plasmid, pMR1 (Rajini Rani and Mahadevan, 1992). The pMR1 plasmid conferred inducible resistance to both Hg^{2+} and phenylmercury acetate (Rajini Rani and Mahadevan, 1994). The mercury resistance determinant of pMR1 had no DNA homology with the mercury resistance region of Tn*501* (Rajini Rani and Mahadevan, 1994). The mercury resistance determinant isolated from *P. putrefaciens* is carried on the incompatibility group J plasmid, pMERPH, which can be transfered to *E. coli* by conjugation (Peters *et al.*, 1991). The pMERPH plasmid expresses inducible, narrow-spectrum mercury resistance and shows distinct differences from the resistant features encoded by transposable elements (Peters *et al.*, 1991).

4.2. pPT23D, pPSR1

Copper compounds are commonly used in agriculture as antimicrobial agents for plants. Plasmid pPT23D, conferring inducible copper resistance, was isolated from the phytopathogen *Pseudomonas syringae* (Bender and Cooksey, 1986). The components of the *copABCD* operon function in copper uptake and exclusion by binding copper in periplasm and inner and outer membranes (Mellano and Cooksey, 1988). Two regulatory genes, *copR* and *copS*, which are similar to known two-component regulatory systems, control the expression of *copABCD* operon (Mills *et al.*, 1993). The amino acid sequence of CopR shares strong similarity with PhoB, an activator protein of phosphate metabolism, and OmpR, a regulatory protein of osmolarity control. The predicted amino acid sequence of CopS indicates that CopS is related to the sensor components that contain the histidine kinase autophosphorylation site. Strains of *Pseudomonas syringae* pv. *syringae* were found to be resistant to copper and streptomycin. These determinants were specified by a 68-kb conjugative plasmid pPSR1 (Sundin and Bender, 1993). The copper resistance genes on pPSR1 had no DNA homology with that of pTP23D.

4.3. pCMS1

Organophosphorus compounds are used as agricultural and domestic pesticides. Plasmid pCMS1 was isolated from a strain *Pseudomonas diminuta* MG, which constitutively hydrolyzes a broad spectrum of organophosphorus agents. The gene (*opd*) encoding organophosphorus phosphotriesterase was cloned and sequenced (McDaniel *et al.*, 1988). The predicted size of the gene product is 35 kDa, but the partially purified enzyme has a molecular weight of approximately 65 kDa, suggesting that the active enzyme is dimeric. An interesting feature of the enzyme is that most of the activity is associated with the membrane fractions.

4.4. pMG2

Plasmid-encoded enhanced UV resistance may provide a selective advantage to the host in certain environments, such as exposure to solar radiation or to DNA-damaging chemicals. Plasmid pMG2 protects *P. aeruginosa* against ultraviolet light and gamma irradiation (Lehrbach *et al.*, 1977a). The plasmid also enhances UV-induced mutagenesis. The regulatory mechanisms operating on plasmid-encoded UV resistance genes have not yet been clarified. Because the expression of DNA repair genes of pMG2 requires a host *recA* gene product (Lehrbach *et al.*, 1977a and 1977b), it would be controlled by a system similar to the *lexA-recA* system of *E. coli* (Kokjohn and Miller, 1994). Both structural analogues of *recA* and *lexA* have been identified in *P. aeruginosa* (Garriga *et al.*, 1992; Miller and Kokjohn, 1990).

5. PLASMIDS FOR BIOSYNTHESIS

P. syringae pathovars are frequently harbored in indigenous plasmids, and some of them are important for plant–bacteria interactions.

5.1. Indole-3-Acetic Acid (IAA) Production

The plant pathogen *P. syringae* causes the production of galls on olive and oleander. The tumorous gall formation depends on the bacterial synthesis of the plant hormone indole-3-acetic acid (IAA). Strains that have lost the ability to produce IAA are avirulent. The genetic determinants for IAA synthesis are located on a plasmid in oleander isolates. The 52-kb plasmid pIAA1 (Palm *et al.*, 1989) and the 72-kb plasmid pIAA2 (Soby *et al.*, 1994) were isolated from *P. syringae* pv. *savastanoi* and specify the IAA synthesis genes (*iaaM* and *iaaH*). IAA is produced from tryptophan by tryptophan 2-monooxygenase, the *iaaM* gene product, and by indoleacetamide hydrolase, the *iaaH* gene product. Certain isolates of *P. syringae* convert IAA to IAA-lysine. The gene for IAA-lysine synthetase (*iaaL*) was also located on pIAA1 (Glass and Kosuge, 1986). Since IAA-lysine is one-third as active as IAA in growth-promoting activity and is resistant to degradation, it may function as a storage form of IAA in gall tissue (Cohen and Bandurski, 1982).

5.2. Ethylene-Forming Enzyme

Ethylene, a plant hormone, is produced by plants and microorganisms, including plant pathogens. There are two biosynthetic pathways in the production of ethylene by microorganisms, *via* 2-keto-4-methylthiobutyric acid by an NADH:Fe(III)EDTA oxidoreductase and *via* 2-oxoglutarate by an ethylene-forming enzyme (Fukuda *et al.*, 1992a; Nagahama *et al.*, 1991; Ogawa *et al.*, 1990). The gene coding ethylene-forming enzyme of *P. syringae* pv. *phaseolicola* PK2 was located in an indigenous plasmid, designated pPSP1 (Nagahama *et al.*, 1994). The gene for the ethylene-forming enzyme of pPSP1 was cloned and expressed in *E. coli* (Fukuda *et al.*, 1992b). Nucleotide sequence analysis of the gene revealed an open reading frame encoding 350 amino acids, which contains the sequences for the 2-oxoglutarate-binding site and the iron-binding site.

5.3. Coronatine Production

Coronatine (COR) is a non-host-specific phytotoxin which induces chlorosis and hypertrophy of plant tissues and is produced by several pathovars of *P. syringae* (Gnanamanickam *et al.*, 1982). Although the mode of action of COR in the virulence of *P. syringae* is not well understood, it may be involved in ethylene synthesis (Kenyon and Turner, 1992). The COR synthesis genes in *P. syringae* pv. glycinea PG4180 and *P. syringae* pv. tomato reside on a 90-kb plasmid designated p4189A and a 95 to 103-kb plasmid designated pPT23A, respectively (Bender *et al.*, 1989; Young *et al.*, 1992). The synthetic pathway of COR has been determined. COR is produced by coupling coronafacic acid (CFA) with coronamic acid (CMA). Organization of the COR biosynthetic cluster, which encodes CFA and CMA synthesis, and the coupling region catalyzing amide bond formation between CFA and CMA was identified (Bender *et al.*, 1993).

5.4. Syringolide Production

A virulence gene D (*avrD*) in *P. syringae* pv. tomato PT23 specifies the production of syringolides, which are C-glycosidic elicitors of plant defense reactions (Keen and Buzzell, 1992; Midland *et al.*, 1993; Smith *et al.*, 1993). Plasmid pPT23B (83 kb) in this strain carries a *avrD* gene (Kobayashi *et al.*, 1989). Location of the *avrD* operon, a partition (*par*) locus, and the plasmid origin of replication (*oriV*) have been mapped (Murillo *et al.*, 1994).

5.5. Malonate Assimilation

Malonate is a competitive inhibitor of succinate dehydrogenase, but some species of *Pseudomonas* can utilize malonate as a sole carbon source. A broad-host-range R-plasmid (60 kb) involved in malonate assimilation was isolated from *P. fluorescens* and was designated pPSF1 (Kim and Kim, 1994). Malonate decarboxylase, a key enzyme in malonate assimilation, is specified by the gene on pPSF1. Plasmid pPSF1 also encodes resistance to ampicillin, kanamycin, and streptomycin and is transmissible to *E. coli* by conjugation. The transformed *E. coli* strains grow on a medium containing malonate as a sole carbon source.

6. DEGRADATIVE PLASMIDS

Pseudomonads show unique metabolic versatility. Some catabolic activities are specified by genes on plasmids, designated degradative plasmids (Gunsalus and Yen, 1981; Timmis *et al.*, 1985). These plasmids occur naturally and are either transmissible or nontransmissible. The

degradative plasmids specify a set of genes involved in the biodegradation of organic compounds, such as aliphatic and aromatic hydrocarbons, alkaloids, and chlorinated aromatics (Chakrabarty, 1976; Frantz
and Chakrabarty, 1986; Haas, 1983). The existence of degradative plasmids with relaxed substrate specificity of enzymes and the presence of
Pseudomonas species widespread in the environment give pseudomonads
a variety of important functions. This chapter provides the characteristics of some individual plasmids with respect to metabolic pathway,
physical mapping, and regulation of gene expression. Table II summarizes the properties of some degradative plasmids. Because a number of
plasmids determined the same or similar phenotypes, only those most
extensively studied are listed.

6.1. CAM

P. putida PpG1 was originally isolated from an enrichment culture
with D-camphor as the carbon source and was shown to carry a plasmid
termed CAM (Rheinwald *et al.*, 1973). The CAM plasmid (about 500 kb)
belongs to incompatibility group P2 and is responsible for early steps of
the D-camphor degradative pathway for converting camphor to isobutylate. The catabolic pathway controlled by the CAM plasmid is the conversion of D-camphor to 5-*exo*-hydroxycamphor by a monooxygenase system, then the conversion to 2,5-diketocamphane by 5-*exo*-hydroxycam-
phor dehydroxygenase. The monooxygenase system consists of three
gene products: *camA* encoding NADH-putidaredoxin reductase, *camB*
encoding putidaredoxin, and *camC* encoding cytochrome P-450cam (Koga
et al., 1989; Unger *et al.*, 1986). The enzyme 5-*exo*-hydroxy camphor
dehydroxygenase is encoded by *camD* (Koga *et al.*, 1989). These genes are
clustered as an operon *camDCAB* and are under negative control by the
regulatory gene *camR* which is autorepressed (Aramaki *et al.*, 1993; Fujita
et al., 1993; Koga *et al.*, 1986). Purified CamR protein forms a homodimer
with a molecular mass of 40 kDa (Aramaki *et al.*, 1995). The binding of
CamR to a specific DNA region of *cam* operon was inhibited by D-camphor, D-3-bromocamphor, adamantane, 2-adamantanone, 5-*exo*-hydro-
xycamphor, and 2,5-diketocamphane (Aramaki *et al.*, 1995). Although the
CAM plasmid contains some isobutylate degradative genes, the complete
set of isobutylate degradative genes is present on a *P. putida* chromosome.

6.2. OCT

The OCT plasmid (about 500 kb) belongs to incompatibility group
P2 and is involved in the degradation of *n*-alkanes, such as octane, hexane,

and decane (Chakrabarty *et al.*, 1973). The entire *n*-alkane degradative pathway is encoded by two transcription units, *alkST* and *alkBFGHJKL*, located on the OCT plasmid and the *alcA* and *aldA* genes on the chromosome (Fig. 1). The first step in converting aliphatic medium-chain-length hydrocarbons to alcohols is catalyzed by alkane monooxygenase (AlkB, AlkG, and AlkT), the conversion to aldehydes is catalyzed by plasmid-encoded alcohol dehydrogenase (AlkJ) and by chromosomal alcohol dehydrogenase (AlcA), and the conversion to fatty acids is catalyzed by aldehyde dehydrogenases, plasmid-encoded AlkH or chromosomal AldA (Kok *et al.*, 1989; Owen *et al.*, 1984; Shapiro *et al.*, 1984). Expression of the *alkBFGHJKL* operon is positively regulated by the *trans*-acting regulatory gene *alkS*, which encodes the 99-kDa protein, whereas the chromosomally encoded alcohol dehydrogenase is partly constitutive (Eggink *et al.*, 1988; Owen, 1986). The second cistron (*alkT*), which encodes the 48-kDa protein, is a component of the alkane monooxygenase complex. AlkB and AlkJ are 41-kDa and 59-kDa membrane-bond proteins respectively. AlkK is a 60-kDa protein of acyl-CoA synthetase located in the cytoplasm, and AlkL is a 25-kDa protein found in the outer membrane whose function is unknown (van Beilen *et al.*,

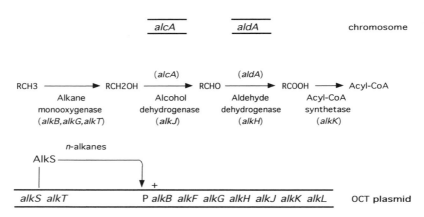

Figure 1. The *n*-alkane degradative pathway and the gene organization of OCT plasmid and chromosome. The degradative pathway is encoded by two operons on the OCT plasmid, *alkST* and *alkBFGHJKL*, and two genes on the chromosome, *alcA* and *aldA*. The *alkB*, *alkG*, and *alkT* genes encode alkane hydroxylase, rubredoxin, and rubredoxin reductase, respectively. The *alkH* and *alkF* encode NAD-dependent aldehyde dehydrogenase and nonfunctional rubredoxin. The *alkJ*, *alkK*, and *alkL* encode alcohol dehydrogenase, acyl-CoA synthetase, and an outer membrane protein. The gene product of *alkS* activates the expression of *alkBFGHJKL* operon in the presence of *n*-alkanes (+).

1992). The OCT plasmid also contains genes for some or all of the early D-lysine metabolic enzymes, such as a D-lysine membrane carrier, lysine racemase, and Δ^1-piperidine-2-carboxylate reductase (Cao *et al.*, 1993).

6.3. TOL(pWW0)

The TOL plasmid represents a group of plasmids that specify degradation of hydrocarbons such as *m*- or *p*-xylene, toluene, 1,2,4-trimethylbenzene, 3-ethyltoluene, and their corresponding alcohols and acid derivatives. The natural host of TOL is *P. putida* mt-2, which was isolated in 1959 by K. Hosokawa in Japan from an enrichment culture supplied with *m*-toluate. The *P. putida* TOL plasmid pWW0 is self-transmissible and belongs to the incompatibilitiy group P9 (Williams and Murray, 1974). The size of pWW0 is 117 kb, in which aboiut a 40-kb consecutive region is needed for the catabolic pathway (Duggleby *et al.*, 1977; Worsey and Williams, 1975). All of the xylene/toluene (*xyl*) genes are located in the 56-kb transposon Tn4651 (Tsuda and Iino, 1988).

The matabolic pathway and the genetic organization of the TOL plasmid (Fig. 2) have been described (Abril and Ramos, 1993; Downing and Broda, 1979; Harayama and Rekik, 1990; Harayama *et al.*, 1985 and 1989; Inouye *et al.*, 1987a and 1987b; Marques and Ramos, 1993; Mermod *et al.*, 1987; Ramos *et al.*, 1986 and 1987). The TOL plasmid contains two operons. The upper pathway operon encodes enzymes for the oxidation of xylene/toluene to toluate/benzoate, and the lower pathway operon encodes enzymes for further degradation of these compounds to substrates for the tricarboxylic acid cycle. Their expression is controlled by two regulatory genes, *xylR* and *xylS*. In the current model,the gene expression for the metabolism of TOL plasmid by xylene is regarded as a cascade amplification system involving induction of a regulatory protein in response to pathway metabolites (Inouye *et al.*, 1987a and 1987b; Ramos *et al.*, 1987). The *xylR* gene plays a key role in this entire regulatory system, and its expression is autogenously regulated (Inouye *et al.*, 1987a). The *xylR* gene product (XylR, 67kDa), which belongs to the NtrC family of regulators of a two-component system, activates the expression of the upper pathway operon and the *xylS* gene in the presence of xylene (Inouye *et al.*, 1988). Both promoters, OP1 (Pu) and Ps, depend on RpoN (σ^{54}) (Inouye *et al.*, 1989; Kohler *et al.*, 1989). In addition, the activation of OP1 requires the integration host factor (IHF) (Perez–Martin *et al.*, 1994). On the other hand, IHF negatively controls the expression of *xylS* (Gomada *et al.*, 1994; Holtel *et al.*, 1992 and 1995). An intrinsic DNA bend structure in the promoter-operator region of Ps plays an important role for the activation of the *xylS* gene.

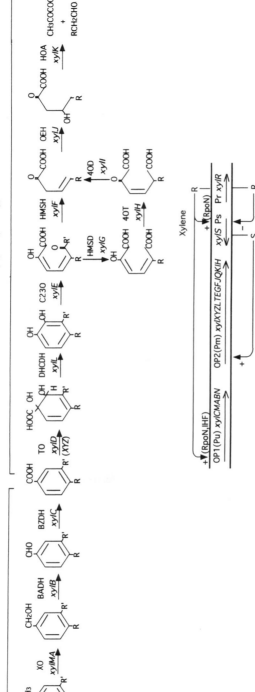

Figure 2. The xylene/toluene degradative pathway and the gene organization of TOL plasmid. The structure is (R=R'=H, toluene), (R=H;R'=CH₃, m-xylene), (R=CH₃;R'=H, p-xylene), (R=H;R'=C₂H₅, 3-ethyltoluene), (R=R'=CH₃, 1,3,4- trimethylbenzene), and corresponding derivatives, respectively. Enzyme and gene abbreviations: XO(xylMA), xylene oxygenase subunits; BADH(xylB), benzyl alcohol dehydrogenase; BZDH(xylC), benzaldehyde dehydrogenase; TO(xylD/xylXYZ), toluate dioxygenase subunits; DHCDH(xylL), dihydrocyclohexadiene carboxylate dehydrogenase; C23O(xylE), catechol 2,3-diosygenase; HMSH(xylF), hydroxymuconic semialdehyde hydrolase; HMSD(xylG), hydroxymuconic semialdehyde dehydrogenase; 4OT(xylH), 4-oxalocrotonate tautomerase; 4OD, 4-oxalocrotonate decarboxylase; OEH(xylJ), 2-oxopent-4-enoate hydratase; HOA(xylK), 2-oxo-4-hydroxypentanoate aldolase. xylCMABN and xylXYZLTEGFJQKIH are operons for the upper pathway and the lower (meta) pathway enzymes of the degradation, respectively. xylS and xylR are regulatory genes. The functions of the xylT and xylQ genes are not known. OP1(Pu), OP2(Pm), Ps, and Pr indicate the operator-promoter regions, respectively. +, activation of transcription; −, repression of transcription. In the presence of xylene, R activates xylXYZLTEGFJQKIH in the absence or presence of toluate. R represses its own expression. The induced S activates both xylS and xylCMABN.

(Gomada *et al.*, 1994). Based on these results, the formation of a DNA loop between the upstream regulatory sequence (URS) located about 100 bp upstream from each promoter and the promoter sequence facilitates the interaction of XylR and RpoN-containing RNA polymerase for the transcriptional activation (Abril *et al.*, 1991; de Lorenzo *et al.*, 1991; Gomada *et al.*, 1992; Inouye *et al.*, 1990). The *xylS* gene product (XylS, 37kDa) is induced in a large quantity and activates the lower pathway operon (Inouye *et al.*, 1986 and 1987b). In the latter process, toluate, which is a metabolite of xylene, greatly enhances the activation of the lower pathway operon (Kessler *et al.*, 1993, 1994a, and 1994b).

The nucleotide sequences of the regulatory genes *xylS* and *xylR*, the structural genes *xylA*, *xylM*, *xylE*, *xylG*, *xylF*, and *xylJ*, and the promoter-operator regions of two operons and two regulatory genes have been determined (Horn *et al.*, 1991; Inouye *et al.*, 1984a, 1984b, 1985, 1986 and 1988; Mermod *et al.*, 1984; Nakai *et al.*, 1983; Suzuki *et al.*, 1991). Xylene monoosygenase (XO), the first enzyme of the pathway, has a broad substrate specificity and oxidizes *m*- and *p*-xylenes, toluene, (methyl)benzyl alcohols, (methyl) benzaldehydes, and indole (Harayama *et al.*, 1989). The XylR protein also exhibits a very broad effector specificity and recognizes the pathway substrates, mono- and disubstituted methyl-, and ethyl-, and chlorotoluenes, (methyl)benzyl alcohols, and *p*-chlorobenzaldehyde (Abril *et al.*, 1989). These degradative mechanisms and the conjugative transfer of TOL plasmid are used to contruct hybrid strains capable of growing in the presence of new compounds (Brinkmann and Reineke, 1992; Duque *et al.*, 1992; Lehrback *et al.*, 1984).

Many different types of naturally occuring plasmids control the degradation of xylene/toluene and related compounds. For example, a 120-MDa plasmid pDK1 specifies toluene, *m*- and *p*-xylenes, and pseudocumene (Kunz and Chapman, 1981), and a 150-MDa plasmid pKJ1 encodes both xylene/toluene degradation and resistance to streptomycin and sulfonamide (Yano and Nishi, 1980). *P. putida* MT15 and MT53 strains contain a 250-kb TOL plasmid pWW15 and a 105-kb TOL plasmid pWW53, which encode toluene/xylene catabolism, respectively (Keil *et al.*, 1985 and 1987). The ability of *P. putida* TMB to degrade xylene, toluene, and 1,2,4-trimethylbenzene is encoded by plasmid pGB (85 kb) (Bestetti and Galli, 1987). The 1,2,4-trichlorobenzene-degrading *P.* sp. strain P51 contains a plasmid pP51 (Kasberg *et al.*, 1995). The pVI150 catabolic plasmid of *P.* sp. strain CF600 carries the *dmp* system encoding the enzymes for the degradation of (methyl)phenols and the regulatory gene *dmpR*. The expression of *dmp* operon is RpoN-dependent, and the DmpR protein is also a member of the NtrC family (Pavel *et al.*, 1994).

6.4. NAH(NAH7) and SAL(SAL1)

The 83-kb NAH7 plasmid from soil bacterium *P. putida* is self-transmissible and encodes enzymes for metabolizing naphthalene or salicylate as the sole carbon and energy source (Yen andGunsalus, 1982). The catabolic genes for metabolism are organized in two operons (Fig. 3); the *nah* operon (*nahA-F*), which encodes six enzymes for metabolizing naphthalene to salicylate, and the *sal* operon (*nahG-M*), which encodes eight enzymes for metabolizing a salicylate to pyruvate and acetaldehyde (Eaton, 1994; Simon *et al.*, 1993; Yen and Gunsalus, 1982). The *nahR* gene encodes a 34-kDa polypeptide which activates transcription of the two operons in response to the inducer salicylate (Schell and Sukordhaman, 1989). Two operons are separated by a 7-kb segment and are transcribed in the same direction. The *nahR* gene is located between two operons and is transcribed in the opposite direction. Promoter sequences similar to the consensus sequence of *E. coli* promoters at the -35 and -10 regions were found in both operons and the *nahR* gene (Schell, 1986). There is another highly conserved region (-82 to -47) of both the *nah* and *sal* promoter regions. The NahR protein binds these regions and activates the transcription of two operons (Schell, 1986). The amino acid sequence of the *nahR* gene deduced indicates that NahR is a member of the LysR family of prokaryotic transcription activators, including NodD, IlvY, and AmpR (Schell and Sukordhaman *et al.*, 1989). Mutagenetic experiments on the *nahR* gene demonstrated that DNA binding activity is clustered in an N-terminal helix-turn-helix motif (aa residues

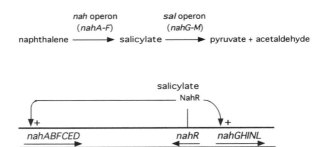

Figure 3. The naphthalene and salicylate degradative pathway and the gene organization of NAH7 plasmid. The *nah* operon (*nahA-F*) encodes enzymes for metabolizing naphthalene to salicylate, and the *sal* operon (*nahG-M*) encodes enzymes for metabolizing salicylate to tricarboxylic acid cycle intermediates. The regulatory gene *nahR* of the NAH7 plasmid encodes a protein which activates both *nah* and *sal* operons in the presence of salicylate.

23–45) or a C-terminal region (aa residues 239–291) and that a domain (aa residues 126–206) is involved in inducer-responsible transcription activation (Schell *et al.*, 1990).

SAL1 plasmid was isolated from strains which grow on salicylate, but not naphthalene (Chakrabarty, 1972). Transposon mutagenesis, electron microscope heteroduplex analysis, and restriction mapanalysis suggest that the SAL1 plasmid is derived from the NAH7 plasmid by DNA substitution in the *nahA* gene region (Yen *et al.*, 1983).

Plasmids pKA1, pKA2, and pKA3, approximately 100kb in size, were isolated from *Pseudomonas* strains, which were recovered from a mixed slurry treatment reactor inoculum for the soil of a manufactured gas plant contaminated with polynuclear aromatic hydrocarbons. Among them, pKA1, a NAH7-like plasmid isolated from *P. fluorescens* 5R, is responsible for the mineralizing anthracene, phenanthrene, and naphthalene (Menn *et al.*, 1993; Sanseverino *et al.*, 1993).

6.5. pHMT112

P. putida ML2 contains a 112-kb catabolic plasmid, pHMT112, which carries genes involved in catabolizing benzene via an *ortho*- or β-ketoadipate pathway. The benzene dioxygenase genes (*bedC1C2BA*) and *cis*-benzene dihydrodiol dehydrogenase gene (*bedD*) are located on the plasmid (Tan and Fong, 1993). The benzene bioxygenase catalyzes oxidation of benzene to *cis*-1,2-dihydroxy-cyclohexa-3,5-diene. The enzyme is a multicomponent complex comprising a flavoprotein reductase encoded by *bedA* gene, a ferredoxin encoded by *bedB*, and iron-sulfur proteins (ISPα and ISPβ) encoded by *bedC1* and *bedC2*, respectively (Tan *et al.*, 1993 and 1994). The amino acid sequences of the benzene dioxygenase complex of pHMT112 deduced show 83%–89% homology with the chromosomally encoded benzene dioxygenase and toluene dioxygenase (Tan *et al.*, 1994).

7. CLONING VECTORS

Host-vector systems have been developed for gene cloning in the metabolically versatile pseudomonads because the organisms can use various organic chemicals as carbon and nitrogen sources and participate in parasitic interactions with plants and animals. The restriction-negative and/or recombination-deficient mutants of *Pseudomonas* suitable for gene cloning (Früh *et al.*, 1983), such as *P. aeruginosa* PAO1162 (Dunn and Holloway, 1971) and *P. putida* KT2440 (Bagdasarian *et al.*,

1981), have been obtained. Many types of cloning vectors, including expression vectors and promoter probe vectors, have been constructed and used (Bagdasarian and Timmis, 1982; Davison *et al.*, 1990; Farinha and Kropinski, 1989; Mermod *et al.*, 1986a; Morales *et al.*, 1991; Neito *et al.*, 1990). The majority of these cloning vectors are broad-host-range plasmids belonging to incompatiblity groups P1, P4/Q, and W. They propagate in a wide range of gram-negative bacteria. Some alternatives to antibiotic-resistant genes as selective markers have been studied, for example, the herbicide-resistant gene of *Pseudomonas* sp. (Fitzgibbon and Braymer, 1990). Specific-purpose vectors of *Pseudomonas* strains have been reviewed (de Lorenzo and Timmis, 1992). Here, some properties of typical plasmids used to construct cloning vectors are described.

7.1. Incompatibility Group P4/Q

The most useful cloning vectors are based on the broad-host-range, nonconjugative (Tra⁻Mob⁺), incompatibility group P4/Q plasmids, including RSF1010, Rlb679, and R300B (Barth and Grinter, 1974; Gautier and Bonewald, 1980; Guerry *et al.*, 1974). The copy number per cell is about 15–20 (Frey *et al.*, 1992; Harring *et al.*, 1985; Morales *et al.*, 1990). Three essential replication genes in RSF1010 (*repA*, *repB* and *repC*) and *oriV* have been identified (Scholz *et al.*, 1989). Because this group of plasmids replicates in a wide range of gram-negative bacterial species and has a small size (8.7 kb), it has been used as a basic replicon to develop DNA cloning vectors.

Plasmid pKT231 (13 kb) is one of the extensively used vectors derived from RSF1010 (Bagdasarian *et al.*, 1981), in particular the *Pseudomonas* species (Grund and Gunsalus, 1983; Timmis, 1981). The plasmid specifies resistance to kanamycin and streptomycin and contains 11 unique restriction sites. Among them, five sites can be used for insertional inactivation of kanamycin resistance, while another four sites can be used for streptomycin resistance. Most pKT231-like vectors are mobilized by conjugation from bacteria, when a donor strain carries a conjugative plasmid that can supply the conjugal transfer function (Bagdasarian *et al.*, 1982). RFS1010-based cosmid vectors, such as pMMB34, have also been constructed (Frey *et al.*, 1983).

Vectors for promoter analysis in *Pseudomonas* are also based on RSF1010 and related plasmids. The vectors contain reporter genes, such as promoterless *lacZ* (β-galactosidase) and *phoA* (alkaline phosphatase) from *E. coli*, *luxAB* (luciferase) genes from *Vibrio harveyi*, the *xylE* (catechol 2,3-dioxygenase) gene from *P. putida* TOL plasmid, and antibiotic-resistant genes, *tetA* and *cat* (de Lorenzo and Timmis, 1992; Farinha and

Kropinski, 1989 and 1990; Konyencsni and Deretic, 1988; Ronald *et al.*, 1990).

Regulated-expression vectors have been developed from RSF1010 derivative. These expression systems are based on the *lacI^q* gene-encoding repressor protein and the hybrid *lac-trp* or *tac-lac* promoter of *E. coli* (Morales *et al.*, 1990), the *trpI* gene-encoding activator protein, and the *trpBA* promoter of *P. putida* (Olekhnovich and Fomichev, 1994), the *xylS* gene-encoding activator protein and the promoter sequence of lower pathway operon from *P. putida* TOL plasmid (Mermod *et al.*, 1986b), and the *nahR* gene encoding activator protein and the *nahG* promoter (Yen, 1991). Each promoter activity is induced by isopropyl-β-D-thio-galactopyranoside (IPTG), indoleglycerol phosphate, *m*-toluate, and salicylate, respectively.

7.2. Incompatibility Group P1

The plasmids belonging to incompatibility group P1 can be transferred to a variety of gram-negative bacteria, including *E. coli*. These plasmids confer resistance to the antibiotics ampicillin, tetracycline, and kanamycin. Native RK2, RP4, and RP1 can be used as a vector. Because of its large size (about 56 kb), attempts have been made to reduce the size for routine use. Plasmid pRK290 (Ditta *et al.*, 1980) constructed from RK2 is 20 kb in size and encodes tetracycline resistance. The copy number is about 3–5 copies per chromosome. Some cosmid vectors derived from RK2/RP4, such as pLAFR1 and pCP13, with low copy number, broad-host-range replicons have been used to construct a gene bank (Darzins and Chakrabarty, 1984; Friedman *et al.*, 1982).

7.3. Incompatibility Group W

The conjugative plasmid pSa belongs to the W incompatibility group of enterobacteria (Tait *et al.*, 1982). The 39-kb plasmid pSa carries a 20-kb segment of DNA that encodes replication and conugal trnsfer functions (Close and Kado, 1992; Tait *et al.*, 1982). The region of 9.6 kb of pSa encodes resistance to chloramphenicol, sulfonamides, streptomycin/spectinomycin, and kanamycin/gentamicin/tobramycin. Because pSa in oncogenic strains of *Agrobacterium tumefaciens* interferes with the formation of crown gall tumors, the plasmid has also been used to construct cloning vectors for genetic engineering of *A. tumefaciens* (Chen and Kado, 1994; Leemans *et al.*, 1982).

pSa-derived vector plasmid pGV1122 (10.8 kb) confers tetracycline and streptomycin/spectinomycin resistance and allows cloning into the unique *Bam*HI, *Sal*I, and *Hind*III sites with inactivation of the tetra-cycline resistance gene (Leemans *et al.*, 1982). It also carries ColE1-type origin which results in a high copy number in *E. coli*. Plasmid pGV1124 (10.8 kb) confers chloramphenicol and streptomycin/spectinomycin re-sistance and allows cloning into the unique *Eco*RI site with inactivation of the chloramphenicol resistance gene (Leemans *et al.*, 1982). Because these incompatibility group W replicon-based vectors are compatible with incompatibility group P4-based RSF1010-like vectors and incom-patibility group P1-based RK2-like vectors, they are used to construct bacteria containing multiple plasmids (Jacoby, 1986).

8. CONCLUSIONS AND PERSPECTIVES

Recently, a sensitive and convenient biosensing system for benzene derivatives in the environment has been developed in *E. coli* by fusing a gene of luciferase to the promoter of the *xylS* gene and introducing the regulatory gene *xylR* of TOL plasmid (Kobatake *et al.*, 1995). The ex-pression of luciferase was induced in the presence 5 μM *m*-xylene. Thus, increasing information about plasmids in *Pseudomonas* has been used for many applications.

Toluene is highly toxic to most microorganisms at 0.1%(v/v) concen-tration. Toxicity is apparently caused by disruption of the cytoplasmic membrane (Sikkema *et al.*, 1995). Inoue and Horikoshi (1989) isolated a toluene-resistant bacterium from Japanese soil after enrichment culture in a nutrient medium containing toluene. One strain, IH-2000, was identified as a strain of *P. putida*. *P. putida* IH-2000 grows in media containing more than 50% (v/v) toluene or high concentrations of cyclo-hexane, *p*-xylene, styrene, and heptanol, but not benzene. The strain cannot utilize toluene as a carbon source. Bacterial strains resistant to those toxic hydrocarbons are considered useful for biotransformation and bioremediation because many hydrophobic solvents are environ-mental pollutants. *P. putida* Idaho was isolated by selecting bacterial growth in a organic–aqueous system at 5 to 50%(v/v). The strain utilizes toluene, *m*-xylene, *p*-xylene, 1,2,4-trimethylbenzene, and 3-ethyltoluene as growth substrates (Cruden *et al.*, 1992). No plasmid DNA was found in *P. putida* Idaho. However, the strain contains DNA regions homologous to the TOL plasmid. The organism degrades toluene, *p*- and *m*-xylenes by the same pathway specified by *P. putida* containing TOL plasmid. *P. putida* DOT-T1, which grows on a minimal medium with 1%(v/v) toluene as the sole carbon source, was isolated from water collected at a wastewater treatment plant (Ramos *et al.*, 1995). The strain grows in the presence of 90%(v/v) toluene, and the solvent tolerance is inducible. The isolate could not use *m*- and *p*-xylenes as the sole carbon source. However, the catabolic potential was expanded to include *m*- and *p*-xylenes by transfer of the TOL plasmid pWWO-Km. In fact, construction of hybrid strains of *Pseudomomas* has been attempted by redesigning the metabolic pathway. These observations indicate that the plasmid–host combinations give rise to great diversity of bacterial strains capable of degrading xenobiotics.

Keil *et al.*, (1987) have suggested that the upper pathway operon, lower pathway operon, and regulatory genes of TOL plasmid may have been inserted as separate genetic entities into TOL plasmid backbones because of the finding of the structurally related TOL-like metabolic pathway on plasmids and chromosomes. Kok *et al.* (1992) have postulated that the alkane and xylene upper pathways originated from a different organism and have only recently transferred to *P. putida.* These features suggest that the rapid evolution of new metabolic pathways for the degradation of synthetic compounds in response to environmental changes occurs in nature. As observed with antibiotics that development of antibiotics stimulates the evolution of resistant genes, new manufactured substrates would promote genetic engineering in *Pseudomonas* cells *in vivo*.

REFERENCES

Abril, M. A., and Ramos, J. L., 1993, Physical organization of the upper pathway operon promoter of the *Pseudomonas* TOL plasmid. Sequence and postional requirement for XylR-dependent activaton of transcription, *Mol. Gen. Genet.* **239:**281–288.

Abril, M. A., Michan, C., Timmis, K. N., and Ramos, J. L., 1989, Regulator and enzyme specificities of the TOL plasmid-encoded upper pathway for degradation of aromatic hydrocarbons and expansion of the substrate range of the pathway, *J. Bacteriol.* **171:**6782–6790.

Abril, M. A., Buck, M., and Ramos, J. L., 1991, Activation of the *Pseudomonas* TOL plasmid upper pathway operon. Identification of binding sites for the positive regulator XylR and for integration host factor protein, *J. Biol. Chem.* **266:**15832–15838.

Aramaki, H., Sagara, Y., Hosoi, M., and Horiuchi, T., 1993, Evidence for autoregulation of *camR*, which encodes a repressor for the cytochrome P-450cam hydrocylase operon on the *Pseudomonas putida* CAM plasmid, *J. Bacteriol.* **175:**7828–7833.

Aramaki, H., Sagara, Y., Kabata, H.. Shimamoto, N., and Horiuchi, T., 1995, Purification and characterization of a *cam* repressor (CamR) for the cytochrome P-450cam hydroxylase operon on the *Pseudomonas putida* CAM plasmid, *J. Bacteriol.* **177:**3120–3217.

Ayres, E. K., Thomson, V. J., Merino, G., Balderes, D., and Figurski, D. H., 1993, Precise deletions in large bacterial genomes by vector-mediated excision (VEX). The *trfA* gene of promiscuous plasmid RK2 is essential for replication in several gram-negative hosts. *J. Mol. Biol.* **230:**174–185.

Bagdasarian, M., and Timmis, K. N., 1982, Host:vector systems for gene cloning in *Pseudomonas, Curr. Top. Microbiol. Immunol.* **96:**47–67.

Bagdasarian, M., Lurz, R., Rückert, B., Franklin, F. C. H., Bagdasarian, M. M., Frey, J., and Timmis, K. N., 1981, Specific purpose cloning vectors. II Broad host range, high copy number, RSF1010-derived vectors, and a host: vector system for gene cloning in *Pseudomonas, Gene* **16**:237–243.

Bagdasarian, M., Bagdasarian, M. M., Lurz, R., Nordheim, A., Frey, A., and Timmis, K. N., 1982, Molecular and functional analysis of the broad host range plasmid RSF1010 and construction of vectors for gene cloning in Gram-negative bacteria, in: *Bacterial Drug Resistance* (S. Mitsuhashi, ed.) Japan Scientific Society Press, Tokyo, pp. 183–197.

Barth, P. T. and Grinter, N. J., 1974, Comparison of the deoxyribonucleic acid molecular weights and homologies of plasmids conferring linked resistance to streptomycin and sulfonamides, *J. Bacteriol.* **120**:618–630.

Bender, C. L., and Cooksey, D. A., 1986, Indigenous plasmids in *Pseudomonas syringae* pv. tomato: conjugative transfer and role in copper resistance, *J. Bacteriol.* **165**:534–541.

Bender, C. L., Malvick, D. K., and Mitchell, R. E., 1989, Plasmid-mediated production of the hytotoxin coronatine in *Pseudomonas syringae* pv. *tomato, J. Bacteriol.* **171**:807–812.

Bender, C. L., Liyanage, H.,Palmer, D., Ullrich, M., Young, S., and Mitchell, R., 1993, Characterizatoin of the genes controlling the biosynthesis of the polyketide phytotoxin coronatine including conjugation between coronafacic and coronamic acid, *Gene* **133**:31–38.

Bestetti, G., and Galli, E., 1987, Characterization of a novel TOL-like plasmid from *Pseudomonas putida* involved in 1,2,4-trimethylbenzene degradation, *J. Bacteriol.* **169**:1780–1783.

Brinkmann, U., and Reineke, W., 1992, Degradation of chlorotoluene by in vivo construction hybrid strains: problems of enzyme specificity, induction and prevention of meta-pathway, *FEMS Microbiol. Lett.* **75**:81–87.

Brown, N. L., Misra, T. K., Winnies, J. N., Schmidt, A., and Silver, S., 1985, The nucleotide sequences of the mercuric resistance operons of plasmid R100 and transposon Tn*501*: further evidencee for *mer* genes which enhance the activity of the mercuric ion detoxification system, *Mol. Gen. Genet.* **202**:143–151.

Brown, N. L., Lund, P. A., and Ni'Bhriain, N., 1989, Mercury resistance in bacteria, in: *Genetic bacterial diversity* (D. A. Hopwood and K. F. Chater, eds.) Academic Press, London, pp. 175–195.

Bryan, L. E., 1984, Aminoglycoside resistance, in: *Antimicrobial Drug Resistance* (L. E. Bryan, ed.) Academic Press, New York, pp. 241–227.

Burkardt, H.–J., Riess, G., and Puhler, A., 1979, Relationship of group P1 plasmid revealed by heroduplex experiments: RP1, RP4, R68, and RK2 are identical, *J. Gen. Microbiol.* **114**:341–348.

Cao, X., Kolonay, J., Jr., Saxton, K. A., and Hartline, R. A., 1993, The OCT plasmid encodes D-lysine membrane transport and catabolic enzymes in *Pseudomonas putida*, *Plasmid* **30**:83–89.

Chakrabarty, A. M., 1972, Genetic basis of the biodegradation of salicylate in *Pseudomonas, J. Bacteriol.* **112**:815–823.

Chakrabarty, A. M., 1976, Plasmids in *Pseudomonas, Annu. Rev. Genet.* **10**:7–30.

Chakrabarty, A. M., Chou, G., and Gunsalus, I. C., 1973, Genetic regulation of octane dissimulation plasmid in *Pseudomonas, Proc. Natl. Acad. Sci. USA* **70**:1137–1140.

Chen, C. Y., and Kado, C. I., 1994, Inhibition of *Agrobacterium tumefaciens* oncogenicity by the *osa* gene of pSa, *J. Bacteriol.* **176**:5697–5703.

Clark, D. L., Weiss, A. A., and Silver, S., 1977, Mercury and organomercurial resistance determined by plasmids in *Pseudomonas, J. Bacteriol.* **132**:186–196.

Close, S. M., and Kado, C. I., 1991, The *osa* gene of pSa encodes a 21.1-kilodalton protein that suppresses *Agrobacterium tumefaciens* oncogenicity, *J. Bacteriol.* **173**:5449–5456.

Close, S. M., and Kado, C. I., 1992, A gene near the plasmid pSa origin of replication encodes a nuclease, *Mol. Microbiol.* **6**:521–527.

Cohen, J. D., and Bandurski, P. S., 1982, Chemistry and physiology of the bound auxins, *Annu. Rev. Plant Physiol.* **33**:403–430.

Cruden, D. L., Wolfram, J. H., Rogers, R. D., and Gibson, D. T., 1992, Physiological properties of a *Pseudomonas strain* which grows wih *p*-xylene in a two-phase (organic-aqueous) medium, *Appl. Env. Microbiol.* **58**:2723–2729.

Darzins, A., and Chakrabarty, A. M., 1984, Cloning of genes controlling alginate biosynthesis from a mucoid cystic fibrosis isolte of *Pseudomonas aeruginosa*, *J. Bacteriol.* **159**:9–18.

Datta, N., Hedges, R. W., Shaw, E. J., Syker, R. B., and Richmond, M. W., 1971, Properties of an R factor from *Pseudomonas aeruginosa*, *J. Bacteriol.* **108**:1244–1249.

Davison, J., Brunel, F., Kaniga, K., and Chevalier, N., 1990, Recombinant DNA vectors for *Pseudomonas*, in: *Pseudomonas: Biotransformations, Pathogenesis, and Evolving Biotechnology* (S. Silver, A. M. Chakrabarty, B. Iglewski, and S. Kaplan, eds.) American Society for Microbiology, Washington, D. C., pp. 242–251.

de Lorenzo, V., and Timmis, K. N., 1992, Specialized host-vector systems for the engineering of *pseudomonas* strins destined for environmental release, in: *Pseudomonas: Molecular Biology and Biotechnology* (E. Galli, S. Silver, and B. Witholt, eds.) American Society for Microbiology, Washington, D. C., pp. 415–428.

de Lorenzo, V., Herrero, M., Metzke, M., and Timmis, K. N. 1991, An upstream XylR- and IHF-induced nucleoprotein complex regulates the sigma 54-dependent Pu promoter of TOL plasmid, *EMBO J.* **10**:1159–1167.

Ditta, G., Stanfields, S., Corbin, D., and Helinski, D. R., 1980, Borad host range DNA cloning system for gram-negative bacteria: Construction of a gene bank of *Rhizobium meliloti*, *Proc. Natl. Acad. Sci. USA* **77**:7347–7351.

Downing, R., and Broda, P., 1979, A cleavage map of the TOL plasmid of *Pseudomonas putida* mt-2, *Mol. Gen. Genet.* **177**:189–191.

Duggleby, C. J., Bayley, S. A., Worsey, M. J., Williams, P. A., and Broda, P., 1977, Molecular sizes and relationship of TOL plasmids in *Pseudomonas*, *J. Bacteriol.* **130**:1274–1280.

Dunn, N. W., and Holloway, B. W., 1971, Pleiotropy of *p*-fluorophenyl-alanine resistance and antibiotic hypersensitive mutants of *Pseudomonas aeruginosa*, *Genet. Res.* **18**:185–197.

Duque, E., Ramos–Gonzales, M. I., Delgado, A., Contreras, A., Molin, S., and Ramos, J. L., 1992, Genetically engineered *Pseudomonas* strains for mineralization of aromatics: Survival, performance, gene transfer, and biological containment, in: *Pseudomonas: Molecular Biology and Biotechnology* (E. Galli, S. Silver, and B. Witholt, eds.) American Society for Microbiology, Washington, D. C., pp. 429–437.

Eaton, R. W., 1994, Organization and evolution of naphthalene catabolic pathways: Sequence of the DNA encoding 2-hydroxychromene-2-carboxylate isomerase and *trans-o*-hydroxybenzylidenepyruvate hydratase-aldolase from the NAH7 plasmid, *J. Bacteriol* **176**:7757–7762.

Eggink, G., Engel, H, Meijer, W. G., Otten, J., Kingma, J., and Witholt, B., 1988, Alkane utilizatoin in *Pseudomonas oleovaorans*. Structure and function of the regulatory locus *alkR*, *J. Biol. Chem.* **263**:13400–13405.

Farinha, M. A., and Kropinski, A. M., 1989, Construction of broad host range vectors for general cloning and promoter selection in *Pseudomonas* and *Escherichia coli*, *Gene* **77**:205–210.

Farinha, M. A., and Kropinski, A. M., 1990, Construction of broad-host-range plasmid vectors for easy visible selection and anlysis of promoters, *J. Bacteriol.* **172**:3496–3499.

Fitzgibbon, J. E., and Braymer, H. D., 1990, Cloning of a gene from *Pseudomonas* sp. strain PG2982 conferring increased glyphosate resistance, *Appl. Environ. Microbiol*, **56**:3382–3388.

Foster, T. J., and Ginnity, F., 1985, Some mercurial resistance plasmids from different incompatibility groups specify *merR* regulatory functions that both repress and induce the *mer* operon of plasmid R100, *J. Bacteriol.* **162**:773–776.

Frantz, B., and Chakrabarty, A. M., 1986, Degradative Plasmids in *Pseudomonas*, in: *The Bacteria, Vol. X, The Biology of Pseudomonas* (J. K. Sokatch ed.) Academic Press, New York, pp. 295–323.

Frey, J., Bagdasarian, M., Feiss, D., Franklin, F. C. H., and Deshusses, J., 1983, Stable cosmid vectors that enable the introduction of cloned fragments into a wide range of gram-negative bacteria, *Gene* **24**:299–308.

Frey, J., Bagdasarian, M. M., and Bagdasarian, M., 1992, Replication and copy number control of the broad-host-range plasmid RSF1010, *Gene* **113**:101–106.

Friedman, A. M., Long, S. R., Brown, S. E., Buikema, W. J., and Ausbel, F. M., 1982, Construction of a broad host range cosmid vector and its use in the gentic analysis of *Rhizobium* mutants, *Gene* **18**:289–296.

Früh, R., Watson, J. M., and Haas, D., 1983, Construction of recombination-deficient strains of *Pseudomonas aeruginosa*, *Mol. Gen. Genet.* **191**:334–337.

Fujita, M., Aramaki, H., Horiuchi, T., and Amemura, A., 1993, Transcription of the *cam* operon and *camR* genes in *Pseudomonas putida* PpG1, *J. Bacteriol.* **175**:6953–6958.

Fukuda, H., Ogawa, T., Tazaki, M., Nagahama, K., Fujii, T., Tanase, S., and Morino, Y., 1992a, Two reactions are simultaneously catalyzed by a single enzyme: The arginine-dependent simultaneous formation of two products, ethylene and succinate, from 2-oxoglutarate by an enzyme from *Pseudomonas syringae*, *Biochem. Biophys. Res. Commun.* **188**:483–489.

Fukuda, H., Ogawa, T., Ishihara, K., Fujii, T., Nagahama, K., Omata, T., Inoue, Y., Tanase, S., and Morino, Y., 1992b, Molecular cloning in *Escherichia coli*, expression and nucleotide sequence of the gene for the ethylene-forming enzyme of *Pseudomonas syringae* pv. *phaseolicola* PK2, *Biochem. Biophys. Res. Commun.* **188**:826–832.

Garriga, S., Calero, S., and Barbe, J., 1992, Nucleotide sequence analysis and comparison of the *lexA* genes from *Salmonella typhimurium*, *Erwinia carotovora*, *Pseudomonas aeruginosa* and *Pseudomonas putida*, *Mol. Gen. Genet.* **235**:125–135.

Gautier, F., and Bonewald, R., 1980, The use of plasmid R1162 and derivatives for gene cloning in the methanol-utilizing *Pseudomonas* AM1, *Mol. Gen. genet.* **178**:375–380.

Glass, N. L., and Kosuge, T., 1986, Cloning of the gene for indoleacetic acid-lysine synthetase from *Pseudomonas syringae* subsp. *savastanoi*, *J. Bacteriol.* **166**:598–603.

Gnanamanickam, S. S., Starratt, A. N., and Ward, E. W. B., 1982, Coronatine production in vitro and in vivo and its relation to symptom development in bacterial blight of soybean, *Can. J. Bot.* **60**:645–650.

Gomada, M., Inouye, S., Imaishi, H., Nakazawa, A., and Nakazawa, T., 1992, Analysis of an upstream regulatory sequence required for activation of the regulatory gene *xylS* in xylene metabolism directed by the TOL plasmid of *Pseudomonas putida*, *Mol. Gen. Genet.* **233**:419–426.

Gomada, M., Imaishi, H., Miura, K., Inouye, S., Nakazawa, T., and Nakazawa, A., 1994, Analysis of DNA bend structure of promoter regulatory regions of xylene-metabolizing genes on the *Pseudomonas* TOL plasmid, *J. Biochem.* **116**:1096–1104.

Grund, A. D., and Gunsalus, I. C., 1983, Cloning of genes for naphthalene metabolism in *Pseudomonas putida, J. Bacteriol.* **156**:89–94.

Guerry, P., Van Embden, J., and Falkow, S., 1974, Molecular nature of two nonconjugative plasmids carrying drug resistance genes, *J. Bacteriol.* **177**:619–630.

Gunsalus, I. C., and Yen, K.–M., 1981, Metabolic plasmid organization and distribution, in: *Molecular Biology, Pathogenicity and Ecology of Bacterial Plasmid* (S. B. Levy, R. C. Clowes, and E. L. Koenig, eds.) Plenum, New York, pp. 499–509.

Haas, D., 1983, Genetic aspects of biodegradation by pseudomonads, *Experientia* **39**:1199–1213.

Hansen, J. B., nad Olsen, R. H., 1978, Isolation of large bacterial plasmids and characterization of the P2 incompatibility group plasmids pMG1 and pMG5, *J. Bacteriol.* **135**:227–238.

Harayama, S., and Rekik, M., 1990, The *meta* cleavage operon of TOL degradative plasmid pWW0 comprises 13 genes, *Mol. Gen. Genet.* **221**:113–120.

Harayama, S., Leppik, R. A., Rekik, M., Mermod, N., Lehrbach, P. R., Reineke, W., and Timmis, K. N., 1985, Gene order of the TOL catabolic plasmid upper pathway operon and oxidation of both toluene and benzyl alcohol by the *xylA* product, *J. Bacteriol.* **167**:455–461.

Harayama, S., Rekik, M., Wubbolts, M., Rose, K., Leppik, R. A., and Timmis, K. N., 1989, Characterization of five genes in the upper-pathway operon of TOL plasmid pWW0 from *Pseudomonas putida* and identification of the gene products, *J. Bacteriol.* **171**:5048–5055.

Harring, V., Scholz, P., Scherzinger, E., Frey, J., Hatfull, F., Willets, N. W., and Bagdasarian, M., 1985, Protein RepC is involved in copy number control of the broad host range plasmid RSF1010, *Proc. Natl. Acad. Sci. USA* **82**:6090–6094.

Hedstrom, R. C., Crider, B. P., and Eagon, R. G., 1982, Comparison of kinetics of active tetracycline uptake and active tetracycline efflux in sensitive and RP4-containing *Pseudomonas pitida, J. Baceriol.* **152**:255–259.

Higashi, A., Sakai, H., Honda, Y., Tanaka, K., Miao, D. M., Nakamura, T., Taguchi, Y., Komano, T., and Bagdasarian, M., 1994, Functional featues of *oriV* of the broad host range plasmid RSF1010 in *Pseudomonas aeruginosa. Plasmid* **31**:196–200.

Holtel, A., Timmis, K. N., and Ramos, J. L., 1992, Upstream binding sequences of the XylR activator protein and integration host factor in the *xylS* gene promoter region of the *Pseudomonas* TOL plasmid, *Nucl. Acids Res.* **20**:1755–1762.

Holtel, A., Goldenberg, D., Giladi, H., Oppenheim, A. B., and Timmis, K. N., 1995, Involvement of IHF protein in expression of the Ps promoter of the *Pseudomonas putida* TOL plasmid, *J. Bacteriol.* **177**:3312–3315.

Honda, Y., Sakai, H., Hiasa, H., Tanaka, K., Komano, T., and Bagdasarian, M., 1991, Functional division and reconstruction of a plasmid replication origin: Molecular dissection of the *oriV* of the broad-host-range plasmid RSF1010, *Proc. Natl. Acad. Sci. USA* **88**:179–183.

Horn, J. M., Harayama, S., and Timmis, K. N., 1991, DNA sequence determination of the TOL plasmid (pWWO) *xylGFJ* genes of *Pseudomonas putida*: Implications for the evolution of aromatic catabolism, *Mol. Microbiol.* **5**:2459–2474.

Ingram, L. C., Richmond, M. H., and Syker, R. B., 1973, Molecular characterization of the R factors implicated in the carbenicillin resistance of a sequence of *Pseudomonas aeruginosa* strains isolated from burns. *Antimicrob. Agents Chemother.* **3**:279–288.

Inoue, A., and Horikoshi, 1989, A *Pseudomonas* thrives in high concentrations of toluene, *Nature(London)*, **338**:264–266.

Inouye, S., Ebina, Y., Nakazawa, A., and Nakazawa, T., 1984a, Nucleotide sequence surrounding transcription initiation site of *xylABC* operon on TOL plasmid of *Pseudomonas putida*, *Proc. Natl. Acad. Sci. USA* **81**:1688–1691.

Inouye, S., Nakazawa, A., and Nakazawa, T., 1984b, Nucleotide sequence of the promoter region of the *xylDEGF* operon on TOL plasmid of *Pseudomonas putida*. *Gene* **29**:323–330.

Inouye, S., Nakazawa A., and Nakazawa, T., 1985, Determination of the transcription initiation site and identification of the protein product of the regulatory gene *xylR* for *xyl* operons on the TOL plasmid, *J. Bacteriol.* **163**:863–869.

Inouye, S., Nakazawa, A., and Nakazawa, T., 1986, Nucleotide sequence of the regulatory gene *xylS* on the *Pseudomonas putida* TOL plasmid and identification of the protein product, *Gene* **44**:235–242.

Inouye, S., Nakazawa, A., and Nakazawa, T., 1987a, Expression of the regulatory gene *xylS* on the TOL plasmid is poistively controlled by the *xylR* gene product, *Proc. Natl. Acad. Sci. USA* **84**:5182–5186.

Inouye, S., Nakazawa, A., and Nakazawa, T., 1987b, Overproduction of the *xylS* gene product and activation of the *xylDLEGF* operon on the TOL plasmid, *J. Bacteriol.* **169**:3587–3592.

Inouye, S., Nakazawa, A., and Nakazawa, T., 1988, Nucleotide sequence of the regulatory gene *xylR* of the TOL plasmid from *Pseudomonas putida*, *Gene* **66**:301–306.

Inouye, S., Yamada, M., Nakazawa, A., and Nakazawa, T., 1989. Cloning and sequence analysis of the *ntrA* (*rpoN*) gene of *Pseudomonas putida*, *Gene* **85**:145–152.

Inouye, S., Gomasa, M., Sangodkar, U. M., Nakazawa, A., and Nakazawa, T., 1990, Upstream regulatory sequence for transcriptional activator XylR in the first operon of xylene metabolism on the TOL plasmid, *J. Mol. Biol.* **216**:251–260.

Ireland, C. R., 1983, Detailed restriction enzyme map of crown gall-suppressive IncW plasmid pSa, showing ends of deletion causing chloramphenicol sensitivity, *J. Bacteriol.* **155**:722–727.

Iyobe, S., tsunoda, M., and Mituhashi, S., 1994, Cloning and expression in Enterobacteriaceae of the extended-spectrum *beta*-lactamase gene from a *Pseudomonas aeruginosa* plasmid, *FEMS Microbiol. Lett.* **121**:175–180.

Jacoby, G. A., 1984, Resistance plasmids of *Pseudomonas aeruginosa*, in: *Antimicrobial Drug Resistance* (L. E. Bryan, ed.) Academic Press, New York, pp. 497–514.

Jacoby, G. A., 1986, Resistance of *Pseudomonas*, in: *The Bacteria, Vol.X, The Biology of Pseudomonas* (J. K. Sokatch ed.) Academic Press, New York, pp. 265–292.

Jacoby, G. A., Jacob, A. E., and Hedges, R. W., 1976, Recombination between plasmids of incompatibility groups P1 and P2, *J. Bacteriol.* **127**:1278–1285.

Jacoby, G. A., Sutton, L., Knobel, L., Mammen, P., 1983, Properties of IncP-2 plamids of *Pseudomonas* spp. *Antimicrob. Agents Chemother.* **24**:168–175.

Kao, J. C., Perry, K. L., and Kado, C. I., 1982, Indoleacetic acid complementation and its relation to host range specifying genes on the Ti plasmid of *Agrobacterium tumefaciens*, *Mol. Gen. genet.* **188**:425–432.

Kasberg, T., Daubaras, D. L., Chakrabarty, A. M., Kinzelt, D., and Reineke, W., 1995, Evidence that operons *tcb*, *tfd*, and *clc* encode maleylacetate reductase, the fourth enzyme of the modified ortho pathway, *J. Bacteriol.* **177**:3885–3889.

Keen, N. T., and Buzzell, R. I., 1991, New disease resistance genes in soybean against *Pseudomonas syringae* pv. *glycinea*: Evidence that one of them interacts with a bacterial elicitor, *Theor. Appl. Genet.* **81**:133–138.

Keil, H., Lebens, M. R., and Williams, P. A., 1985, TOL plasmid pWW15 contains two nonhomologous, independently regulated catechol 2,3-dioxygenase genes, *J. Bacteriol.* **63**:248–255.

Keil, H., Saint, C. M., and Williams, P. A., 1987, Gene organization of the first catabolic operon of TOL plasmid pWW53: Production of indigo by the *xylA* gene product, *J. Bacteriol.* **169:**764–770.

Kenyon, J. S., and Turner, J. R., 1992, The stimulation of ethylene synthesis in *Nicotiana tabacum* leaves by the phytotoxin coronatine, *Plant Physiol.* **100:**219–224.

Kessler, B., de Lorenzo, V., and Timmis, K. N., 1993, Identification of a *cis*-acting sequence within the Pm promoter of the TOL plasmid which confers XylS-mediated responsiveness to substituted benzoates, *J. Mol. Biol.* **230:**699–703.

Kessler, B., Herrero, M., Timmis, K. N., and de Lorenzo, V., 1994a, Genetic evidence that the XylS regulator of the *Pseudomonas* TOL *meta* operon controls the Pm promoter through weak DNA-protein interactions, *J. Bacteriol.* **176:**3171–3176.

Kessler, B., Timmis, K. N., and de Lorenzo, V., 1994b, The organization of the Pm promoter of the TOL plasmid reflects the structure of its cognate activator protein XylS, *Mol. Gen. Genet.* **244:**596–605.

Kim, Y. S., and Kim, E. J., 1994, A plasmid responsible for malonate assimilation in *Pseudomonas fluorescens*, *Plasmid*, **32:**219–221.

Kobayashi, D. Y., Tamaki, S. J., and Keen, N. T., 1989, Cloned avirulence genes from the tomato pathogen *Pseudomonas syringae* pv. *tomato* confer cultivar specificity on soybeans, *Proc. Natl. Acad. Sci. USA* **86:**157–161.

Kobatake, E., Niimi, T., Haruyama, T., Ikariyama, Y., and Aizawa, M., 1995, Biosensing of benzene derivatives in the environment by luminescent *Escherichia coli*, *Biosensors & Bioelectronics*, **10:**601–605.

Koga, H., Aramaki, H., Yamaguchi, E., Takeuchi, K., Horiuchi, T., and Gunsalus, I. C., 1986, *camR*, a negative regulator locus of the cytochrome P-450cam hydroxylase operon, *J. Bacteriol.* **166:**1089–1095.

Koga, H., Yamaguchi, E., Matsunaga, K., Aramaki, H., and Horiuchi, T., 1989, Cloning and nucleotide sequences of NADH-putidaredoxin reductase gene (*camA*) and putidaredoxin gene (*camB*) involved in cytochrome P-450cam hydroxylase of *Pseudomonas putida*, *J. Biochem.* **106:**831–836.

Kohler, T., Haayama, S., Ramos, J. L., and Timmis, K. N., 1989, Involvement of *Pseudomonas putida* RpoN sigma factor in regulation of varous metabolic functions, *J. Bacteriol.* **171:**4326–4333.

Kok, M., Oldenhuis, R., van der Linden, M. P., Raatjes, P., Kingma, J., van Lelyveld, P. H., and Witholt, B., 1989, The *Pseudomonas oleovorans* alkane hydroxylase gene, *J. Biol. Chem.* **264:**5435–5441.

Kok, M., Shaw, J. P., and Harayama, S., 1992, Comparison of two hydrocarbon monooxygenases of *Pseudomonas putida*, in: *Pseudomonas: Molecular Biology and Biotechnology* (E. Galli, S. Silver, and B. Witholt, eds.) American Society for Microbiology, Washington, D. C., pp. 214–222.

Kokjohn, T. A., and Miller, R. V., 1994, IncN plasmids mediate UV resistance and error-prone repair in *Pseudomonas aeruginosa* PAO, *Microbiol.* **140:**43–48.

Konyencsni, W., and Deretic, V., 1988, Borad host range plasmid and M13 bacteriophage-derived vectors for promoter analysis in *Escherichia coli* and *Pseudomonas aeruginosa*, *Gene* **74:**357–386.

Kornacki, J. A., West, A. H., and Firshein, W., 1984, Proteins encoded by the *trans*-acting replication and maintenance regions of broad host range plasmid RK2, *Plasmid* **11:**48–57.

Krishnapillai, V., Nash, J., and Lanka, E., 1984, Insertion mutations in the promiscuous IncP-1 plasmid R18 which affect its host range between *Pseudomonas* species, *Plasmid* **12:**170–180.

Kues, U., and Stahl, U., 1989, Replication of plasmids in gram-negative bacteria, *Microbiol. Rev.* **53**:491–516.

Kunz, D. A., and Chapman, P. J., 1981, Isolation and characterization of spontaneously occurring TOL plasmid mutants of *Pseudomonas putida* HS1, *J. Bacteriol.* **146**:952–964.

Lanka, E., Kröger, M., and Fürste, J. P., 1984, Plamid RP4 encodes two forms of a DNA primase, *Mol. Gen. Genet.* **194**:65–72.

Leemans, J., Langenakens, J., De Greve, H., Deblaere, R., Van Montague, M., and Schell, J., 1982, Broad-host-range cloning vectors derived from the W-plasmid Sa, *Gene* **19**:361– 364.

Lehrbach, P., Kung, A. H. C., and Lee, B. T. O., 1977a, Loss of ultraviolet light-induced mutability in *Pseudomonas aeruginosa* carrying mutant R plasmids, *J. Gen. Microbiol.* **101**:135–141.

Lehrbach, P., Kung, A. H. C., Lee, B. T. O., and Jacoby, G. A., 1977b, Plasmid modification of radiation and chemical mutagen sensitivity in *Pseudomonas aeruginosa*. *J. Gen. Microbiol.* **98**:167–176.

Lehrbach, P. R., Zeyer, J., Reineke, W., Knackmuss, H. J., and Timmis, K. N., 1984, Enzyme recruitment *in vitro*: Use of cloned genes to extend the range of haloaromatics degraded by *Pseudomonas* sp. strain B13, *J. Bacteriol.* **158**:1025–1032.

Li, X. Z., Livermore, D. M., and Nikaido, H., 1994a, Role of efflux pump(s) in intrinsic resistance of *Pseudomonas aeruginosa*; resistance to tetracycline, chloramphenicol, and norfloxacin, *Antimicrob. Agents Chemother.* **38**:1732–1741.

Li, X. Z., Ma, D., Livermore, D. M., and Nikaido, H., 1994b, Role of efflux pump(s) in intrinsic resistance of *Pseudomonas aeruginosa*: Active efflux as a contributing factor to *beta*-lactam resistance, *Antimicrob. Agents Chemother.* **38**:1742–1752.

Loper, J. E., and Kado, C. I., 1979, Host range conferred by the virulence-specifying plasmid of *Abrobacterium tumefaciens*, *J. Bacteriol.* **139**:591–596.

Marques, S., and Ramos, J. L., 1993, Transcriptional control of the *Pseudomonas putida* TOL plasmid catabolic pathways, *Mol. Microbiol.* **9**:923–929.

Matsuhashi, Y., Yagisawa, M., Kondo, S., Takeuchi, T., and Umezawa, H., 1975, Aminoglycoside 3′-phosphotransferases I and II in *Pseudomonas aeruginosa*, *J. Antibiot.* **28**:442–447.

Matthew, M., 1979, Plasmid-mediated β-lactamase of gram-negative bacteria: Properties and distribution, *J. Antimicrob. Chemother.* **5**:349–358.

McCombie, W. R., Hansen, J. B., Zylstra, G. J., Maurer, B., and Olsen, R. H., 1983, *Pseudomonas* streptomycin resistance transposon associated with R-plasmid mobilization, *J. Bacteriol.* **155**:40–48.

McDaniel, C. S., Harper, L. L., and Wild, J. R., 1988, Cloning and sequencing of a plasmid-borne gene (*opd*) encoding a phosphotriesterase, *J. Bacteriol.* **170**:2306–2311.

Mellano, M. A., and Cooksey, D. A., 1988, Induction of the copper resistance operon from *Pseudomonas syringae* pv. *tomato*, *J. Bacteriol.* **170**:4399–4401.

Menn, F. M., Applegate, B. M., and Sayler, G. S., 1993, NAH plasmid-mediated catabolism of anthracene and phenanthrene to naphthoic acids, *Appl. Env. Microbiol.* **59**:1938–1942.

Mermod, N., Lehrbach, P. R., Reineke, W., and Timmis, K. N., 1984, Transcription of the TOL plasmid toluate catabolic pathway operon of *Pseudomonas putida* is determined by a pair of coordinately and positively regulated overlapping promoters, *EMBO J.* **3**:2461–2466.

Mermod, M., Lehrbach, P. R., Don, R. H., and Timmis, K. N., 1986a, Gene cloning and manipulation in *Pseudomonas*, in: *The Bacteria, Vol.X, The Biology of Pseudomonas* (J. K. Sokatch, ed.) Academic Press, New York, pp. 325–355.

Mermod, N., Ramos, J. L., Lehrbach, P. R., and Timmis, K. N., 1986b, Vectors for regulated expression of cloned genes in a wide range of gram-negative bacteria, *J. Bacteriol.* **167**:447–454.

Mermod, N., Ramos, J. L., Bairoch, A., and Timmis, K. N., 1987, The *xylS* gene positive regulator of TOL plasmid pWW0: Identification, sequence analysis, and overproduction leading to constitutive expression of *meta* cleavage operon, *Mol. Gen. Genet.* **207**:349–354.

Miao, D. M., Sakai, H., Tanaka, K., Honda, Y., Komano, T., and Bagdasarian, M., 1995, Functional distinction among structural subsections in the specific priming signal for DNA replication of the broad-host-range plasmid RSF1010, *Biosci. Biotechnol. Biochem.* **59**:920–921.

Midland, S. L., Keen, N. T., Sims, J. J., Midland, M. M., Stayton, M. M., Burton, V., Smith, M. J., Mazzola, E. P., Graham, K. J., and Clardy, J., 1993, The structures of syringolides 1 and 2, novel C-glycosidic elicitors from *Pseudomonas syringae* pv. *tomato*, *J. Org. Chem.* **58**:2940–2945.

Miller, R. V., and Kokjohn, T. A., 1990, General microbiology of *recA*: Environmental and evolutionary significance, *Annu. Rev. Microbiol.* **44**:265–294.

Mills, S. D., Jaslavich, C. A., and Cooksey, D. A., 1993, A two-component regulatory system required for copper-inducible expression of the copper resistance operon of *Pseudomonas syringae*, *J. Bacteriol.* **175**:1656–1664.

Morales, V., Bagdasarian, M. M., and Bagdasarian, M., 1990, Promiscuous plasmids of the IncQ group: Mode of replication and use for gene cloning in gram-negative bacteria, in: *Pseudomonas: Biotransformations, Pathogenesis, and Evolving Biotechnology* (S. Silver, A. M. Chakrabarty, B. Iglewski, and S. Kaplan, eds.) American Society for Microbiology, Washington, D. C., pp. 229–241.

Morales, V., Bäckman, A., and Bagdasarian, M., 1991, A series of wide host range low copy number vectors that allow direct screening for recombinants, *Gene* **97**:39–47.

Murillo, J., Shen, H., Gerhold, D., Sharma, A., Cooksey, D. A., and Keen, N. T., 1994, Characterization of pPT23B, the plasmid involved in syringolide production by *Pseudomonas syringae* pv. *tomato* PT23, *Plasmid* **31**:275–287.

Nagahama, K., Ogawa, T., Fujii, T., Tazaki, M., Tanase, S., Morino, Y., and Fukuda, H., 1991, Purification and properties of an ethylene-forming enzyme from *Pseudomonas syringae*, *J. Gen. Microbiol.* **137**:2281–2286.

Nagahama, K., Yoshino, K., Matsuoka, M., Sato, M., Tanase, S., Ogawa, T., and Fukuda, H., 1994, Ethylene production by strains of the plant-pathogenic bacterium *Pseudomona syringae* depends upon the presence of indigenous plasmids carrying homologous genes for the ethylene-forming enzyme, *Microbiology* **140**:2309–2313.

Nakai, C., Kagamiyama, H., Nozaki, M., Nakazawa, T., Inouye, S., Ebina, Y., and Nakazawa, A., 1983, Complete nucleotide sequence of the metapyrocatechase gene on the TOL plasmid of *Pseudomonas putida* mt-2, *J. Biol. Chem.* **258**:2923–2928.

Nakazawa, T., Inouye, S., and Nakazawa, A., 1980, Physical and functional mapping of RP4-TOL plasmid recombinants: Analysis of insertion and deletion mutants, *J. Bacteriol.* **144**:222–231.

Neito, C., Fernandez–Tresguerres, E., Sanchez, N., Vicente, M., and Diaz, R., 1990, Cloning vectors from a naturally occurring plasmid of *Pseudomonas savastanoi*, specifically tailored for genetic manipulations in *Pseudomonas*, *Gene* **87**:145–149.

Nikaido, H., 1992, Nonspecific and specific permeation channels of the *Pseudomonas aeruginosa* outer membrane, in: *Pseudomonas: Molecular Biology and Biotechnology* (E. Galli, S. Silver, and B. Witholt, eds.) American Society for Microbiology, Washington, D. C., pp. 146–153.

Ogawa, T., Takahashi, M., Fujii, T., Tazaki, M., and Fukuda, H., 1990, The role of NADH:Fe(III)EDTA oxidoreductase in ethylene formation from 2-keto-methylthiolbutyrate, *J. Ferment. Bioeng.* **93:**177–181.

Olekhnovich, I.N., and Fomichev, Y. K., 1994, Controlled-expression shuttle vector for pseudomonads based on the *trpIBA* genes of *Pseudomonas putida, Gene* **140:**63–65.

Owen, D. J., 1986, Molecular cloning and characterization of sequences from the regulatory cluster of the *Pseudomonas plasmid alk* system, *Mol. Gen. Genet.* **203:**64–72.

Owen, D. J., Eggink, G., Hauer, B., Kok, M., McBeth, D. L., Yang, Y. L., and Shapiro, J. A., 1984, Physical structure, gentic content, and expression of the *alkBAC* operon, *Mol. Gen. Genet.* **197:**373–383.

Palleroni, N. J., Kunisawa, R., Contopoulou, R., and Doudoroff, M., 1973, Nucleic acid homologies in the genus *Pseudomonas, Int. J. Syst. Bacteriol.* **23:**333–339.

Palm, C. J., Gaffney, T., and Kosuge, T., 1989, Contranscription of genes encoding indoleacetic acid production in *Pseudomonas syringae* subsp. *savastanoi, J. Bacteriol.* **171:**1002–1009.

Pavel, H., Forsman, M., and Shingler, V., 1994, An aromatic effector specificity mutant of the transcriptional regulator DmpR overcomes the growth constraints of *Pseudomonas* sp. strain CF600 on *para*-substituted methylphenols, *J. Bacteriol.* **176:**7550–7557.

Perez–Martin, J., Timmis, K. N., and de Lorenzo, V., 1994, Coregulation by bent DNA. Functional substitutions of the integration host factor site at sigma 54-dependent promoter Pu of the upper TOL operon by intrinsically curved sequences, *J. Biol. Chem.* **269:**22657–22662.

Peters, S. E., Hobman, J. L., Strike, P., and Ritchie, D. A., 1991, Novel mercury resistance determinants carried by IncJ plasmids pMERPH and R391, *Mol. Gen. Genet.* **228:**294–299.

Quinn, J. P., 1992, Intrinsic antibiotic resistance in *Pseudomonas aeruginosa,* in: *Pseudomonas: Molecular Biology and Biotechnology* (E. Galli, S. Silver, and B. Witholt, eds.) American Society for Microbiology, Washington, D. C., pp. 154–160.

Rajini Rani, D. B., and Mahadevan, A., 1992, Plasmid mediated metal and antibiotic resistance in marine *Pseudomonas, Biometals* **5:**73–80.

Rajini Rani, D. B., and Mahadevan, A., 1994, Cloning and expression of the mercury resistance genes of marin *Pseudomonas* sp. strain MR1 plasmid pMR1 in *Escherichia coli, Res. Microbiol.* **145:**121–127.

Ramos, J. L., Stolz, A., Reineke, W., and Timmis, K. N., 1986, Altered effector specificities in regulators of gene expression: TOL plasmid *xylS* mutants and their use to engineer expansion of the range of aromatics degraded by bacteria, *Proc. Natl. Acad. Sci. USA* **83:**8467–8471.

Ramos, J. L., Mermod, N., and Timmis, K. N., 1987, Regulatory circuits controlling transcription of TOL plasmid operon encoding *meta*-cleavage pathway for degradation of alkylbenzoates by *Pseudomonas, Mol. Microbiol.* **1:**293–300.

Ramos, J. L., Duque, E., Huertas, M.–J. and Haïdour, A., 1995, Isolation and expansion of the catabolic potential of a *Pseudomonas putida* strain able to grow in the presence of high concentrations of aromatic hydrocarbons, *J. Bacteriol.* **177:**3911–3916.

Reimmann, C., and Haas, D., 1986, IS21 insertion in the *trfA* replication control gene of chromosomally integrated plasmid RP1: A property of stable *Pseudomonas aeruginosa* Hfr strains. *Mol. Gen. Genet.* **203:**511–519.

Rheinwald, J. G., Chakrabarty, A. M., and Gunsalus, I. C., 1973, A transmissible plasmid contolling camphor oxidation in *Pseudomonas putida, Proc. Natl. Acad. Sci. USA* **70:**885–889.

Ronald, S. L., Kropinski, A. M., and Farinha, M. A., 1990, Construction of broad-host-range vectors for the selection of divergent promoters, *Gene* **90:**145–148.

Rosenberg, C., Casse–Delbart. F., Dusha. I., David, M., and Boiucher, C., 1982, Megaplasmids in the plant-associated bacteria *Rhizobium meliloti* and *Pseudomonas solanacearum, J. Bacteriol.* **150:**402–406.

Sanseverino, J., Applegate, B. M., King, J. M., and Sayler, G. S., 1993, Plasmid-mediated mineralization of naphthalene, phenanthrene, and anthracene, *Appl. Env. Microbiol.* **59:**1931–1937.

Schell, M. A., 1986, Homology between nucleotide sequences of promoter regions of *nah* and *sal* operons of NAH7 plasmid of *Pseudomonas putida, Proc. Natl. Acad. Sci. USA* **83:**369–373.

Schell, M. A., and Sukordhaman, M., 1989, Evidence that the transcription activator encoded by the *Pseudomonas putida nahR* gene is evolutionarily related to the transcription activators encoded by the *Rhizobium nodD* genes, *J. Bacteriol.* **171:**1952–1959.

Schell, M. A., Brown, P. H., and Raju, S., 1990, Use of saturation mutagenesis to localize probable functional domains in the NahR protein, a LysR-type transcription activator, *J. Biol. Chem.* **265:**3844–3850.

Scherzinger, E., Haring, V., Lurz, R., and Otto, S., 1991, Plasmid RSF1010 DNA replication in vitro promoted by purified RSF1010 RepA, RepB, and RepC proteins, *Nucl. Acids Res.* **19:**1203–1211.

Scholz, P., Haring, V., Wittmann–Liebold, B., Ashman, K., Bagdasarian, M., and Scherzinger, E., 1989, Complete nucleotide sequence and gene organization of the broad-host-range plasmid RSF1010, *Gene* **75**(2):271–288.

Shapiro, J. A., Charbit, A., Benson, S., Caruso, M., Laux, R., Meyer, R., and Banuett, F., 1981, Perspectives for genetic engineering of hydrocarbon oxidizing bacteria, in: *Trends in the Biology of Fermentations for Fuels and Chemicals* (A. Hollaender, ed.), Plenum, New York, pp. 243–272.

Shapiro, J. A., Owen, D. J., Kok, M., and Eggink, G., 1984, *Pseudomonas* hydrocarbon oxidation, in: *Genetic Control of Environmental Pollutants* (G. S. Omenn and A. Hollaender, eds.) Plenum, New York, pp. 229–238.

Shingler, V., and Thomas, C. M., 1984, Transcription of the *trfA* region of broad host range plasmid RK2 is regulated by *trfB* and *korB, Mol. Gen. Genet.* **195:**523–529.

Sikkema, J., de Bont, J. A. M., and Poolman, B., 1995, Mechanisms of membrane toxicity of hydrocarbons, *Microbio. Rev.* **59:**201–222.

Simon, M. J., Osslund, T. D., Saunders, R., Ensley, B. D., Suggs, S., Harcourt, A., Suen, W.–C., Gibson, D. T., and Zylstra, G. J., 1993, Sequences of genes encoding naphthalene dioxygenase in *Pseudomonas putida* strain G7 and NCIB 9816-4, *Gene* **127:**31–37.

Smith, M. J., Mazzola, E. P., Sims, J. J., Midland, S. L., Keen, N. T., Burton, V., and Stayton, M. M., 1993, The syringolides: Bacterial C-glycosyl lipids that trigger plant disease resistance, *Tetrahedon Lett.* **34:**223–226.

Soby, S., Kirkpatrick, B., and Kosuge, T., 1994, Characterization of high-frequency deletions in the *iaa*-containing plasmid, pIAA2, of *Pseudomonas syringae* pv. *savastanoi, Plasmid* **31:**21–30.

Summers, A. O., and Lewis, E., 1973, Volatilization of mercuric chloride by mercury-resistant plasmid-bearing strains of *Escherichia coli, Staphylococcus aureus,* and *Pseudomonas aeruginosa, J. Bacteriol.* **113:**1070–1072.

Sundin, G. W., and Bender, C. L., 1993, Ecological and genetic analysis of copper and streptomycin resistance in *Pseudomonas syringae* pv. *syringae, Appl. Envioron. Microbiol.* **59:**1018–1024.

Suzuki, M., Hayakawa, T., Shaw, J. P., Rekik, M., and Harayama, S., 1991, Primary structure of xylene monooxygenase: Similarities to and differences fromthe alkane hydroxylation system, *J. Bacteriol.* **173**:1690–1695.

Tait, R. C., Rempel, H., Rodriguez, R. L., and Kado, C. I., 1985, The aminoglycoside-resistance operon of the plasmid pSa: Nucleotide sequence of the streptomycin-spectinomycin resistance gene, *Gene* **36**:97–104.

Tait, R. C., Close, T. J., Rodriguez, R. L., and Kado, C. I., 1982, Isolation of the origin of replication of the IncW-group plasmid pSa, *Gene* **20**:39–49.

Tan, H. M., and Fong, K. P., 1993, Molecular analysis of the plasmid-borne *bed* gene cluster from *Pseudomonas putida* ML2 and cloning of the *cis*-benzene dihydrodiol dehydrogenase gene, *Can. J. Microbiol.* **39**:357–362.

Tan, H. M., Tang, H. Y., Joannou, C. L., Abdel–Wahab, N. H., and Mason, J. R., 1993, The *Pseudomonas putida* ML2 plasmid-encoded genes for benzene dioxygenase are unusual in codon usage and low in G+C content, *Gene* **130**:33–39.

Tan, H. M., Joannou, C. L., Cooper, C. E., Butler, C. S., Cammack, R., and Mason, J. R., 1994, The effect of ferredoxin (BED) overexpression on benzene dioxygenase activity in *Pseudomonas putida* ML2. *J. Bacteriol.* **176**:2507–2512.

Timmis, K. N., 1981, Gene manipulation *in vitro*, in: *Genetics as a Tool in Microbiology* (S. W. Glover and D. A. Hopwood, eds.) Cambridge University Press, Cambridge, England, pp. 49–109.

Timmis, K. N., Lehrbach, P. R., Harqayama, S., Don, R. H., Mermod, N., Bas, S., Leppik, R., Weightman, A. J., Reineke, W., and Knackmuss, H.–J., 1985, Analysis and manipulations of plasmid-encoded pathways for the catabolism of aromatic compounds by soil bacteria, in: *Plasmids in Bacteria* (D. Helinski, S. N. Cohen, D. Clewell, D. Jackson, and A. Hollaender, eds.) Plenum, New York, pp. 719–739.

Tsuda, M., and Iino, T., 1988, Identification and characterization of Tn4653, a transposon covering the toluene transposon Tn4651 on TOL plasmid pWW0, *Mol. Gen. Genet.* **213**:72–77.

Unger, B. P., Gunsalus, I. C., and Sligar, S. G., 1986, Nucleotide sequence of the *Pseudomonas putida* cytochrome P-450cam gene and its expression in *Escherichia coli*, *J. Biol. Chem.* **261**:1158–1163.

Valentine, C. R. I., and Kado, C. I., 1989, Molecular genetics of the IncW plasmids, in: *Promiscuous Plasmids of Gram-Negative Bacteria* (C. M. Thomas, ed.) Academic Press, London, pp. 125–163.

Valentine, C. R., Heinrich, M. J., Chissoe, S. L., and Roe, B. A., 1994, DNA sequence of direct repeats of the *sulI* gene of plasmid pSa, *Plasmid* **32**:222–227.

van Beilen, J. B., Eggink, G., Enequist, H., Box, R., and Witholt, B., 1992, DNA sequence determination and functional characterization of the OCT-plasmid-encoded *alkJKL* genes of *Pseudomonas oleovorans*, *Mol. Microbiol.* **6**:3121–3136.

Watanabe, T., Furuse, C., and Sakaizumi, S., 1968, Transduction of various R-factors by phage P_1 in *Escherichia coli* and by phage P_{22} in *Salmonella typhimurium*, *J. Bacteriol.* **96**:1791–1796.

Watanabe, M., Iyobe, M., Inoue, M., and Mitsuhashi, S., 1991, Transferable imipenem resistance in *Pseudomonas aeruginosa*, *Antimicrob. Agents Chemother.* **35**:147–151.

Williams, P. A., and Murray, K., 1974, Metabolism of benzoate and the methylbenzoates by *Pseudomonas putida* (*arvilla*) mt-2: Evidence for the existence of a TOL plasmid, *J. Bacteriol.* **120**:416–423.

Worsey, M. J., and Williams, P. A., 1975, Metabolism of toluene and xylenes by *Pseudomonas putida* (*arvilla*) mt-2: Evidence for a new function of the TOL plasmid, *J. Bacteriol.* **124**:7–13.

Yano, K., and Nishi, T., 1980, pKJ1, a naturally occurring conjugative plasmid coding for toluene degradation and resistance to streptomycin and sulfonamides, *J. Bacteriol.* **143:**552–560.

Yen, K. M., 1991, Construction of cloning cartridges for development of expression vectors in gram-negative bacteria, *J. Bacteriol.* **173:**5328–5335.

Yen, K.–M., and Gunsalus, I. C., 1982, Plasmid gene organization: Naphthalene/salicylate oxidation. *Proc. Natl. Acad. Sci. USA* **79:**874–878.

Yen, K.–M., Sullivan, M., and Gunsalus, I. C., 1983, Electron microscope heteroduplex mapping of naphthalene oxidation gene on the NAH7 and SAL1 plasmids, *Plasmid* **9:**105–111.

Young, S. A., Park, S. K., Rodgers, C., Mitchell, R. E., and Bender, C. L., 1992, Physical and functional characterization of the gene cluster encoding the polyketide phytotoxin coronatine in *Pseudomonas syringae* pv. glycinea, *J. Bacteriol.* **174:**1837–1843.

Carbohydrate Catabolism in *Pseudomonas aeruginosa*

2

LOUISE M. TEMPLE, ANDREW E. SAGE,
HERBERT P. SCHWEIZER, and
PAUL V. PHIBBS, JR.

1. INTRODUCTION

The goal of this review is to update the reader on recent data elucidating the physiology and genetics of glycolytic pathways in *P. aeruginosa*, the most thoroughly investigated member of the pseudomonads. Glycolytic pathways in this organism have several unique features. Lacking phosphofructokinase, *P. aeruginosa* metabolizes three- and six-carbon sugars via a central cycle which includes the Entner–Doudoroff pathway (EDP) enzymes, rather than utilizing the fermentation pathway of Embden–Meyerhoff–Parnas (EMP) (Entner and Doudoroff, 1952; Kersters and DeLey, 1968). Another unique physiological feature is that a product of the EDP, glyceraldehyde 3-phosphate, is largely recycled through the central cycle, rather than continuing to pyuvate via the lower EMP pathway (Banerjee, 1989; Phibbs, 1988). Thus, the latter enzymes in *P. aeruginosa* seem to serve gluconeogenic rather than the more usual catabolic functions in other organisms. Whereas the metabolism of glucose is preferred by *Escherichia coli*, *P. aeruginosa* utilizes succinate and other tricarboxylic acid cycle intermediates before glucose (Anderson and Wood, 1969; Belvins *et al.*, 1975; Hylemon and Phibbs, 1972; Midgley

LOUISE M. TEMPLE • Department of Biology, Drew University, Madison, New Jersey 07940.   ANDREW E. SAGE • WorldWide Microbiology Group, Millipore Corp., Bedford, Massachusetts 01730.   HERBERT P. SCHWEIZER • Department of Microbiology, Colorado State University College of Veterinary Medicine, Fort Collins, Colorado 80523.   PAUL V. PHIBBS, JR. • Department of Microbiology and Immunology, East Carolina University School of Medicine, Greenville, North Carolina 27858.

Pseudomonas, edited by Montie. Plenum Press, New York, 1998.

and Dawes, 1973; and Tiwari and Campbell, 1969). In addition, this organism lacks an oxidative hexose monophosphate pathway (Phibbs, 1988). Therefore, the regulation of carbon flow in this organism is interesting if for no other reason than it is different from the well-studied *Enterobacteriaceae* and eukaryotic cells. In a more applied way, understanding the regulation of these pathways has important medical implications because the resulting carbon pools are precursors for alginate, the viscous exopolysaccharide known to be a virulence factor in lung infections of cystic fibrosis patients (Banerjee *et al.*, 1983, 1985; and Govan *et al.*, 1988).

The utilization of glucose, gluconate, mannitol, fructose, and glycerol are considered in this chapter. Each of these compounds is transported and metabolized by peripheral pathways which feed into a central cycle including the EDP enzymes (Fig. 1). Mutants lacking enzymes in the peripheral and central metabolic pathways have enabled the elucidation of the physiological role played by each pathway. The mutant strains have made possible the chormosomal mapping of approximately 30 of the known structural and regulatory genes for the transport and metabolism of the C_3 and C_6 carbon compounds listed above (Table I). In addition, complementation of the enzyme deficiencies in these mutants has facilitated the cloning of chromosomal fragments containing structural and regulatory genes specifying these metabolic processes. Chimeric plasmids from these clonings have revealed previously unmapped genes, which have been subsequently mutated using homologous recombination with an interrupted or incomplete copy of the gene on a suicide plasmid. Nucleotide sequence analyses and manipulations of the cloned chromosomal fragments, including mRNA analyses, have revealed novel genetic regulatory mechanisms. Much of the recent work has depended on new molecular genetic techniques and a number of biological reagents developed specifically for *Pseudomonas* (detailed in other chapters of this volume). The data resulting from these studies have enabled us to begin unraveling the complex regulation of the enzymes in these pathways.

In preparing this chapter, we have built on the comprehensive review of Lessie and Phibbs (1984) and the minireview of Phibbs (1988) for analysis of work in earlier years. In addition, the article summarizing phenotypes of mutants in glucose metabolism (Fraenkel, 1986). has been helpful. The work in this review represents efforts primarily from the laboratories of Dr. Paul Phibbs (East Carolina University School of Medicine), Dr. Herbert Schweizer (Colorado State University), and Dr. Elizabeth Worobec (University of Manitoba).

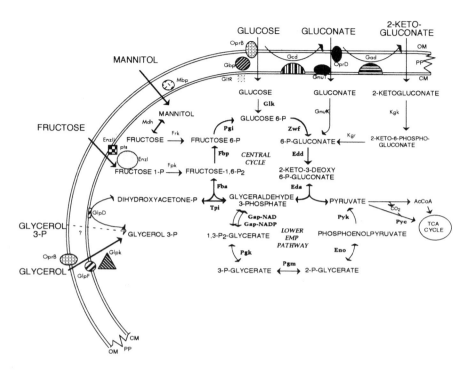

Figure 1. Pathways of carbohydrate metabolism in *Pseudomonas aeruginosa*. Abbreviations: OM, outer membrane; CM, cytoplasmic membrane; PP, periplasm; Gad, gluconate dehydrogenase; Gcd, glucose dehydrogenase; Gbp, glucose-binding protein; Mbp, mannitol-binding protein; Kgk, 2-ketogluconate kinase; Gnuk, gluconate kinase; GnuT, gluconate permease; Kgr, 2-keto-6-phosophogluconate reductase; Glk, glucokinase; Zwf, glucose 6-phosphate dehydrogenase; Edd, 6-phosphogluconate (Entner–Duodoroff) dehydratase; Eda, 2-keto-3-deoxy-6-phosphogluconate aldolase; Pyc, pyruvate carboxylase; Pyk, pyruvate kinase; Eno, enolase; Pgm, phosphoglucomutase; Pgk, 3-phosphoglycerate kinase; Gap, glyceraldehyde 3-phosphate dehydrogenase; Tpi, triose phosphate isomerase; Pgm, phosphoglucoisomerase; Fba, fructose 1,-6-bisphosphate aldolase; Fbp, fructose 1,6-bisphosphatase; Pgi, phosphoglucoisomerase; Frk, fructokinase; Mdh, mannitol dehydrogenase; Fpk, fructose 1-phosphate kinase; GlpD, glycerol 3-phosphate dehydrogenase; GlpK, glycerol kinase.

2. GLUCOSE UTILIZATION VIA PHOSPHORYLATION

Extracellular glucose can be brought directly into the cell by an active transport system and phosphorylated intracellularly by gluco-

Table I. Genetic Loci and Proteins Specifying Carbohydrate Catabolism of *P. aeruginosa*[a]

Gene	Protein	Molecular weight	Mutant phenotype	Protein function	Map location
			Glucose utilization		
oprB	OprB porin	48,000[b,c]	Unknown	Glucose, glycerol channel	ND[d]
gbp	Glucose-binding protein	44,000[b]	Unknown	Periplasmic glucose binding	ND
gltR	Response regulator	27,000[a]	Glucose negative	Two-component response	39·
				Regulator	
glk	Glucokinase	36,500[c]	Glucose negative	Enzymatic[e]	39
			Gluconate and 2-ketogluconate utilization		
kgu-11*			2KG negative		48
gnuK	Gluconokinase	18,000[c]	Gluconate negative	Enzymatic	48
gnuT	Gluconate permease	48,000[c]	No gluconate uptake	Transport	48
gcd	Glucose dehydrogenase	—	Unknown	Enzymatic	ND
gad	Gluconate dehydrogenase	138,000[b]	Unknown	Enzymatic	ND
gnuR	Gluconate repressor	—	Gluconate operon constitutive	Regulatory	48
oprD	Porin D	45,000[b,c]	Unknown	Gluconate channel	60
			Fructose utilization		
ptsI*	Enzyme I	72,000[b]	No fructose uptake	Enzymatic	34
ptsII*	Enzyme II	ND	No fructose uptake	Transport of fructose?	34
ptsRI		ND	PTS system not induced		34
			Mannitol utilization		
mbp	Mannose-binding protein	37,000[b]	Unknown	Periplasmic mannose binding	

Gene	Protein	MW	Function	Phenotype	Ref.
mtuD	Mannitol dehydrogenase		Enzymatic	Mdh negative	48
mtuR				Pathway not induced	48
mtu9002				Mdh and Fk negative	48
mtu-19				Mdh negative; No mannitol transport	48
Glycerol and glycerol, 3-phosphate utilization					
glpD	G3P dehydrogenase	56,000[b,c]	Enzymatic	Glycerol and G3P Negative	38–41,30*f*
glpM		12,000[c]		Unknown	30
glpF	Glycerol diffusion facilitator	29,000[c]	Glycerol facilitator	No glycerol transport	30
glpK	Glycerol kinase	120,000[b]	Enzymatic	Glycerol negative	30
glpR	Glycerol repressor	28,000[c]	Glp repressor	*glp* constitutive	30
glpR2		25,000[c]			50–55
oprB	Porin B	48,000	Porin protein	Unknown	
Central pathway					
fba	Fructose bisphosphate aldolase		Enzymatic	Glucose and mannitol negative	ND, linked to *pgk*
fbp	Fructose bisphosphatase		Enzymatic	Slow growth on glucose	ND
pgi	Phosphoglucoisomerase	190,000[b]	Enzymatic	Mannitol negative	1
zwf	Glucose, 6-phosphate dehydrogenase	54,000[c]	Enzymatic	Mannitol negative; No aerobic glucose utilization	39

(continued)

Table I. (*Continued*)

Gene	Protein	Molecular weight	Mutant phenotype	Protein function	Map location
edd	6-Phosphogluconate Dehydratase	65,000[c]	Glucose and mannitol negative	Enzymatic	39
hexR	*hex* regulon repressor	120,000[b]	*hex* regulon constitutive	Regulatory	39
eda	KDPG aldolase	24,000[c]	Glucose and mannitol negative	Enzymatic	39
			Lower EMP		
*gap-*NADP	G3P dehydrogenase	315,000[b]	Unknown	Enzymatic in gluconeogenesis	ND
*gap-*NAD	G3P dehydrogenase	140,000[b] 36,000[c]	Unknown	Enzymatic Regulatory?	39
pgk	Phosphoglycerokinase		Glutamate, succinate, lactate negative	Enzymatic	ND, linked to *fba*
eno	Enolase			Enzymatic	
pyc	Pyruvate carboxylase		Lactate, pyruvate, glycerol negative, 6-C compound negative	Enzymatic	35

[a] For references, see text.
[b] M_r determined from purified protein.
[c] Molecular weight predicted from DNA coding sequence.
[d] ND, not determined.
[e] For reaction catalyzed, see text.
[f] Two locations reported.

kinase to produce the central metabolite 6-phosphogluconate. This route of metabolism is obligatory during anaerobic catabolism (Hunt and Phibbs, 1981, 1983) and predominates when glucose concentrations are low (Whiting *et al.*, 1976a). Metabolism of glucose by this route is subject to catabolite repression control.

2.1. Transport

Extracellular glucose induces a specific glucose transport system, which is regulated independently of subsequent enzymes in the dissimilatory pathway (Hylemon and Phibbs, 1972). The glucose transport system includes at least three components which have been identified: an outer membrane protein (OprB), a periplasmic binding protein (GBP), and a putative environmental response regulatory protein (GltR). As yet unidentified, a permease must serve as a membrane carrier protein to facilitate the passage of the molecule across the inner membrane.

2.1.1. Porin B

OprB was identified by sequence homology and antigenic cross-reactivity with the same protein from *P. putida* (Hancock and Carey, 1980; Saravolac *et al.*, 1991; Trias *et al.*, 1988; and Wylie and Worobec, 1993). OprB has been purified and the gene cloned and sequenced (predicted M_r = 47,597, Wylie and Worobec, 1994, 1995). Studies showed that OprB forms large diameter, anion-selective pores with a specific glucose-binding site and has physical properties consistent with other known porins. Thus, OprB apparently binds and channels glucose through the outer membrane and into the periplasm (Wylie and Worobec, 1994, 1995). However, studies showed that OprB, rather than acting strictly as a glucose-selective porin, facilitates the diffusion of a wide range of carbohydrates and is more appropriately described as a carbohydrate-selective porin or channel (see below, in glycerol uptake).

2.1.2. Glucose-Binding Protein

The second component is a glucose-specific, periplasmic-binding protein (GBP) of M_r = 44,000 (Stinson *et al.* 1977). GBP has been purified and characterized and has been shown to bind glucose specifically at a K_m of 8μM, but its role in glucose uptake remains unclear. It has been postulated that GBP acts to transfer the glucose directly across the periplasm from OprB to the transport machinery of the cytoplasmic membrane. At low glucose concentrations, approaching its K_m, GBP activity appears to be the driving force behind glucose transport into the cytoplasm (Wylie and Worobec, 1993).

2.1.3. Glucose Transport Regulator

Genetic analysis of glucose transport was first carried out by Cuskey and Phibbs (1985) who constructed mutants defective in glucose transport by selecting glucose-negative derivatives of a strain lacking the oxidative pathway of glucose utilization. These mutant strains (PFB360 and 362) showed no detectable glucose uptake when grown in the presence of glucose. The glucose uptake locus (originally *glcT*, then *gltB*, since redesignated *gltR*) (Sage *et al.*, 1996) was shown in prototrophic reduction tests to be tightly linked with the *zwf* and *edd* loci, located at minutes 38–40 on the revised *P. aeruginosa* chromosome map. These results were confirmed with transductional analyses using phage F116L and G101.

Cuskey *et al.* (1985) used strain PFB362 to select a chimeric plasmid containing a 6.0-kb chromosomal DNA fragment which restored the ability of this strain to grow on glucose (Fig. 2). Subsequently, *gltR* was subcloned onto a 1.1-kb *Sal*I fragment which restored both glucose binding and uptake to *gltR* mutants PFB360 and PFB362. In addition, osmotic shock fluids or purified GBP from the parental strain PAO1 restored glucose binding and uptake and chemotactic responsiveness to glucose, to the mutant bacteria (Sly *et al.*, 1993). These observations seemed to indicate that the gene for *P. aeruginosa* glucose-binding protein is located on the 1.1-kb *Sal*I fragment, but the authors also stated that the data do not rule out the presence of a regulatory gene controlling expression of the glucose transport machinery (Sly *et al.*, 1993).

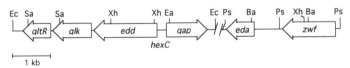

Figure 2. Map and genomic organization of the region containing the glucose phosphorylative pathway enzyme genes, central pathway enzyme genes, and regulatory loci in *P. aeruginosa*. Hatch marks represent a distance of less than 11 kb on the PAO1 chromosome. Only selected restriction enzyme cleavage sites are marked: Ba, *Bam*HI; Ea, *Eag*I; Ec, *Eco*RI; Ps, *Pst*I; Sa, *Sal*I; Xh, *Xho*I. Other abbreviations: *gltR*, glucose transport regulatory protein; *glk*, glucokinase; *edd*, 6-phosphogluconate (Entner–Doudoroff) dehydratase; *gap*, glyceraldehyde 3-phosphate dehydrogenase; *eda*, 2-keto-3-deoxy-6-phosphogluconate aldolase; *zwf*, glucose 6-phosphate dehydrogenase; *hexC*, regulatory locus. See text for references.

Expecting to find the gene for GBP or another transport function, Sage and Phibbs performed sequence analysis of the region. These data revealed an ORF of 729 bp specifying a polypeptide of 242 amino acid residues (M_r = 27,381 daltons), much too small to encode GBP (Sage *et al.* 1993, 1996). A second unexpected observation was that the ORF is not fully contained on the 1.1-kb fragment, which nonetheless can complement the defect of the mutants. Even though the ORF extends 120 bases beyond the cloned fragment, apparently the truncated gene product is still functional. Comparison of the predicted peptide sequences with Genbank did not show homologies with known sugar-binding proteins. Also, the predicted peptide failed to exhibit any significant homology with the N-terminal sequences of purified GBP of *P. aeruginosa* or the predicted peptide sequence of the recently characterized glucose-binding protein gene of *P. putida* (Sage *et al.*, 1996). Instead, the predicted protein exhibited 33 to 46% identity to the two-component, environmentally responsive regulator proteins, with greatest similarity to the OmpR subfamily: OmpR of *E. coli* and *Salmonella typhimurium*, PhoB of *P. aeruginosa*, VirG of *Agrobacter tumefaciens*, and VanR of *Enterococcus faecium* (Sage *et al.*, 1996). Of special interest in the context of this article, AgmR of *P. aeruginosa* (Schweizer *et al.*, 1995, see below), another two-component response regulator involved in regulating glycerol metabolism in *P. aeruginosa*, exhibits some sequence identity with GltR (21.6% identity, Sage *et al.*, 1996). N-termini of response regulators are the effector domains wherein phosphorylation by the sensor histidine kinase occurs, whereas DNA recognition and binding functions are carried by the C-termini (Parkinson and Kofold, 1992). Residues Asp-13, Asp-56, and Lys-108, highly conserved in the effector domains across a number of response regulators, were found in GltR. These residues function in the phosphorylation of the effector protein. No putative ATP-binding or helix-turn-helix DNA binding motifs were identified in the GltR sequence (Sage *et al.*, 1996, Parkinson, 1993; Parkinson and Kofold, 1992).

A strain lacking an intact copy of the *gltR* open reading frame was constructed by recombining an internal fragment of the ORF carried on a nonreplicating suicide vector with the wild-type gene in a recipient strain blocked in the oxidative pathway. The resulting antibiotic-resistant transconjugant fails to utilize glucose as a sole carbon source and cannot take up glucose (Sage *et al.*, 1996). However, fructose, mannitol, or glycerol still support growth, indicating that this mutation is a specific defect in glucose metabolism. The defined mutant has all of the phenotypic traits of the chemically induced glucose negative strains PFB360 and PFB362. Revertants, which had lost the recombined plasmid, recovered the capacity to take up and utilize glucose. These results indicated that *gltR* encodes a two-component response regulator required for glucose uptake (Sage *et al.*, 1996).

Sensor histidine kinase genes are often juxtaposed with the associated response regulator gene and may also be cotranscribed. Therefore, analysis of 250 bp downstream of *gltR* was carried out in search of an ORF encoding a putative histidine kinase. Examination of the six possible reading frames failed to reveal any ORFs with homology to sensor histidine kinases. This does not rule out the possibility that the coding region for the expected kinase occurs downstream. The sensor regions of these molecules which occur at the N-termini, have little similarity to each other, apparently providing the specificity for each system. Thus more extensive sequencing may be necessary to reveal a gene identifiable by sequence homologies. Alternatively, the kinase gene may occur elsewhere in the genome. Examples of nonjuxtaposed genes for the two-component systems have been documented, and such may be the case with *gltR*. Significantly, the *agmR* gene in glycerol metabolism of *P. aeruginosa* (see later in this article) does not appear to be in an operon with the gene for its sensory component (Schweizer *et al.*, 1995).

2.2. Phosphorylation

Glucokinase catalyzes the intracellular, ATP-dependent phosphorylation of glucose to glucose 6-phosphate. Mutants lacking glucokinase (Glk) activity cannot utilize glucose as a sole carbon source when the alternative oxidative route is blocked. Glk activity is expressed coordinately with at least four other glycolytic enzymes: Zwf, Edd, Eda, and Gap-NAD (Fig. 1). Activity is elicited in response to glucose, gluconate, mannitol, fructose, and glycerol and is repressed by TCA cycle intermediates, such as succinate.

The chimeric plasmid which complements the glucose transport mutation (see above) also restores a glucose-positive phenotype and Glk activity to the Glk-deficient strain PRP444 (Cuskey *et al.*, 1985). The *glk* locus was subsequently localized to a chromosomal fragment which contained a 996 bp ORF (Fig. 2) shown to complement the defect in PRP444 (Sage, unpublished data). The ORF predicted a polypeptide of 331 amino acids, exhibiting 32% identity with glucokinase from *Zymomonas mobilis*, and contains a potential ATP-binding site (Sage *et al.*, 1993). Upstream of the beginning of the ORF were several potential promoter sequences similar to known *Pseudomonas* sigma consensus sequences.

However, the significance of these sequences, if any, awaits further analysis. Preliminary evidence indicates that this gene may be expressed from its own promoter and also may be cotranscribed with *edd*, the gene for which lies immediately upstream of *glk* (Sage and Phibbs, unpublished).

3. OXIDATIVE PATHWAY OF GLUCOSE UTILIZATION

Extracellular glucose may be metabolized by one of two oxidative routes and enter the central cycle as 6-phosphogluconate (Fig. 1). These pathways are predominant during oxidative growth; in fact, gluconate is reported to cause repression of the glucose transport system leading to the phosphorylative route of metabolism (Whiting *et al.*, 1976a,b). The first step leading to either oxidative pathway involves the oxidation of extracellular glucose to gluconate by an NAD(P)-independent dehydrogenase (Gcd) which is membrane-bound (Matsushita, 1980). Loss of glucose dehydrogenase activity under anaerobic conditions may result from lack of expression of its cofactor, pyrroloquinoline quinone (Duine *et al.*, 1979; Goosen *et al.*, 1987). Transport of gluconate has been associated with an outer membrane porin protein, OprD, $M_r = 45,000$ (Huang *et al.*, 1992; Huang and Hancock, 1993), the gene for which has been mapped at minute 60 on the *P. aeruginosa* chromosome map (Liao *et al.*, 1996). Gluconate can be subsequently transported directly into the cell via an active transport system involving gluconate permease (GnuT), then phosphorylated by ATP-dependent gluconokinase, as demonstrated in membrane vesicles (Stinnett and Eagon, 1973). An alternative fate of gluconate is further oxidation, by a second membrane-bound dehydrogenase, to 2-ketogluconate (2KG). Gluconate dehydrogenase (Gad) has a monomeric molecular weight of 138,000, which includes a cytochrome c1 and a covalently bound flavin (Matsushita, 1979). 2KG is then internalized by active transport and converted in two steps (catalyzed by ATP-dependent 2KG kinase and NAD(P)H-dependent 2KGP reductase) to the central metabolite, 6-phosphogluconate.

Because of the close proximity of the gluconate and mannitol markers, a cosmid with cloned chromosomal DNA containing mannitol utilization genes (see below) was tested and found to complement the gluconate pathway defects in PRP658 and PRP660 (Wallace, 1989). Genes for both gluconokinase (*gnuK*) and the gluconate permease (*gnuT*) were cloned together on a 3 kb *Sal*I DNA fragment (Hager *et al.*, 1997). Both activities were still inducible and could also be expressed individually on smaller fragments, suggesting that the genes are separately expressed. Two open reading frames, separated by 96 bp, encode proteins of 18 and 48 kd, corresponding to *gnuK* and *gnuT*, respectively. The 3 kb fragment also contained a gluconate regulatory protein *gnuR*, divergently expressed from *gnuK*. Deduced amino acid sequences of the *P. aeruginosa* gluconate regulatory protein, gluconokinase, and gluconate permease showed 37–40% identity with homologs from *Escherichia coli* and *Bacillus* (Hager *et al.*, 1997).

In conjugation experiments utilizing the mobilizing plasmid R68.45, these markers were mapped at minutes 46–48 on the *P. aeruginosa* chromosome, relative to other known markers, in the following order: *catA1 ben-4 mtu-9002 gnuK1 kgu-11 tyu-9030*. Transductional analysis using phage F116L demonstrated cotransduction of the *gnuK1* locus at a frequency of 80% with *tynB1* and showed the close relationship between two other mannitol utilization markers, *mtu-4* and *mtu-2* and both *gnuK1* and *gnuT1* (Wallace, 1989). The *kgu-11* marker was not cotransducible with either *gnuK1* or other markers (Wallace and Phibbs, 1988). No cloned chromosomal fragment, which complements the *kgu-11* marker (2-KG utilization), has been identified to date.

Because of the close proximity of the gluconate and mannitol markers, a cosmid with cloned chromosomal DNA containing mannitol utilization genes (see below) was tested and it was found to complement the gluconate pathway defects in PRP658 and PRP660 (Wallace, 1989). A 3.1-kb *Sal*I subclone restored the ability to grow on gluconate as a sole carbon source to both strains. Although gluconate transport was restored to wild-type induced levels, the activity of gluconokinase in the partial diploid was only 25% of wild-type, suggesting that sequences needed for regulation were missing from the chimeric plasmid. Alternatively, a truncated version of the kinase, resulting from deletion of a portion of the gene during the cloning process, may have been only partially active. Subsequent subcloning of the 3.1-kb plasmid localized the transport function to a 1.8-kb *Sal*I-*Eco*RI fragment and the kinase function (still at reduced levels) to a 1.3-kb *Sal*I-*Eco*RI fragment (Wallace, 1989).

4. FRUCTOSE UTILIZATION

Fructose is brought into the cell via a PEP-dependent phosphotransferase system producing intracellular fructose 1-phosphate (Baumann and Baumann, 1975; Durham and Phebbs, 1982; Roehl and Phebbs,

1982; Sawyer *et al.*, 1977; Van Dijken and Quayle, 1977). This vectorial transfer is unique among carbohydrate uptake mechanisms in this organism. Other carbon sources are accumulated by active transport. The phosphotransferase system appears to be comprised of only two components, a soluble enzyme of 72,000 M_r (EnzI) and a membrane-associated enzyme II. The Hpr-like low molecular weight phosphate-carrier protein found in other phosphotransferase systems (PTS) is lacking in *P. aeruginosa*. Intracellular fructose 1-phosphate is subsequently converted to fructose 1,6-bisphosphate by 1-phosphofructokinase, which is co-induced with both components of the fructose 1-phosphotransferase system during growth on fructose.

Fructose 1,6-bisphosphate is metabolized primarily via the central cycle to pyruvate and glyceraldehyde 3-phosphate (Fig. 1). A secondary route of catabolism produces triose phosphate via the lower EMP reactions. The relative importance of fructose dissimilation via the central cycle or EMP was discerned from glucose positive revertants of double mutants lacking pyruvate kinase or enolase activity (Phibbs, 1988). The original Zwf-mutants grew at approximately 20% of the wild-type rate on fructose, whereas the double mutants (Zwf-, Pyk-, or Zwf-, Eno-) did not grow on fructose. Zwf+ revertants exhibited normal growth rates on glucose. Furthermore, strains lacking Pgi, Edd, or Eda also grew very slowly on fructose, confirming the minor contribution of the EMP pathway in metabolizing fructose.

Three classes of mutants in the PTS system were obtained. *pts*I and *pts*II strains lost the complete activity of a single component and retained the low activity of the other, and *pts*R strains lost the activity of both components. All *pts* mutants expressed 1-phosphofructokinase at greatly reduced levels. The patterns of expression indicate that fructose 1-phosphate may serve as the physiological inducer of both the PTS system and 1-phosphofructokinase. The *pts* mutations are clustered at minute 34 on the *P. aeruginosa* chromosome (Roehl and Phibbs, 1982). To our knowledge, no fructose utilization genes have been cloned.

5. MANNITOL UTILIZATION

Exogenous mannitol is converted to intracellular fructose 6-phosphate via a mannitol active transport system, mannitol dehydrogenase, and fructokinase, which are all induced specifically in the presence of mannitol (Phibbs and Eagon, 1970). *P. aeruginosa* is naturally deficient in the glycolytic 6-phosphoglycerokinase (Tiwari and Campbell, 1969).

Fructose 6-phosphate is obligatorily metabolized to glyceraldehyde 3-phosphate and pyruvate via the central cycle. Mutants lacking Pgi, Zwf, Edd, or Eda cannot grow on mannitol (Phibbs *et al.*, 1978). The active transport system is mediated by a periplasmic-binding protein (M_r = 37,000, pI = 8.3) specific for mannitol (Eisenberg and Phibbs, 1982). Nearly 100 independently isolated mutants defective in the peripheral mannitol utilization pathway have been isolated (*mtu* for mannitol utilization). These strains have lost mannitol dehydrogenase activity alone or have lost Mdh together with loss of the transport and/or the kinase. To date, no strains have been isolated which solely affect either mannitol transport or fructokinase (Roehl and Phibbs, 1981). Quantitative two-factor transductions (prototrophic reduction tests) show that the *mtu* mutations are tightly linked on the genome at minute 48 (Roehl and Phibbs, 1981). The natural deficiency of *P. putida* for the entire mannitol peripheral pathway provides a means of identifying an R'-plasmid (vector R68.45) containing inducible wild-type *P. aeruginosa* genes for the peripheral pathway proteins (Phibbs *et al.*, 1987). In addition, a 35-kb chromosomal fragment which complements the *mtu* mutations in *P. aeruginosa* was cloned into a small cosmid vector (Dail and Phibbs, unpublished).

6. GLYCEROL AND GLYCEROL 3-PHOSPHATE UTILIZATION

The role of *P. aeruginosa* mucoid strains in pulmonary infections of cystic fibrosis patients has sparked interest in the uptake and metabolism of glycerol and indeed in carbohydrate metabolism generally in this organism. As an opportunistic pathogen, this organism plays an important role in the mortality of the disease, primarily because it can infect the lungs and develop an atypical mucoid form which produces copious amounts of alginate, a viscous exopolysaccharide and an important virulence factor (May *et al.*, 1991). Phosphatidylcholine, a major component of lung surfactant, is degraded by *P. aeruginosa* phospholipase C (Terry *et al.*, 1992) and extracellular lipases or esterases to glycerol and fatty acids. Phosphatidylcholine is used by *P. aeruginosa* as a sole source of carbon, nitrogen, and phosphorus for growth (Terry *et al.*, 1991, 1992), and growth on this nutrient leads to the emergence of mucoid subpopulations and elevated levels of alginate. Furthermore, glycerol is a particularly good carbon source for alginate synthesis by bronchial and cystic fibrosis isolates of *P. aeruginosa* and readily promotes the appearance of mucoid variants of nonmucoid bacteria when supplied at high concentrations (Terry *et al.*, 1991). Thus, elucidation of the regulated pathways of glycerol metabolism in *P. aeruginosa* is vitally important to understanding disease causation by this organism in cystic fibrosis.

The primary route of metabolism of glycerol in *P. aeruginosa* is via six carbon intermediates in the central cycle (see below). In the peripheral pathway, glycerol enters the cell via the glycerol diffusion facilitator (encoded by *glpF*). It is immediately phosphorylated by glycerol kinase (*glpK*) and trapped intracellularly as *sn*-glycerol-3-phosphate (glycerol-P) which is subsequently oxidized to dihydroxyacetone phosphate (DHAP) by membrane-associated glycerol-P dehydrogenase (*glpD*). Entry into the central cycle is by the action of fructose 1,6-bisphosphate aldolase to form fructose 1,6-bisphosphate from DHAP and its isomer, glyceraldehyde 3-phosphate. Alternatively, 20% or less of the glycerol carbons are shunted to pyruvate through the action of five EMP enzymes. This secondary pathway permits slow growth on glycerol of mutants lacking central cycle enzymes, such as Zwf, Edd, or Eda (the identical situation exists with fructose utilization; see above). Although Cuskey and Phibbs (1985) defined a putative positive regulatory locus (*glpR2*), recent genetic studies indicate that the glycerol catabolism genes are members of a *glp* regulon that is negatively regulated by the Glp repressor, the product of *glpR* (Schweizer and Po, 1996).

6.1. Glycerol and Glycerol-P Uptake

Until very recently, lack of specific uptake mutants and non-metabolizable substrate analogs has hindered progress in understanding the mechanism of glycerol and glycerol-P uptake. Three studies of glycerol and glycerol-P transport have been reported using wild-type *P. aeruginosa* strains (Siegel and Phibbs, 1979; Tsay *et al.*, 1971; Williams *et al.*, 1994), but it was not until very recently that detailed genetic analyses and gene cloning helped to unravel the nature of glycerol and glycerol-P uptake and to solve some of the apparent contradictions reported previously (Schweizer, 1991; Schweizer and Po, 1994; Schweizer *et al.*, 1997).

6.2. Glycerol Transport

Using log phase *P. aeruginosa* PAO1 cells, Siegel and Phibbs (1979) reported saturation kinetics for the uptake of radiolabeled glycerol with an apparent K_m of 13 μM. The uptake system was specific for glycerol and was not inhibited by glycerol-P. Expression of glycerol uptake activity is subjected to strong catabolite repression control by succinate or malate. Certain metabolic poisons or enzymes inhibitors (arsenate, azide, cyanide, iodoacetate, and dinitrophenol, but not *N,N*-dicyclohexylcarbodiimide or chloromercuribenzoate) inhibit the initial rate of glycerol uptake. Energy dependence of the uptake system is also implicated, using the glucose dehydrogenase (Gcd) negative strain PAO311 and measuring glycerol uptake in the presence of various inducers of Gcd expression. In these experiments, substrates which induce Gcd synthesis (suggested by Eagon, 1971, as an energy transducer via glucose oxidation) also greatly stimulate glycerol uptake in PAO1 but not in PAO311. These results taken together suggest an energy requirement for glycerol uptake. However, the data do not rule out the possibility that intracellular phosphorylation is an energy source driving the uptake system.

Utilizing strain PAO1, Tsay *et al.* (1971) detected binding of ^{14}C glycerol by osmotic shock fluids. This finding and the apparent shock-sensitivity of glycerol transport suggest a periplasmic glycerol-binding protein. This notion was corroborated in a subsequent study by Siegel and Phibbs (1979) who showed that cellular transport of glycerol and binding of glycerol by shock fluid is inhibited by sodium azide and *N*-ethylmaleimide.

Using a continuous culture with glycerol limitation, Williams *et al.* (1994) report glycerol uptake with a K_m of 19 μM in *P. aeruginosa* strain NM48. However, neither change in the rate of glycerol uptake in the presence of *P*- trifluoromethoxyphenylhydraxone (a powerful energy uncoupling agent) nor the involvement of a periplasmic binding protein was observed. In contrast, the experiments presented in this study suggest a very rapid phase of glycerol uptake rather than binding and that this uptake is most likely caused by facilitated diffusion. Induction of glycerol uptake is paralleled by simultaneous production of glycerol kinase and a 48,000-M_r outer membrane protein, both of which are absent in noninduced cultures. Production of this outer membrane protein in cells grown under glucose limitation and N-terminal sequence information suggests its identity with the glucose porin OprB of *P. aeruginosa* (Wylie *et al.*, 1993). OprB has been characterized as a porin capable of facilitating the diffusion of several carbohydrates, including glycerol (Wylie and Worobec, 1995). Expression of OprB is apparently regulated separately from glycerol uptake, glycerol-P dehydrogenase, and glycerol kinase activities.

Cloning of the glycerol facilitator structural gene *glpF* and its close association with the glycerol kinase structural gene *glpK* (Fig. 3 and see below), showed that glycerol transport in *P. aeruginosa* is indeed mediated by a facilitated diffusion system which is structurally and functionally similar to those found in other bacteria (Schweizer *et al.*, 1997). Analysis of the deduced amino acid sequence of the glycerol facilitator indicated that it is a protein with extremely low polarity which explains

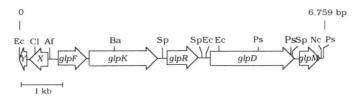

Figure 3. Map and genomic organization of the region containing the glycerol catabolic gene cluster of *P. aeruginosa*. Only selected restriction enzyme cleavage sites are marked: Af, *Afl*II; Ba, *Bam*HI; Cl, *Cla*I; Ec, *Eco*RI; Nc, *Nco*I; Ps, *Pst*I; Sp, *Sph*I. Other abbreviations: *glp*D (*sn*-glycerol-3-phosphate dehydrogenase); *glp*F and *glp*K (encoding membrane-associated glycerol diffusion facilitator and glycerol kinase, respectively); *glp*M (membrane protein affecting alginate synthesis); *glp*R (*glp* regulatory gene); X and Y, orfX and orfY, encoding a transcriptional regulator and carbohydrate kinase, respectively, or unknown functions. See text for references.

its aberrant mobility on SDS PAGE (observed molecular mass of 25 kDa versus a calculated mass of 28,897) by the binding of more SDS by the abundant hydrophobic amino acids, as previously noted for *E. coli* GlpF. Hydropathic analyses of the GlpF primary sequence and predictions of possible transmembranous segments further indicated that it probably is an integral membranous protein with six potential membrane-spanning helices (Weissenborn and Larson, 1992; Schweizer *et al.*, 1997). It has previously been pointed out that this design is unique among bacterial plasma membrane transporters, which usually contain 12 transmembranous helices (6 + 6 design) (Nikaido and Saier, 1992). Signature analysis confirmed the previous conclusions that GlpF belongs to the family of major intrinsic membranous proteins that normally produce large, open, aqueous channels presumably at the interface of four subunits (Nikaido and Saier, 1992). Although some properties of the glycerol facilitator protein are similar to those of channel-type transporters, its function must be somewhat different to avoid leakage of solutes from the cytoplasm. As is true with other plasma membrane transporters, the channel is likely to be gated so that it opens only when the proper ligand interacts with its binding site. Alternatively, gating could also be achieved by close physical interaction with glycerol kinase, which is required for efficient glycerol phosphorylation (Voegel *et al.*, 1993). Interaction with glycerol kinase also insures cytoplasmic trapping of glycerol in the form of glycerol-P, and phosphorylation also might serve as an energy source for driving the uptake system. A plasmid-borne *glpFK* deletion was constructed and inserted into the chromosome. The resulting mutant failed to grow on 10 mM glycerol and completely lacked glycerol transport activity but transported and grew normally on glycerol-P. In contrast, mutants solely deficient in glycerol uptake grow at high (millimolar) concentrations of glycerol (Schweizer and Po, 1994), indicating that at high concentrations glycerol passes freely through the cytoplasmic membrane and that the facilitator is required only for glycerol transport at low substrate concentrations.

6.3. Uptake of Glycerol-P

Earlier studies by Siegel and Phibbs (1979) indicate that in contrast to glycerol transport, uptake of glycerol-P is slower, is not saturable, and is dramatically inhibited by unlabeled glycerol. In addition, the glycerol-P transport system is subject to strong catabolite repression control, is inhibited by the same inhibitors as glycerol uptake, and is energy-dependent.

The relationship between the uptake of glycerol and glycerol-P has not been fully elucidated although recent genetic evidence indicates the presence of two separate, coinducible transport systems for glycerol and glycerol-P (Schweizer and Po, 1994; Schweizer *et al.*, 1997). A mutant lacking high affinity glycerol transport and glycerol-P dehydrogenase could not grow on either glycerol or glycerol-P (Schweizer and Po, 1994). However, millimolar concentrations of glycerol, which allow free diffusion of glycerol across the cell envelope independent of GlpF, induced the transport of glycerol-P, suggesting a separate, inducible, glycerol-P transport system. Isolation of a *glpFK* deletion strain further supports the existence of a separate glycerol-P transport system because this mutant transported and grew normally on glycerol-P (Schweizer *et al.*, 1996). Furthermore, completion of the nucleotide sequence of the region containing the glycerol catabolic genes *glpF*, *glpK* and *glpD* and their negative regulatory gene *glpR*, indicated that the gene(s) responsible for glycerol-P uptake are physically separated from the hitherto identified *glp* genes (Fig. 3).

6.4. Glycerol-P Dehydrogenase

The gene encoding glycerol-P dehydrogenase (*glpD*) was cloned by complementation of *glpD* mutant PAO104 (Schweizer and Po, 1994) utilizing the bacteriophage mini-D3112-based in vivo cloning method (Darzins and Casabadan, 1989). The nucleotide sequence predicted a protein with a deduced molecular mass of 56,150 daltons, which was verified by expression in a T7 polymerase/T7 promoter system. The predicted amino acid sequence of GlpD revealed homologies with several bacterial flavin-binding proteins. The flavin-binding domain close to the

N-terminus of GlpD exhibits 72% identity to the corresponding domain of the *E. coli* GlpD protein. Overall, GlpD from *P. aeruginosa* is 56% identical and 69% similar to the GlpD from *E. coli*. Comparison of the *P. aeruginosa*, *E. coli*, and *Bacillus subtilis* GlpD sequences revealed a conserved region spanning amino acid residues 250 to 400 which has significant homology with regions found in the triose phosphate isomerase sequences of *E. coli* and *Saccharomyces cerevisiae*. This conserved region may encompass, at least in part, the glycerol-P-binding site (Austin and Larson, 1991; Schweizer and Po, 1994). It is interesting to note that although the GlpD primary sequence does not contain any pronounced hydrophobic regions, glycerol-P dehydrogenase is tightly membrane-associated (Schweizer and Po, 1994).

A defined *glpD* mutant was constructed by inserting a tetracycline-resistance-encoding cassette into a restriction site approximately midway through the gene and returning the mutated sequences into the corresponding chromosomal region into wild-type PAO1 using a *sacB*-based gene replacement procedure (Schweizer, 1992; Schweizer and Hoang, 1995). The resulting *glpD* strain PAO151 no longer grew on glycerol and glycerol-P but transported these carbon sources normally. A mucoid strain derived from *glpD* mutant PAO151 by inactivation of *mucB* (Martin *et al.*, 1993) exhibited much reduced ability to produce alginate from various carbon sources. Because such a pleiotropic phenotype for a *glpD* mutant is not expected, the observed effects are most likely caused by the polar effects of *glpD* insertional mutations on the expression of the downstream gene *glpM* (see below). These data indicate that the transcription of *glpD* and therefore *glpM* may be required for efficient alginate synthesis from various carbon sources.

6.5. The GlpM Protein

During mutational analysis of *glpD*, strains unaffected in glycerol metabolism were isolated which contained insertions downstream of *glpD*. Similarly to the *glpD mucB* mutants, in a *mucB* genetic background these mutants were pleiotropically defective in alginate biosynthesis (Schweizer *et al.*, 1995). The insertions were located in the *glpM* gene (acronym for membrane-associated *glp* gene product) that encoded a protein of 109 amino acids with a predicted molecular weight of 12,039. The *glpD-glpM* intergenic region contains no recognizable promoter sequence, and although it contains palindromic sequences capable of forming stable stem-loop structures, experimental evidence indicates that these structures are efficient transcription terminators in vivo. Thus, *glpM* may be part of a transcriptional unit with *glpD*. Codon preference analysis and expression of a protein of the expected size in an *E. coli* T7 expression system indicates in vivo expression of the GlpM protein. Several lines of evidence indicate that GlpM is a membrane-associated protein: (1) a Kyte–Doolittle analysis shows that GlpM is a hydrophobic protein with four putative transmembrane domains and (2) a GlpM-LacZ hybrid protein was expressed and localized to the membrane fraction of *P. aeruginosa* PAO1 harboring the gene fusion plasmid.

To assess the physiological function of GlpM, a defined mutant in the *glpM* locus was constructed by insertional inactivation of *glpM* with a gentamicin-resistance-encoding cassette and crossing the mutation into wild-type PAO1. The resulting mutant PAO206 grew normally on glycerol and expresses glycerol-P dehydrogenase and glycerol transport in the same manner as PAO1, suggesting that *glpM* and its product are not required for function of the peripheral glycerol pathway. However, as with the *glpD mucB* strains described above, a *mucB* derivative of PAO206 (PAO206M) was greatly impaired in alginate biosynthesis from various carbon sources. Since PAO206M could not synthesize alginate from mannitol, it does not seem that this mutant merely affects carbon flux through its metabolic pathways, but rather that it seems to affect alginate biosynthesis more directly (Schweizer *et al.*, 1995). The relationship of this putative membrane protein to glycerol metabolism in particular remains to be elucidated.

6.6. Glycerol Kinase

This enzyme catalyzes the ATP-dependent, irreversible phosphorylation of glycerol to glycerol-P. The K_m reported for glycerol of the enzyme in crude extracts is 19 μM (Williams *et al.*, 1994) which compares to 10–40 μM for the purified enzyme (McCowen *et al.*, 1987). The M_r of the native enzyme is 120,000 daltons (McCowen *et al.*, 1987) and the M_r of the denatured form reported is between 51,300 (Williams *et al.*, 1994) and 56,000 (Schweizer *et al.*, 1997), indicating that the enzyme probably exists as a homodimer. Williams *et al.* (1994) noted that 12% of kinase activity is associated with the membrane fraction of *P. aeruginosa*. Thus, it may be activated by interaction with the glycerol facilitator permease as demonstrated for the *E. coli* enzyme (Voegel *et al.*, 1993). Glycerol kinase is encoded by *glpK*, the promoter-distal gene of the *glpFK* operon (Schweizer *et al.*, 1997). The calculated molecular mass of GlpK is 56,063. Analysis of the glycerol kinase amino acid sequence revealed the conserved nature of this protein. It is very similar in size and amino acid composition to the corresponding proteins from *E. coli* (81% identity; 88% similarity) and to a putative GlpK homologue from the genome sequence of *Haemophilus influenzae* (72% identity; 86% similarity). A comparably high degree of similarity was also found with GlpK from other gram-negative and gram-positive bacteria and with eukaryotic glycerol kinases. When compared to other bacterial glycerol kinases, the *P. aeruginosa* protein contains all of the conserved motifs including the N-terminal ATP binding site, the two signatures of the FGGY family of carbohydrate kinases, and the amino acid residues involved in glycerol binding (Hurley *et al.*, 1993; Schweizer *et al.*, 1997). As in *E. coli* glycerol kinase (Hurley *et al.*, 1993), the functional domains, i.e., the domains involved in ATP and glycerol binding, are located in the N-terminal half of the protein. This suggests that as in *E. coli*, the C-terminal half of the protein might constitute a regulatory domain that participates in allosteric protein–protein interactions whose nature remains to be elucidated. Division into separate domains may explain why *glpK* insertion mutants that synthesize truncated GlpK proteins retain the ability to grow slowly on glycerol and still exhibit low levels of glycerol transport (Schweizer *et al.*, 1997). The truncated glycerol kinases synthesized by these strains can phosphorylate the glycerol passing through the facilitator channel and trap it as glycerol-P in the cytoplasm.

6.7. Regulatory Loci

6.7.1. AgmR

The observation (Cuskey and Phibbs, 1985; McCowen *et al.*, 1981) that a glycerol negative mutant (PRP406) lacked activity of transport, kinase, and dehydrogenase led to the hypothesis that the *P. aeruginosa glp* genes are positively regulated and that PRP406 is defective in a gene (termed *glpR2*) encoding a positive regulator. Schweizer (1991) utilized this mutant to clone a chromosomal DNA fragment that complemented this defect. It was determined that a single gene, *agmR* (acronym for activation of glycerol metabolism), which defined a protein and has a molecular mass of 24,422 daltons, was sufficient to restore the ability of PRP406 to grow on glycerol. It was shown that the AgmR protein belongs to the family of bacterial response regulatory proteins, and the presence of a putative C-terminal helix-turn-helix domain indicates that it may be a DNA-binding protein. Surprisingly, although AgmR-expressing plasmids restored the ability of PRP406 to grow on glycerol, expression of glycerol transport and glycerol-P dehydrogenase activities was not restored. Furthermore, *agmR* insertion mutants in a wild-type *P. aeruginosa* genetic background had no detectable *glp* phenotype. Therefore, the reasons for AgmR-mediated *glpR8* complementation activity remain unknown but is most likely caused by some sort of suppresser activity by overproduced AgmR protein.

6.7.2. GlpR

A regulatory gene, termed *glpR*, was identified between the *glpK* and *glpD* genes (Fig. 3) which encodes a protein of 251 amino acids 59% identical to the Glp repressor from *E. coli* and expressed as a 28-kDa protein in a T7 expression system (Schweizer and Po, 1996). Inactivation of chromosomal *glpR* by insertion of a tetracycline-resistance-encoding cassette followed by gene replacement (Schweizer and Hoang, 1995) resulted in constitutive expression of GlpF-mediated glycerol transport activity and *glpD'-xylE*$^+$ activity. These activities were strongly repressed after introducing a multicopy plasmid containing the *glpR* gene. The same plasmid also efficiently repressed expression of a chromosomal *glpT-lacZ*$^+$ transcriptional fusion in an *E. coli glpR* mutant. Analysis of the *glpD* and *glpF* upstream regions identified conserved palindromic sequences up to 70% identical to the *E. coli glp* operator consensus sequence. The results suggest that the operons of the *glp* regulon in *P. aeruginosa* are negatively regulated by the action of a *glp* repressor, which is similar structurally and functionally to the GlpR protein of *E. coli*.

6.8. Chromosomal Mapping of Glycerol Metabolism Genes

Transductional analysis using phage F116L was employed to map the genetic defects on the glycerol negative mutants PRP406 (*glpR2*) and PRP408 (*glpD8*) (Cuskey and Phibbs, 1985). In this study, the *glpD8* marker was cotransducible at low frequencies with the hexose catabolism gene cluster, namely, *edd, eda, zwf*, and at higher frequencies with the *amiR* and *phe-2* markers, all of which are localized in the 38–41 minute region of the *P. aeruginosa* chromosome map. In the same study, *glpR2* could not be linked with any known markers tested, indicating that *glpR2* and *glpD8* are not closely linked loci. Using chromosomal macro-restriction fragment analysis, Liao *et al.* (1996) physically mapped *glpD* and therefore the closely linked *glpF, glpK, glpR*, and *glpM* genes to the same chromosomal *Dpn*I and *Spe*I fragments in the 30-min region of the chromosome as *algD*, encoding the key enzyme GDP-mannose dehydrogenase of the alginate biosynthetic pathway. Presently it is not clear why these separate studies place the *glpD* region in different regions of the *P. aeruginosa* chromosome. It should also be mentioned that *agmR*, the gene complementing the *glpR2* defect of strain PRP406 by hybridization, was mapped to the *Spe*I fragment D in the 50–55 min region of the *P. aeruginosa* PAO1 chromosome and was linked by R68.45-mediated conjugational crosses to the *catA*, *gcu*-1, and *cys*-54 markers (Holloway *et al.*, 1994; Schweizer, unpublished data).

7. CENTRAL CYCLE

A group of six enzymes (Fba, Fbp, Pgi, Zwf, Edd, and Eda) catalyze reactions through which six-carbon and certain three-carbon compounds are catabolized (Fig. 1). These enzymes, related genetic loci, and the role in various metabolic processes are considered in turn. The first three enzymes are constitutively produced, whereas the last three are expressed during growth on glycerol, fructose, mannitol, glucose, and gluconate.

7.1. Fructose 1,6-Bisphosphate Aldolase

Fructose 1,6-bisphosphate aldolase (Fba) catalyzes the reversible cleavage of fructose 1,6-bisphosphate to glyceraldehyde 3-phosphate and dihydroxyacetone (which are freely interchangeable via triose phosphate isomerase). Mutants lacking Fba cannot grow on glucose, gluconate, or glycerol and also cannot utilize gluconeogenic substrates, such as glutamate, succinate, or lactate (Banerjee *et al.*, 1987). A chimeric plasmid containing *P. aeruginosa* chromosomal DNA can complement the deficient strains. The gene encoding 3-phosphoglycerate kinase is also present on the chromosomal insert containing the presumed Fba gene. To our knowledge, these loci have not been mapped on the *P. aeruginosa* chromosome.

7.2. Fructose 1,6-Bisphosphatase

Fructose 1,6-bisphosphatase (Fbp) irreversibly dephosphorylates fructose 1,6-bisphosphate to yield fructose 6-phosphate. In *P. aeruginosa*, Fbp functions in metabolizing glycerol and fructose and in recycling glyceraldehyde 3-phosphate (see below), whereas the enzyme serves largely in gluconeogenic processes in other organisms. Efforts to isolate mutants completely lacking Fbp activity have been unsuccessful despite numerous attempts, suggesting that the gene may be essential. Mutants possessing 10–30% of parental enzymic activities were constructed by Banerjee (1989). These strains grow poorly on gluconate and exhibit low levels of Fba activity. Fructose 1,6-bisphosphate and triose phosphates accumulate in amounts higher than parental strains during the growth of these strains, presumably because of lowered activity of Fba and Fbp. The ability of these strains to produce alginate implicated Fbp as essential in alginate biosynthesis. Several chimeric plasmids containing chromosomal fragments controlling alginate synthesis or Fba activity did not complement the Fbp/Fba deficiency. Taken together, these observations indicate that the Fbp/Fba deficient strains contain a polar or regulatory mutation.

7.3. Phosphoglucoisomerase

Phosphoglucoisomerase (Pgi) catalyzes the reversible interchange between fructose 6-phosphate and glucose 6-phosphate. Mutants in this locus fail to grow on mannitol as a sole carbon source and grow at greatly reduced rates on fructose and glycerol (Phibbs *et al.*, 1978). Wild-type growth rates on lactate are strongly inhibited by mannitol, suggesting that high levels of an intermediate compound are toxic to the cells. Three independent mutant strains lacking Pgi activity were used to map the *pgi* locus to minute 1 of the *P. aeruginosa* chromosome, between the *car* and *ilv* loci (Calligeros *et al.*, 1996).

7.4. Glucose 6-Phosphate Dehydrogenase

Glucose 6-phosphate dehydrogenase (Zwf), which catalyzes the NAD(P)-dependent, irreversible oxidation of glucose 6-phosphate to 6-phosphogluconate, is strictly coinduced with 6-PG dehydratase and KDPG aldolase which catalyze the two subsequent steps in dissimilation of glucose. Mutants lacking Zwf activity cannot grow on mannitol or glucose (when the oxidative glucose pathway is blocked) and grow very slowly on fructose or glycerol (Phibbs *et al.*, 1978). The *zwf* locus was originally mapped in the 39-minute region of the *P. aeruginosa* chromosome along with loci for the two Entner–Doudoroff pathway enzymes (Roehl *et al.*, 1983) and later cloned with the structural gene for Eda on a chromosomal fragment (Fig. 2, Temple, *et al.*, 1990). It was shown that expression of *zwf* is regulated at the level of transcription. Sequence analysis revealed an ORF that specified a protein of $M_r = 53,900$, which is 52% identical to glucose, 6-phosphate dehydrogenase from *E. coli* (Ma *et al.*, 1998).

7.5. Entner–Doudoroff Dehydratase

Entner–Doudoroff dehydratase (Edd) catalyzes the irreversible dehydration of 6-phosphogluconate to 2-keto-3-deoxy-6-phosphogluconate.

The *edd* locus was mapped with the *zwf* and *eda* loci to minute 39 on the *P. aeruginosa* chromosome (Roehl *et al.*, 1983). The structural gene was cloned on a 6.0-kb *Eco*RI chromosomal fragment complementing the *gltR* (formerly *gltB*) defect (Fig. 2, Cuskey and Phibbs, 1985). In addition to *edd*, this fragment also contains the genes for glucokinase and NAD-specific glyceraldehyde 3-phosphate dehydrogenase (Temple *et al.*, 1990, 1994). Subcloning and sequencing of the region revealed an ORF of 1833 bp potentially encoding a 610 amino acid protein with a predicted M_r of 65,140, which compares favorably to the 63,000 dalton protein (Edd) from *Zymomonas mobilis*. The amino acid sequence exhibits similarity of 51–76% with Edd sequences from *Z. mobilis* and *E. coli* and with dihydroxy acid dehydratase from *E. coli* (Temple *et al.*, 1994). A potential ribosome binding site is upstream of the ORF for Edd. The start of transcription was mapped 87 bp upstream of the ATG translation start sequence, and a message of approximately 2.0 kb was visualized. However, no terminatorlike sequences were apparent in the nucleotide sequence downstream of the ORF. The ORF for glucokinase begins just 101 bases downstream from the stop of *edd*, and the occasionally detected longer transcript from the *edd* promoter (see above) suggests that *edd* may be cotranscribed with glucokinase (Sage and Phibbs, unpublished).

7.6. Entner–Doudoroff Aldolase

Entner–Doudoroff aldolase (Eda) catalyzes the reversible cleavage of 2-keto-3-deoxy-6-phosphogluconate to glyceraldehyde 3-phosphate and pyruvate. The mutant specifically lacking Eda activity displays the same growth phenotype as those lacking Edd, that is, inability to utilize glucose, gluconate or mannitol and very slow growth on fructose or glycerol. A chromosomal fragment complementing the defect in PAO1838 was cloned along with the structural gene for Zwf (Temple *et al.*, 1990, Fig. 2). Nucleotide sequencing revealed an ORF predicting a protein of $M_r = 36,500$ with 43% identity to Eda from *E. coli* (Hager and Phibbs, unpublished data). A gap of approximately 725 nucleotides occurs between the genes, indicating that *eda* may be transcribed separately from *zwf* (Phibbs *et al.*, unpublished data).

A cosmid vector useful for cloning large pieces of DNA in *Pseudomonas* (Wolff and Phibbs, 1986) was used to isolate a *P. aeruginosa* chromosomal fragment which contains *gltR*, *glk*, *edd*, *gap*(NAD), *eda*, and *zwf* (in that order). The insert in this clone was 30–40 kb, thus showing the close proximity of this group of carbohydrate catabolism genes (Phibbs *et al.*, unpublished data).

8. LOWER EMBDEN–MEYERHOFF–PARNAS (EMP) PATHWAY

P. aeruginosa lacks a complete EMP pathway, because it is naturally deficient in 6-phosphofructokinase (Tiwari and Campbell, 1969). However, the five EMP pathway enzymes necessary to metabolize glyceraldehyde 3-phosphate to pyruvate are all present. Of these five, pyruvate kinase, enolase, phosphoglycerate mutase, and 3-phosphoglycerate kinase are all expressed constitutively, whereas NAD-dependent glyceraldehyde 3-phosphate dehydrogenase is inducible.

8.1. Glyceraldehyde 3-Phosphate Dehydrogenase

Glyceraldehyde 3-phosphate dehydrogenase (Gap) exists in two forms in *P. aeruginosa*, NAD- and NADP-specific (Rivers and Blevins, 1987). The NADP-dependent form is produced constitutively and appears to be essential during growth on gluconeogenic substrates. The gene for NADP-Gap has been cloned in a cosmid vector and shown to be unlinked on the chromosome to the *glk-zwf-edd-eda* cluster or the *fba* and *pgk* genes (Banerjee *et al.*, 1987).

The NAD-Gap gene was discovered during the sequencing of DNA adjacent to the *edd* locus (Temple *et al.*, 1994, Fig. 2). The 1011-bp ORF extends divergently from the Edd ORF. The start codon is 128 bases from the start codon for *edd*. Extensive (45–60%) sequence identity at the amino acid level was noted between the predicted protein ($M_r = 36{,}238$) and other Gaps, which are widespread in the animal and plant kingdoms. Conserved amino acid residues involved in the catalytic site and the NAD-binding site were found in appropriate locations in the predicted sequence. The NAD-Gap ORF is preceded by a potential Shine–Delgarno sequence, and the start of mRNA transcription was mapped 33 bases upstream of the ATG start site, only nine bases from the mRNA start of *edd*. A sequence perfectly matching the -10 *rpoD*-type and two sequences closely related to the -35 *rpoD*-type consensus sequences of *P. aeruginosa* were found upstream of the transcriptional start site. A sequence exhibiting near perfect homology with the integration host factor (IHF) consensus sequence occurs at bases 80–82 upstream of the transcriptional start site. Additionally, several inverted repeat motifs were located in the upstream region, one of which closely resembled the Hut repressor protein binding site of the histidine utilization system in *P. putida* (Allison and Phillips, 1990; Hu *et al.*, 1989). It was shown that expression of *gap* is controlled at the level of transcription (Temple *et al.*, 1994).

8.2. Phosphoglycerate Kinase

Phosphoglycerate kinase (Pgk) catalyzes the reversible phosphorylation of 3-phosphoglycerate to 1,3-bisphosphoglycerate. A mutant which lacks Pgk activity cannot utilize glucose, gluconate, or glycerol and cannot grow on glutamate, succinate, or lactate, indicating that the enzyme is essential for gluconeogenesis (Banerjee *et al.*, 1987). A chimeric plasmid containing *P. aeruginosa* chromosomal DNA complements the deficiency in Pgk activity. The presumed Pgk gene is closely linked to the Fba gene but unlinked to several other carbohydrate loci (see above).

8.3. Phosphoglycerate Mutase (Pgm), Enolase (Eno), and Pyruvate Kinase (Pyc)

Phosphoglycerate mutase and enolase catalyze reversible reactions producing phosphoenolpyruvate. Pyruvate kinase catalyzes the irreversible dephosphorylation of phosphoenolpyruvate to pyruvate. Mutants lacking enolase or pyrvuate kinase activity showed parental growth rates on glycerol, fructose, and galactose, indicating that dissimilation of carbohydrates through the EDP does not depend on the lower EMP pathway enzymes (Phibbs, unpublished data).

8.4. Pyruvate Carboxylase

Pyruvate carboxylase (Pyc) facilitates the complex anapleurotic reaction funneling pyruvate and CO_2 into the TCA cycle. Enzymic activity is highest during growth on glucose and glycerol and lower during growth on succinate and acetate (Phibbs *et al.*, 1974). Mutants totally lacking Pyc activity cannot grow on lactate, pyruvate, glycerol, or any six-carbon sugar tested, but can utilize acetate, citrate, and succinate as a sole carbon source. These studies showed that Pyc serves a critical physiological role in anapleurotic CO_2 fixation during growth on C_6 and C_3 compounds but is not required for growth on TCA cycle intermediates. The *pyc* locus was mapped to minute 35 on the *P. aeruginosa* chromosome (Roehl *et al.*, 1983).

9. RECYCLING OF GLYCERALDEHYDE 3-PHOSPHATE

A number of studies have shown that glyceraldehyde 3-phosphate is preferentially recycled through the central cycle rather than being catabolized to pyruvate via the EMP enzymes (reviewed in Lessie and Phibbs,

1984; Banerjee, 1989). This recycling, starting with two molecules of glyceraldehyde 3-phosphate and yielding one pyruvate and one glyceraldehyde 3-phosphate, was first proposed in *Xanthomonas phaseoli* (Hochster and Katznelson, 1958). The seven enzymes constituting this pathway (triose phosphate isomerase, Fba, Fbp, Pgi, Zwf, Edd, and Eda), are all present in certain obligate methylotrophs which do not utilize carbohydrates as carbon sources (Colby *et al.*, 1979). In the case of *P. aeruginosa*, these enzymes may serve two functions: both catabolic, feeding pyruvate into the TCA cycle, and anabolic, providing six-carbon intermediates (e.g., fructose 6-phosphate, glucose 6-phosphate) for glycoproteins, glycolipids, and carbohydrate polymers in various cellular functions and structures.

10. REGULATION OF CENTRAL AND LOWER EMP PATHWAYS

Hex Regulon

Activity of at least five enzymes, Glk, Zwf, Edd, Eda, and NAD-Gap, are affected by a locus (*hexC*) present in the divergent promoter region between *edd* and *gap*(NAD). The phenotypic effect of the *hexC* locus when present on a multicopy plasmid in *P. aeruginosa* PAO1 is a two- to ninefold increase in the activities of the *hex* regulon enzymes (Cuskey *et al.*, 1985; Temple *et al.*, 1990, and 1994). The *hexC* effect probably results from titration of a repressor protein when the repressor binding site is present in multiple copies. We have identified a protein (HexR) that binds to the *hexC* region as shown by gel shift assays (Proctor *et al.*, 1997). This binding activity has been purified and has an apparent native MW of approximately 120 kd (Proctor *et al.*, 1997). HexR likely binds an inverted repeat (TGTTGTttttACAACA) centered at -61 nucleotides 5' of the transcriptional start site of *gap*. A similar sequence, (TGTTGTttaattACtACA) is located at -71 nucleotides 5' of the transcriptional start site of *zwf* (Proctor *et al.*, 1997). The binding of HexR to both regions is inhibited by low concentrations of KDPG but not 6-phosphogluconate. These observations indicate the KDPG serves as the physiological inducer of *hexC* repressed genes, in particular, *edd*, *gap*, and *zwf*. The identification of KDPG as the physiological inducer explains why certain mutants lacking Edd activity also lose inducibility of the other *hex* regulon enzymes. Because these mutants cannot produce KDPG, none of the *hexC* regulon enzymes are expressed even when exogenous inducers are present. Current work is ongoing to purify the *hexC* repressor further and identify its other binding sites.

The simplest hypothesis explaining the *hexC* effect is that each promoter in the *hex* regulon contains a binding site for the putative repressor; however, some evidence indicates that each promoter in the regulon is not identical with respect to the *hexC* phenomenon. For example, chromosomal fragments containing the structural genes for Eda and Zwf, each expressed from its own cloned promoter, do not confer the *hexC* effect. HexR may bind with differing affinities to promoters of the genes in the regulon, or there may be other factors that act secondarily (in a positive fashion) at non-*hexC*-containing promoter(s) in the regulon. Because of the nonequivalent nature of the promoters, the *hex* regulon appears to be relatively complex, mechanistically.

Further complexity results from two other sources. First, the *hex* regulon enzymes are subject to catabolite repression control (see below). The *hexC* locus on a multicopy plasmid showed no effect on the repression of the *hex* regulon enzymes, indicating that *hexC* is separate from and recessive to the locus for catabolite repression control (Temple *et al.*, 1990). Secondly, the promoter region between the translational starts of Edd and Gap-NAD genes contains three consensus binding sequences for integration host factor (IHF), a protein known to be involved in DNA bending and in both positive and negative transcriptional control (for review, see Goosen and dePutte, 1995). Purified *E. coli* IHF binds the divergent promoter containing *hexC*, but this binding is unaffected by the presence of IPTG (Proctor *et al.*, 1997a,b). Thus, the IHF binding site and the *hexC* loci are separate. Therefore it appears that a minimum of three control mechanisms are operative within this small divergent promoter region.

In addition to the interesting transcriptional regulation of the *hex* regulon, the physiological and genetic relationships of the five enzymes raise several other questions. First, why is glucokinase stimulated under conditions in which it does not appear to be needed? For example, glucokinase is induced when *P. aeruginosa* is grown on fructose as a sole carbon source, a condition where no intracellular glucose is produced from catabolism of the external carbon source. This phenomenon may result from differential transcription. There is evidence for transcriptional readthrough from the gene for *edd* into *glk* (Sage, unpublished). Because Edd is required for metabolizing fructose via the central cycle, at least some Glk would be produced due to the induction of Edd, if polycistronic transcripts are made. Careful studies of mRNA transcripts under various conditions should resolve this question.

Secondly, why is Gap-NAD produced during growth on glucose, fructose, mannitol, and glycerol, when the substrate for that enzyme is recycled via the central cycle to pyruvate, rather than metabolized by

Gap-NAD? The molecule may serve to prevent toxic buildup of triose phosphates or hexose phosphates. Interesting in this context, however, are examples of Gap-NAD serving functions other than glycolytic catalysis. It has been shown that the eukaryotic homologue of Gap-NAD binds tRNA in a sequence-specific manner and perhaps is a transporter of tRNA from the cytoplasm into the nucleus (Singh and Green, 1993). Competition studies suggest that the tRNA binding domain of Gap-NAD includes the NAD-binding site, so that the binding of tRNA may exclude NAD. Furthermore, a viable Gap-NAD gene is required in yeast even when it is not needed physiologically, for example, when the organism is grown on lactate (MacAllister and Holland, 1985). Gap-NAD has been implicated in several other cellular activities unrelated to glycolysis, such as single-stranded nucleic acid binding, protein phosphorylation, transcriptional regulation, and DNA repair (Karpel and Burchard, 1981; Perucho *et al.*, 1977; Ryazanov, 1985). Which, if any, of these functions the Gap-NAD of *P. aeruginosa* serves remains to be elucidated.

Lastly, what role, if any, is being played by IHF in regulating the *hex* regulon and perhaps other regulatory units? IHF from *P. aeruginosa* has been purified and the genes cloned and sequenced. The predicted protein subunits show over 70% identity with the molecule from *E. coli* (Delic–Atree *et al.*, 1995). It been shown that IHF is required for expressing *algD* in *P. aeruginosa* and the binding site occurs at the 3' end of the gene (Wozniak, 1993). It has been shown that OmpR from *E. coli*, which appears to have homologs in *P. aeruginosa*, interacts with IHF to repress transcription of *ompF*. Binding studies utilizing IHF from *P. aeruginosa* and the several available cloned promoter regions should yield interesting results and help expand the understanding of IHF's regulatory role in this organism.

Thus, the *hex* regulon appears to be a unique combination of genetic regulatory elements, including a repressor protein, transcriptional regulators similar to IHF, and catabolite repression control at a level yet to be determined. The regulon includes three very closely spaced structural genes. Two genes are transcribed from divergent promoters, and two remaining structural genes are located several kilobases distant from the first group. We are continuing to elucidate the *hex* regulon control using mapping, sequencing, transcriptional analyses, and physiological studies.

11. CATABOLITE REPRESSION CONTROL (CRC)

All of the inducible transport systems and enzymes for carbohydrate utilization in *P. aeruginosa* are sensitive to strong catabolite repression control when grown in the presence of acetate or most tricarboxylic acid cycle intermediates (reviewed in Collier *et al.*, 1996). In contrast to the CRC system in *E. coli*, cyclic-AMP is not an effector of CRC of these enzymes or of other inducible enzymes for utilizing noncarbohydrates (Phillips and Mulfinger, 1981; Siegel *et al.*, 1977). A strain which had lost the biphasic growth characteristic of parental *P. aeruginosa* (Wolff *et al.*, 1991) was restored to normal growth when it was transformed with a chimeric plasmid containing a 2-kb chromosomal fragment (MacGregor *et al.*, 1991). The product of the *crc* gene was identified as a protein with $M_r = 30,000$ and expressed from the chimeric plasmid in both *E. coli* and *P. aeruginosa*. The nucleotide sequence of the gene showed homology to the apurinic/apyrimidinic endonuclease family of DNA repair enzymes (MacGregor *et al.*, 1996). The significance of this homology is unknown, as *crc* mutants do not have a DNA repair phenotype. A strain containing *crc* interrupted with a kanamycin resistance gene had lost the normal biphasic growth pattern. It is not known yet whether the *crc* gene product interacts with regulatory regions of CRC-sensitive promoters or catalyzes the production of another compound which is the effector molecule. In one instance (see above), it has been shown that CRC is separate from and dominant over a separate regulatory system, *hexC*. Analyses of promoter sequences controlling the genes regulated by the CRC system are likely to reveal consensus binding sites for the effector molecule(s).

12. CLUSTERING OF GENES FOR GLYCOLYTIC ENZYMES

At least 25 genetic loci specifying carbohydrate catabolism genes have been mapped on the *P. aeruginosa* chromosome (reviewed in Holloway *et al.*, 1994; Liao *et al.*, 1996). Most of these markers were mapped by transductional analysis. More recently, macrorestriction mapping and Southern hybridization procedures have resulted in a combined physical and genetic map (Holloway *et al.*, 1994; Ratnaningsih *et al.*, 1990; Romling *et al.*, 1992). Genes in *P. aeruginosa* that are functionally related are sometimes clustered, a phenomenon first described by Leidigh and Wheelis (1973) and termed "supraoperonic clustering." Twenty of the carbohydrate genes which have been mapped are located within three large clusters: the glycerol metabolism genes at minute 30, the phosphorylative glucose metabolism and EDP enzyme genes at minute 38, and the mannitol and oxidative glucose metabolism genes at minute 48. Thus, the majority of genes specifying carbohydrate catabolism fall within a region which is less than 25% of the chromosome. These observations support the proposal of Holloway and Morgan (1986) that the chromosomal structure of existing pseudomonads has evolved by integration of extrachromosomal genetic material into a smaller ancestral genome.

REFERENCES

Allison, S. L., and Phillips, A. T., 1990, Nucleotide sequence of the gene encoding the repressor for the histidine utilization genes of *Pseudomonas putida*, *J. Bacteriol.* **172**:5470–5476.

Anderson, R. L., and Wood, W. R., 1969, Carbohydrate metabolism in microorganisms, *Ann. Rev. Microbiol.* **23**:539–578.

Austin, D., and Larson, T. J., 1991, Nucleotide sequence of the *glpD* gene encoding aerobic *sn*-glycerol-3-phosphate dehydrogenase of *Escherichia coli* K-12. *J. Bacteriol.* **173**:101–107.

Banerjee, P. C., 1989, Fructose-bisphosphatase-deficient mutants of mucoid *Pseudomonas aeruginosa*, *Folia Microbiologica* **34**:81–86.

Banerjee, P. C., Vanags, R. I., Chakrabarty, A. M., and Maitra, P. K., 1983, Alginic acid synthesis in *Pseudomonas aeruginosa* mutants defective in carbohydrate metabolism, *J. Bacteriol.* **155**:238–245.

Banerjee, P. C., Vanags, R. I., Chakrabarty, A. M., and Maitra, P. K., 1985, Fructose, 1,6-bisphosphate aldolase activity is essential for synthesis of alginate from glucose by *Pseudomonas aeruginosa*, *J. Bacteriol.* **161**:458–460.

Banerjee, P. C., Darzins, A., and Maitra, P. K., 1987, Gluconeogenic mutations in *Pseudomonas aeruginosa* genetic linkage between fructose-bisphosphate aldolase and phosphoglycerate kinase, *J. Gen. Microbiol.* **133**:1099–1108.

Baumann, P., and Baumann, L., 1975, Catabolism of D-fructose and D-ribose by *Pseudomonas doudoroffii*. I. Physiological studies and mutant analysis. *Arch. Microbiol.* **105**(3):225–240.

Blevins, W. T., Feary, T. W., and Phibbs, P. V., Jr., 1975, 6-Phosphogluconate dehydratase deficiency in pleiotropic carbohydrate negative mutant strains of *Pseudomonas aeruginosa*, *J. Bacteriol.* **121**:942–949.

Calligeros, J. E., Matsumoto, H., Gates, J. E., and Phibbs, P. V., Jr., 1996, Characterization and genetic mapping of phosphoglucoisomerase mutations in *Pseudomonas aeruginosa*, *Curr Microbiol.* **33**:347–351.

Colby, J., Dalton, H., and Whittenbury, R. 1979, Biological and biochemical aspects of microbial growth on 1 carbon compounds, *Ann. Rev. Microbiol.* **33**:481–518.

Collier, D. N., Hager, P. W., and Phibbs, P. V., Jr., 1996, Catabolite repression control in the *Pseudomonads*, *Res. Microbiol.* **147**:551–561.

Cuskey, S. M., and Phibbs, P. V., Jr., 1985, Chromosomal mapping of mutations affecting glycerol and glucose catabolism in *Pseudomonas aeruginosa* PAO, *J. Bacteriol.* **162**:872–880.

Cuskey, S. M., Wolff, J. A., Phibbs, P. V., Jr., and Olsen, R. H., 1985, Cloning of Genes specifying carbohydrate catabolism in *Pseudomonas aeruginosa* and *Pseudomonas putida*, *J. Bacteriol.* **162**:865–871.

Darzins, A., and Casabadan, M. J., 1989, *In vivo* cloning of *Pseudomonas aeruginosa* genes with mini-D3112 transposable bacteriophage, *J. Bacteriol.* **171**:3917–3925.

Delic-Attree, I., Toussaint, B., and Vignais, P. M., 1995, Cloning and sequence analysis of the genes coding from the integration host factor (IHF) and HU proteins of *Pseudomonas aeruginosa*, *Gene* **154**:61–64.

Duline, J. A., and Frank Jzn, J., 1981, Quino proteins, a novel class of dehydrogenases, *Trends Biochem. Sci.* **6**:278–280.

Duine, J. A., and Jongejan, J. A., 1989, Quinoproteins, enzymes with pyrrolo-quinoline quinone as cofactor. *Annu. Rev. Biochem.* **58**:403–426.

Durham, D. R., and Phibbs, P. V., Jr., 1982, Fractionation and characterization of the phosphoenolpyruvate:fructose 1-phosphotransferase system from *Pseudomonas aeruginosa*, *J. Bacteriol.* **149**:534–541.

Eagon, R. G., 1971, 2-Deoxyglucose transportation via passive diffusion and its oxidation, not phosphorylation, to 2-deoxygluconic acid by *Pseudomonas aeruginosa*, *Can. J. Biochem.* **49**:606–613.

Eisenberg, R. C., and Phibbs, P. V., Jr., 1982, Characterization of an inducible mannitol-binding protein from *Pseudomonas aeruginosa*, *Curr. Microbiol.* **7**:229–234.

Entner, N., and Doudoroff, M., 1952, Glucose and gluconic acid oxidation of *Pseudomonas saccharophila*, *J. Biol. Chem.* **196**:853–862.

Fraenkel, D. G., 1986, Mutants in glycolysis, *An. Rev. Biochem.* **55**:317–337.

Goosen, N., and van de Putte, P., 1995, The regulation of transcription initiation by integration host factor, *Mol. Microbiol.* **16**:1–7.

Goosen, N., Vermasas, D. A. M., and van de Putte, P., 1987, Cloning of the genes involved in synthesis of coenzyme pyrrolo-quinoline-quinone from *Acenetobacter calcoaceticus*, *J. Bacteriol.* **169**:303–307.

Gottschalk, G., Bender, R., Heath, H. E., and Gaudy, E. T., 1978, Relationship between catabolism of glycerol and metabolism of hexosephosphate derivatives by *Pseudomonas aeruginosa*, *J. Bacteriol.* **136**:638–646.

Govan, J. R. W., 1988, Alginate biosynthesis and other unusual characteristics associated with the pathogenesis of *Pseudomonas aeruginosa* in cystic fibrosis, in: *Bacterial Infections of Respiratory and Gastrointestinal Mucosae*, (E. Griffiths, W. Donachie, and J. Stephen, eds.), IRL Press, Oxford, pp. 67–96.

Hager, P. W., Covert-Rinaldi, A., Wallace, W. H., and Phibbs, P. V., Jr., 1997, Cloning and sequence analysis of the gluconate operon of *Pseudomonas aeruginosa* PAO, Abstracts of the VI International Congress of *Pseudomonas*: Molecular Biology and Biotechnology, pg. 71.

Hancock, R. E., and Carey, A. M., 1980, Protein D1, a glucose-inducible, pore-forming protein from the outer membrane of *Pseudomonas aeruginosa*, *FEMS Microbiol. Lett.* **8**:105–109.

Hochster, R. M., and Katzneleon, H., 1958, On the mechanism of glucose-6-phosphate oxidation in cell-free extracts of *Xanthomonas phaseoli* (XP8), *Can. J. Biochem. Physiol.* **36**:669–689.

Holloway, B. W., and Morgan, A. F., 1986, Genome organization in *Pseudomonas*, *Annu. Rev. Microbiol.* **40**:79–105.

Holloway, B. W., Krishnapillai, V., and Morgan, A. F., 1979, Chromosomal genetics of *Pseudomonas*, *Microbiol. Rev.* **43**:73–102.

Holloway, B. H., Römling, U., and Tümmler, B., 1994, Genomic mapping of *Pseudomonas aeruginosa* PAO, *Microbiol.* **140**:2907–2929.

Hu, L., Allison, S. L., and Phillips, A. T., 1989, Identification of multiple repressor recognition sites in the *hut* system of *Pseudomonas putida*, *J. Bacteriol.* **171**:4189–4195.

Huang, H., and Hancock, R. E. W., 1993, Genetic definition of the substrate selectivity of outer membrane porin protein OprD of *Pseudomonas aeruginosa, J. Bacteriol.* **175:**7793–7800.

Huang, H., Siehnel, R. J., Bellido, F., Rawling, E., and Hancock, R. E., 1992, Analysis of two gene regions involved in the expression of the imipenem-specific, outer membrane porin protein OprD of *Pseudomonas aeruginosa, FEMS Microbiol. Lett.* **76:**267–273.

Hurley, J. H., Faber, H. R., Worthylake, D., Meadow, N. D., Roseman, S., Pettigrew, D. W., and Remington, S. J., 1993, Structure of the regulatory complex of *Escherichia coli* III[Glc] with glycerol kinase, *Science* **259:**673–677.

Hunt, J. C., and Phibbs, P. V., Jr., 1981, Failure of *Pseudomonas aeruginosa* to form membrane-associated glucose dehydrogenase activity during anaerobic growth with nitrate, *Biochem. Biophys. Res. Commun.* **102:**1393–1399.

Hunt, J. C., and Phibbs, P. V., Jr., 1983, Regulation of alternate peripheral pathways of glucose catabolism during aerobic and anaerobic growth of *Pseudomonas aeruginosa, J. Bacteriol.* **154:**793–804.

Hylemon, P. B., and Phibbs, P. V., Jr., 1972, Independent regulation of hexose catabolizing enzymes and glucose transport activity in *Pseudomonas aeruginosa, Biochem. Biophys. Res. Commun.* **48:**1041–1048.

Karpel, R. L., and Burchard, A. C., 1981, A basic isozyme of yeast *Saccharomyces cerevisiae* glyceraldehyde, 3-phosphate dehydrogenase with nucleic acid helix destabilizing activity, *Biochim. Biophys. Acta.* **64:**256–267.

Kersters, K., and DeLey, J., 1968, The occurrence of the Entner–Doudoroff pathway in bacteria, *Antonie van Leewenhoek* **34:**393–408.

Leidigh, B. J., and Wheelis, M. L., 1973, The clustering on the *Pseudomonas putida* chromosome of genes specifying dissimilatory functions. *J. Mol. Evol.* **2**(4):235–242.

Lessie, T. G., and Phibbs, P. V., Jr., 1984, Alternative pathways of carbohydrate utilization in pseudomonads, *Ann. Rev. Microbiol.* **38:**359–387.

Liao, X., Charlebois, I., Ouellet, C., Morency, M.-J., Dewar, K., Lightfoot, J., Foster, J., Siehnel, R., Schweizer, H. P., Lam, J., Hancock, R. E. W., and Levesque, R. C., 1996, Physical mapping of 32 genetic markers on the *Pseudomonas aeruginosa* PAO1 chromosome, *Microbiology* **142:**79–86.

Ma, J. F., Hager, P. W., Howell, M. L., Phebbs, P. V., and Hassett, D. J., 1998, Cloning and characterization of the *Pseudomonas aeruginosa* zwf gene encoding glucose-6-phosphate dehydrogenase, an enzyme important in resistance to methyl viologen (paraquat). *J. Bacteriol.* **180**(7):1741–1749.

MacAlister, L., and Holland, M. J., 1985, Differential expression of the three yeast glyceraldehyde, 3-phosphate dehydrogenase genes, *J. Biol. Chem.* **280:**15013–15018.

MacGregor, C. H., Wolff, J. A., Arora, S. K., and Phibbs, P. V., Jr., 1991, Cloning a catabolite repression control CRC gene from *Pseudomonas aeruginosa*, expression of the gene in *Escherichia coli*, and identification of the gene product in *Pseudomonas aeruginosa, J. Bacteriol.* **173:**7204–7212.

MacGregor, C. H., Arora, S. K., Hager, P. W., Dail, M. B., and Phibbs, P. V., Jr., 1996, The nucleotide sequence of the *Pseudomonas aeruginosa* pyrE-crc-rph region and the purification of the *crc* gene product, *J. Bacteriol.* **178:**5627–5635.

Martin, D. W., Holloway, B. W., and Deretic, V., 1993, Characterization of a locus determining the mucoid status of *Pseudomonas aeruginosa*: AlgU shows sequence similarities with a *Bacillus* sigma factor, *J. Bacteriol.* **175:**1153–1164.

Matsushita, K., Shinagawa, E., Adachi, O., and Ameyama, M., 1979, Membrane-bound D-gluconate dehydrogenase from *Pseudomonas aeruginosa*. Purification and structure of cytochrome-binding form, *J. Biochem.* **85:**1173–1181.

Matsushita, K., Shinagawa, E., Adachi, O., and Ameyama, M., 1979, D-gluconate dehydrogenase from bacteria, 2-keto-D-gluconate yielding, membrane bound, *Methods Enzymol.* **89:**187–193.

Matsushita, K., Shinagawa, E., Adachi, O., and Ameyama, M., 1979, Membrane-bound D-gluconate dehydrogenase from *Pseudomonas aeruginosa*. Its kinetic properties and a reconstitution of gluconate oxidase, *J. Biochem.* **86:**249–256.

Matsushita, K., Yamada, M., Shinagawa, E., Adachi, O., and Ameyama, M., 1980, Membrane-bound respiratory chain of *Pseudomonas aeruginosa* grown aerobically. *J. Bacteriol.* **141**(1):389–392.

May, T. B., Shinabarger, D., Maharaj, R., Kato, J., Chu, L., Devault, J. D., Roychoudthury, S., Zielinkski, N. A., Berry, A., Rothmel, R. K., Misra, T. K., and Chakrabarty, A. M., 1991, Alginate synthesis by *Pseudomonas aeruginosa*: A key pathogenic factor in chronic pulmonary infections of cystic fibrosis patients, *Clin. Microbiol. Rev.* **4:**191–206.

McCowen, S. M., Phibbs, P. V., Jr., and Feary, T. W., 1981, Glycerol catabolism in wild-type and mutant strains of *Pseudomonas aeruginosa, Curr. Microbiol.* **5:**191–196.

McCowen, S. M., Sellers, J. R., and P. V., Phibbs, Jr., 1987, Characterization of fructose, 1,6-diphosphate-insensitive catabolic glycerol kinase of *Pseudomonas aeruginosa, Curr. Microbiol.* **14:**323–327.

Midgley, M., and Dawes, E. A., 1973, The regulation of transport of glucose and methyl alpha glucoside in *Pseudomonas aeruginosa, Biochem. J.* **132:**141–154.

Ng, F. M. W., and Dawes, E. A., 1973, Chemostat studies on the regulation of glucose metabolism in *Pseudomonas aeruginosa* by citrate, *Biochem. J.* **132:**129–140.

Nikaido, H., and Saier, M. H., 1992, Transport proteins in bacteria: Common themes in their design, *Science* **258:**936–942.

O'Brien, R. W., 1975, Enzymatic analysis of the pathways of glucose catabolism and gluconeogenesis in *Pseudomonas citronellosis, Arch. Microbiol.* **103:**71–76.

Olsen, R. H., Debusscher, G., and McCombie, W. R., 1982, Development of broad host-range vectors and gene banks: Self-cloning of the *Pseudomonas aeruginosa* chromosome, *J. Bacteriol.* **150:**60–69.

Parkinson, J. S., 1993, Signal transduction schemes of bacteria, *Cell* **73:**857–871.

Parkinson, J. S., and Kofold, E. C., 1992, Communication modules in bacterial signaling proteins, *Ann. Rev. Genet.* **26:**71–112.

Perucho, M., Salas, J., and Salas, M. L., 1977, Identification of the mammalian DNA binding protein P-8 as glyceraldehyde 3-phosphate dehydrogenase, *Eur. J. Biochem.* **81:**557–562.

Phibbs, P. V., Jr., 1988, Genetic analysis of carbohydrate metabolism in *Pseudomonas*, in: *Microbial Metabolism and the Carbon Cycle*, (S. R. Hagedorn, R. S. Hanson, and D. A. Kunz, eds.), Harwood Academic Publishers, New York, pp. 412–436.

Phibbs, P. V., Jr., and Eagon, R. G., 1970, Transport and phosphorylation of glucose, fructose, and mannitol by *Pseudomonas aeruginosa. Arch. Biochem. Biophys.* **138**(2):470–482.

Phibbs, P. V., Jr., Feary, T. W., and Blevins, W. T., 1974, Pyruvate carboxylase deficiency in pleiotropic carbohydrate negative mutant strains of *Pseudomonas aeruginosa, J. Bacteriol.* **118:**999–1009.

Phibbs, P. V., Jr., McCowen, S. M., Feary, T. W., and Blevins, W. T., 1978, Mannitol and fructose catabolic pathways of *Pseudomonas aeruginosa* carbohydrate negative mutants and pleiotropic effects of certain enzyme deficiencies, *J. Bacteriol.* **133:**717–728.

Phibbs, P. V., Jr., Srivastava, R., Chunfang, Z., and Holloway, B. W., 1987, Expression of the *P. aeruginosa* mannitol utilization genes in *P. putida. Abstr. Ann. Meet. Am. Soc. Microbiol.* **H-25**, p. 143.

Phillips, A. T., and Mulfinger, L. M., 1981, Cyclic adenosine 3',5'-monophosphate levels in *Pseudomonas putida* and *Pseudomonas aeruginosa, J. Bacteriol.* **145**:1286–1292.

Proctor, W. D., Hager, P. W., and Phibbs, P. V., Jr., 1997, Purification and characterization of HexR, a putative repressor protein involved in the regulation of carbohydrate catabolism by *Pseudomonas aeruginosa* PAO1, Abstracts of the VI International Congress of *Pseudomonas*: Molecular Biology and Biotechnology, pg. 148.

Proctor, W. D., Arora, S., Hager, P., and Phibbs, P. V., Jr., 1997, Integration host factor and the putative repressor protein *hexR* bind the *hexC* locus of *Pseudomonas aeruginosa*, *Abstr. Annu. Meet. Am. Soc. Microbiol.* **K-95**, 357.

Ratnaningsih, E., Dharmsthiti, S., Krishnapillai, V., Morgan, A., Sinclair, M., and Holloway, B. W., 1990, A combined physical and genetic map of *Pseudomonas aeruginosa* PAO, *J. Gen. Microbiol.* **136**:2351–2357.

Rivers, D. B., and Blevins, W. T., 1987, Multiple enzyme forms of glyceraldehyde, 3-phosphate dehydrogenase in *Pseudomonas aeruginosa, J. Gen. Microbiol.* **133**:3159–3164.

Roehl, R. A., and Phibbs, P. V., Jr., 1981, Genetic mapping of mutations in the mannitol catabolic pathway of *Pseudomonas aeruginosa, Abstr. Ann. Meet. Am. Soc. Microbiol.* **K70**, 149.

Roehl, R. A., and Phibbs, P. V., Jr., 1982, Characterization and genetic mapping of fructose phosphotransferase mutations in *Pseudomonas aeruginosa, J. Bacteriol.* **149**:897–905.

Roehl, R. A., Feary, T. W., and Phibbs, P. V., Jr., 1983, Clustering of mutations affecting central pathway enzymes of carbohydrate catabolism in *Pseudomonas aeruginosa, J. Bacteriol.* **156**:1123–1129.

Romling, U., Duchene, M., Essar, D. W., Galloway, D., Guidi–Rontani, C., Hill, D., Lazdunski, A., Millet, R. V., Scheifer, K. H., Smith, D. W., Toschka, H. Y., and Tummler, B., 1992, Localization of *alg*, *opr*, *phn*, 4.5S RNA, 6S RNA, *tox*, *trp*, and *xcp* genes, *rrn* operons, and the chromosomal origin on the physical genome may of *Pseudomonas aeruginosa* PAO, *J. Bacteriol.* **174**:327–330.

Ryazanov, A. G., 1985, Glyceraldehyde 3-phosphate dehydrogenase is one of the three major RNA-binding proteins of rabbit reticulocytes, *FEBS Lett.* **182**:131–134.

Sage, A. E., Proctor, W. D., and Phibbs, P. V., Jr., 1996, A two-component response regulator, *gltR*, is required for glucose transport activity in *Pseudomonas aeruginosa, J. Bacteriol.* **178**:6064–6066.

Sage, A., Temple, L. M., Christie, G. E., and Phibbs, P. V., Jr., 1993, Nucleotide sequence and expression of the glucose catabolism and transport genes in *Pseudomonas aeruginosa, Prog. Abstr. Fourth Int. Symp. Pseudomonas: Biotechnology and Molecular biology*, 1993, Vancouver, British Columbia, Canada, p. 105.

Savrolac, E. G., Taylor, N. F., Benz, R., and Hancock, R. E. W., 1991, Purification of glucose-inducible outer membrane protein OprB of *Pseudomonas putida* and reconstitution of glucose-specific pores, *J. Bacteriol.* **173**:4970–4976.

Sawyer, M. H., Baumann, P., Baumann, L., Berman, S. M., Canovas, J. L., and Berman, R. H., 1977, Pathways of D-fructose catabolism in species of *Pseudomonas, Arch. Microbiol.* **112**(1):49–55.

Schweizer, H. P., 1991, The *agmR* gene, an environmentally responsive gene, complements defective *glpR*, which encodes the putative activator for glycerol metabolism in *Pseudomonas aeruginosa, J. Bacteriol,* **173**:6798–6806.

Schweizer, H. P., 1992, Allelic exchange in *Pseudomonas aeruginosa* using novel ColE1-type vectors and a family of cassettes containing a portable *oriT* and the counter-selectable *Bacillus subtilis sacB* marker, *Mol Microbiol.* **6**:1195–1204.

Schweizer, H. P., and Hoang, T., 1995, An improved system for gene replacement and *xylE* fusion analysis in *Pseudomonas aeruginosa, Gene* **158**:15–22.

Schweizer, H. P., and Po, C. 1994, Cloning and characterization of the *sn*-glycerol 3-phosphate dehydrogenase structural gene *glpD* of *Pseudomonas aeruginosa*, *J. Bacteriol.* **176**:2184–2193.

Schweizer, H. P., and Po, C., 1996, Regulation of glycerol metabolism in *Pseudomonas aeruginosa*: Characterization of the *glpR* repressor gene, *J. Bacteriol.* **178**:5215–5221.

Schweizer, H. P., Po, C., and Bacic, M. K., 1995, Identification of *Pseudomonas aeruginosa* *glpM*, whose gene product is required for efficient alginate biosynthesis from various carbon sources, *J. Bacteriol.* **177**:4801–4804.

Schweizer, H. P., Jump, R., and Po, C., 1997, Structure and gene-polypeptide relationships of the region encoding glycerol diffusion facilitator (*glpF*) and glycerol kinase (*glpK*) of *Pseudomonas aeruginosa*, *Microbiol.* **143**:1287–1297.

Siegel, L. S., and Phibbs, P. V., Jr., 1979, Glycerol and L-α-glycerol 3-phosphate uptake by *Pseudomonas aeruginosa*, *Curr. Microbiol.* **2**:251–256.

Siegel, L. S., Hylemon, P. B., and Phibbs, P. V., Jr., 1977, Cyclic adenosine 3',5'-monophosphate levels and activities of adenylate cyclase and cyclic adenosine 3',5'-monophosphate phosphodiesterase in *Pseudomonas* and *Bacteroids*, *J. Bacteriol.* **129**:87–96.

Singh, R., and M. R. Green, 1993, Sequence-specific binding of transfer RNA by glyceraldehyde, 3-phosphate dehydrogenase, *Science* **259**:365–368.

Sly, L. M., Worobec, E. A., Perkins, R. E., and Phibbs, P. V., Jr., 1993, Reconstitution of glucose uptake and chemotaxis in *Pseudomonas aeruginosa* glucose transport defective mutants, *Can. J. Microbiol.* **39**:1079–1083.

Stinnet, J. D., and Eagon, R. G., 1973, Comparison of protein content of cytoplasmic membrane and outer cell wall membrane of *Pseudomonas aeruginosa*, *Abstr. Annu. Meet. Am. Soc. Microbiol.* **73**:182.

Stinnett, J. D., Guymon, L. F., and Eagon, R. G., 1973, A novel technique for the preparation of transport-active membrane vesicles from *Pseudomonas aeruginosa*: Observations on gluconate transport, *Biochem. Biophys. Commun. Res.* **52**:284–290.

Stinson, M. W., Cohen, M. A., and Merrick, J. M., 1977, Purification and properties of the periplasmic glucose-binding protein of *Pseudomonas aeruginosa*, *J. Bacteriol.* **131**:672–681.

Temple, L., Cuskey, S. M., Perkins, R. E., Bass, R. C., Morales, N. M., Christie, G. E., Olsen, R. H., and Phibbs, P. V., Jr., 1990, Analysis of cloned structural and regulatory genes for carbohydrate utilization in *Pseudomonas aeruginosa* PAO, *J. Bacteriol.* **172**:6396–6404.

Temple, L., Sage, A. E., Christie, G. E., and Phibbs, P. V., Jr., 1994, Two genes for carbohydrate catabolism are divergently transcribed from a region of DNA containing the *hexC* locus in *Pseudomonas aeruginosa* PAO1, *J. Bacteriol.* **176**:4700–4709.

Terry, J. M., Pina, S. E., and Mattingly, S. J., 1991, Environmental conditions which influence mucoid conversion in *Pseudomonas aeruginosa* PAO1, *Infect. Immun.* **59**:471–477.

Terry, J. M., Pina, S. E., and Mattingly, S. J., 1992, Role of energy metabolism in conversion of nonmucoid *Pseudomonas aeruginosa* to the mucoid phenotype, *Infect. Immun.* **60**:1329–1335.

Tiwari, N. P., and Campbell, J. R. R., 1969, Enzymatic control of the metabolic activity of *Pseudomonas aeruginosa* grown in glucose or succinate medium, *Biochim. Biophy. Acta* **192**:395–401.

Trias, J., Rosenberg, E. Y., and Nikaido, H., 1988, Specificity of the glucose channel formed by protein D1 of *Pseudomonas aeruginosa*, *Biochim. Biophys. Acta* **938**:493–496.

Tsay, S.-S., Brown, K. K., and Gaudy, E. T., 1971, Transport of glycerol by *Pseudomonas aeruginosa*, *J. Bacteriol.* **108**:82–88.

Van Dijken, J. P., and Quayle, J. R., 1977, Fructose metabolism in four *Pseudomonas* species. *Arch. Microbiol.* **114**(3):281–286.

Voegel, R. T., Sweet, G. D., and Boos, W., 1993, Glycerol kinase of *Escherichia coli* is activated by interaction with the glycerol facilitator, *J. Bacteriol.* **175**:1087–1094.

Wallace, W. H., 1989, Genetic and biochemical analysis of gluconate utilization in *Pseudomonas aeruginosa* PAO1, Ph.D. Thesis, Department of Microbiology and Immunology, East Carolina University School of Medicine, Greenville, NC.

Wallace, W. H., and Phibbs, P. V., Jr., 1988. Chromosomal mapping of mutations affecting the oxidative pathway of glucose catabolism in *Pseudomonas aeruginosa* PAO, *Abstr. Ann. Meet. Am. Soc. Microbiol.* **K-115**, 225.

Weissenborn, D. L., and Larson, T. J., 1992, Structure and regulation of the *glpFK* operon encoding glycerol diffusion facilitator and glycerol kinase of *Escherichia coli* K-12, *J. Biol. Chem.* **267**:6122–6131.

Whiting, P. H., Midgley, M., Dawes, E. A., 1976a, The role of glucose limitation in the regulation of the transport of glucose, gluconate, and 2-oxo-gluconate and of glucose metabolism in *Pseudomonas aeruginosa*, *J. Gen. Microbiol.* **92**:304–310.

Whiting, P. H., Midgley, M., Dawes, E. A., 1976b, The regulation of transport of glucose, gluconate, and 2-oxo-gluconate and of glucose metabolism in *Pseudomonas aeruginosa*, *Biochem. J.* **154**:659–668.

Williams, S. G., Greenwood, J. A., and Jones, C. W., 1994, The effect of nutrient limitation on glycerol uptake and metabolism in continuous cultures of *Pseudomonas aeruginosa*, *Microbiol.* **140**:2961–2969.

Wolff, J. A., and Phibbs, P. V., Jr., 1986, Construction and use of a small cosmid cloning vector that replicates in *Pseudomonas aeruginosa*, *Plasmid* **16**:228.

Wolff, J. A., MacGregor, C. H., Eisenberg, R. C., and Phibbs, P. V., Jr., 1991, Isolation and characterization of catabolite repression control mutants of *Pseudomonas aeruginosa* PAO, *J. Bacteriol.* **173**:4700–4706.

Wozniak, D. J., and Ohman, D. E., 1993, Involvement of the alginate *algT* gene and integration host factor in the regulation of the *Pseudomonas aeruginosa algB* gene, *J. Bacteriol.* **75**:4145–4153.

Wylie, J. L., and Worobec, E. A., 1993, Substrate specificity of the high-affinity glucose transport system of *Pseudomonas aeruginosa*, *Can. J. Microb.* **39**:722–725.

Wylie, J. L., and Worobec, E. A., 1994, Cloning and nucleotide sequence of the *Pseudomonas aeruginosa* glucose-selective OprB porin gene and distribution of OprB within the family *Pseudomonadaceae*, *Eur. J. Biochem.* **22**:505–512.

Wylie, J. L., and Worobec, E. A., 1995, The OprB porin plays a central role in carbohydrate uptake in *Pseudomonas aeruginosa*, *J. Bacteriol.* **17**:3021–3026.

Wylie, J. L., Bernegger-Egli, C., O'Neil, J. D. J., and Worobec, E. A., 1993. Biophysical characterization of OprB, a glucose-inducible protein of *Pseudomonas aeruginosa*, *J. Bioenerg. Biomembr.* **25**:547–556.

Polysaccharides

Lipopolysaccharide and Capsular Polysaccharide

3

ESTELLE J. McGROARTY

1. INTRODUCTION

In the study of polysaccharides from *Pseudomonas*, the two most commonly examined species are *Pseudomonas aeruginosa* and *Pseudomonas syringae*. The fluorescent pseudomonads that are phytopathogenic are collectively classified as *P. syringae* and contain more than 40 distinct pathovars (Young *et al.*, 1978). These pathovars damage many agriculturally important plants (Gvozdyak *et al.*, 1989) and generally possess a narrow host range (Fahy and Loyd, 1983). *Pseudomonas aeruginosa* is commonly found in aquatic environments, and its high adaptive potential is reflected in its importance as an opportunistic animal pathogen. Infections of the respiratory tract, urinary tract, burn wounds, and blood are often lethal (Cross *et al.*, 1983). In patients with cystic fibrosis (CF), *P. aeruginosa* causes chronic pulmonary infections which are resistant to antibiotic therapy, making this pathogen the major cause of morbidity and mortality in these patients (Thomassen *et al.*, 1987). Cell envelope components may be critical for the pathogenicity of these pseudomonads. Because the literature on these structures is more complete, the structures of the envelope polysaccharides from *P. aeruginosa* are presented in detail in this review. The structures of polysaccharides from other species, especially the lipopolysaccharide (LPS) of *P. syringae*, are also presented for comparison.

The cell envelope of *P. aeruginosa* and other pseudomonads is typical of a Gram-negative organism, containing a proximal cell membrane

ESTELLE J. McGROARTY • Department of Biochemistry, Michigan State University, East Lansing, Michigan 48824.

Pseudomonas, edited by Montie. Plenum Press, New York, 1998.

and an intermediate peptidoglycan layer. The distal layer is comprised of an outer membrane, the outer monolayer of which consists of LPS and protein. The LPS layer is exposed on the cell surface and plays a critical role in the interaction of the cell with the environment, including other biological systems, and can be the dominant antigen during bacterial infections. An additional mucoid capsulelike coating can be detected on the surface of many pseudomonad strains. For *P. aeruginosa* this glycocalyx is comprised mainly of neutral and acidic polysaccharides. The polysaccharides in the lipopolysaccharides and the glycocalyx are the most important surface antigens in *P. aeruginosa*. Further, these polysaccharides appear to interfere with antibody-induced killing during infection (Hatano *et al.*, 1995). In this chapter these polysaccharides are considered, and their structure, synthesis, and role on the bacterial surface and in pathogenesis is discussed.

2. LIPOPOLYSACCHARIDE

2.1. Structure

The structures of the LPS from *Pseudomonas* species have been extensively studied because of their importance in host specificity for *P. syringae* and in infection for *P. aeruginosa*. The overall architecture of LPS molecules from *P. aeruginosa* is similar to that of many gram-negative species. They are made up of three regions: a lipid component, termed lipid A, a core oligosaccharide rich in phosphate residues, and a distal immunologically reactive O-polysaccharide chain rich in amino sugars (Lüderitz *et al.*, 1982). Many studies have shown that LPS on the cell surface is heterogeneous in its structure. This heterogeneity is discussed in detail later. The hydrophobic lipid A forms the outer monolayer of the outer membrane and anchors the molecule into the cell wall, making it an integral component of this membrane. The structure of lipid A is discussed in another chapter. Only the general features are highlighted here. This component of LPS is responsible for inducing the endotoxic effects of LPS and participates in initiating a number of pathophysiological responses (Kropinski *et al.*, 1985). The lipid A is comprised of a phosphorylated $\beta 1 \rightarrow 6$ glucosamine disaccharide head group onto which fatty acids are ester- and amide-linked. The fatty acids are saturated hydroxy and nonhydroxy fatty acids (Bhat *et al.*, 1990; Wilkinson, 1983) but differ from those in the lipid A of enteric bacteria. *P. aeruginosa* LPS lacks the typical hydroxytetradecanoic acid but instead contains 3-hydroxydecanoic, 3-hydroxydodecanoic and 2-hydroxydodecanoic acids (Wilkinson, 1983). Complete analysis of the identity and positions of the ester- and amide-bound fatty acids in *P. aeruginosa* LPS has been defined by several groups (Bhat *et al.*, 1990; Karunaratne *et al.*, 1992). It has also been demonstrated that the lipid A disaccharide head group is substituted at the C-1 and C-4′ positions with phosphomonoesters (Karunaratne *et al.*, 1992; Masoud *et al.*, 1994) and that for PAO1 strains, 4-amino-4-deoxyarabinose is bound in substochiometric amounts to the C-4′ through a phosphoester link (Bhat *et al.*, 1990).

The components detected in the core region of LPS are more conserved than the O-polysaccharide substituents when comparing different *P. aeruginosa* strains. Components commonly found in the core oligosaccharide include D-glucose, L-rhamnose, D-galactosamine, L-glycero-D-manno-heptose, 2-keto-3-deoxyoctulosonic (KDO) acid and unusually high concentrations of phosphate (Karunaratne *et al.*, 1992; Kroponski *et al.*, 1985; Peterson *et al.*, 1985; Rivera *et al.*, 1988a). Other noncarbohydrate components are also in this oligosaccharide, including L-alanine and ethanolamine (Wilkinson, 1983). The core region can be divided into two regions, the outer core and inner core. The inner core is comprised of KDO, heptose, phosphate, and probably ethanolamine (Masoud *et al.*, 1995). A recent analysis of the core structure of LPS from O6 serotype indicates the presence of a 7-*O*-carbamoyl heptose (Masoud *et al.*, 1995). The outer core commonly contains D-glucose, L-rhamnose, D-galactosamine and L-alanine, but the detailed chemical structure of this region has been reported for only a few strains (Drewry *et al.*, 1975; Masoud *et al.*, 1994, 1995; Rowe and Meadow, 1983). Generally, the core is attached to the lipid A head group through an acid labile ketosidic linkage between KDO and the C-6′ of Gln II (see Fig. 1). One or two heptose residues are distal to the KDO components and in some strains a branched glucose residue which, along with ethanolamine and high levels of phosphate, comprises the remainder of the inner core. The outer core lacks any phosphate substituents, and structural heterogeneity in this region in isolates from different strains has been determined using monoclonal antibodies as probes. Evidence indicates that different serotypes have similar or identical inner core structures, but the structure of the outer cores diverge (De Kevit and Lam, 1994; Masoud *et al.*, 1994; Yokota *et al.*, 1989, 1992), although the structural differences are not well defined (Yokota *et al.*, 1989). Recently, specific differences in the outer core structure between O5 and O6 serotypes have been reported (Dasgupta *et al.*, 1994). The alanine is reportedly attached only to a portion of the molecules from a single strain, bonded to galactosamine (Kropinski *et al.*, 1985; Lam *et al.*, 1989). Other groups in the core that may vary from molecule to molecule include the content and position of phosphate and ethanolamine (Knirel, 1990).

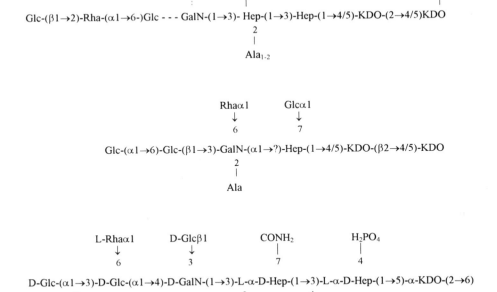

Figure 1. Proposed structures for the core oligosaccharide region of lipopolysaccharide from an O2 strain (top) (Drewry *et al.*, 1975), an O3 strain (center) (Rowe and Meadow, 1983) and an O6 strain (bottom) (Masoud *et al.*, 1995) of *Pseudomonas aeruginosa*. Abbreviations: Glc, glucose; GalN, galactosamine; Hep, L-*glycero*-D-*manno*-heptose; KDO, 2-keto-3-deoxyoctulosonic acid; Rha, rhamnose; EtN, ethanolamine.

The outermost component of LPS, the O-polysaccharide, is composed of repeating oligosaccharide units of usually two to five residues and is linked to the core oligosaccharide (Knirel, 1990). Differences in the structure of the O-polysaccharides from different strains form the basis of the O-serotype classification of gram-negative bacteria including *P. aeruginosa*. The different O-antigen structures induce specific antibodies used in the serotype classification. Different serotype classification systems for *P. aeruginosa* have been described, including the Fisher system containing seven serogroups (Fisher *et al.*, 1969; Horton *et al.*, 1977), the Habs scheme of 12 serotypes (habs, 1957), the Lanyi system which distinguishes 23 serotypes (Lanyi, 1966), the Homma system of 15 serotypes (Homma, 1976, 1982), and the most extensively used serotyping system, the International Antigenic Typing Scheme (IATS), consisting of 17 standard O serotypes (Liu *et al.*, 1983). Recently an additional three serotypes have been reported, bringing the total number of distinct IATS serotypes to 20 (Liu and Wang, 1990). Thus, it appears that for this species different strains can make at least 20 distinct O-polysaccharide chains. The precise chemical structures of 17 of the IATS standard serotypes have been characterized by Knirel and co-workers (1988a).

When surveying the composition of the O-polysaccharide of different serotypes, it can be seen that these structures contain very unusual sugar components not commonly found in other natural polysaccharides. There are several major classes of monosaccharide components. The neutral monosaccharides include D-ribose, D-xylose, D-glucose, and L-rhamnose. Monoamino sugars are commonly found in these polymers, but they consist of very uncommon aminosaccharides and include the deoxyhexosamines, D-quinovosamine D- and L-fucosamine, and their derivatives, such as bacillosamine. Acidic aminosugars are also detected in a number of the O-polysaccharides including 2-amino-2-deoxyuronic acids, 2,3-diamino-2,3-dideoxyuronic acids, and 5,7-diamino-3,5,7,9-tetradeoxynonulosonic acids. Most of the amino groups on the aminosugars are acetylated but several O-polysaccharides are substituted by other acyl groups, such as formyl, acetimidoyl and hydroxybutyryl moieties. Also, many of the sugars are modified with O-acetyl groups (Knirel, 1990).

The structure of the repeating units of O-polysaccharides from *P. aeruginosa* consist of linear tri- or tetrasaccharide repeat units with only one group showing a branched repeat unit (Knirel, 1990; Knirel *et al.*, 1988a). The regularity of the repeat units is disrupted by nonstoichiometric O-acetylation of up to 40% of the repeating units in the O3 serogroup and by amidation of carboxyl groups of uronic acid groups in the O13 serogroup. Also in the O2 serogroup there is partial epimerization at the C-5 of one of 2,3-diamino-2,3-dideoxyuronic acid residues. Another anomaly in the repeat structure of *P. aeruginosa* O-antigens is seen in the O9 serogroup. The repeat units in this group have amidic and glycosidic linkages.

The structure of the O-polysaccharides of *P. syringae* strains has been studied in some detail; the pathovars are reported to be heterogeneous and, on the basis of polysaccharide composition, are divided into 9 serogroups (Pastushenko and Simonovich, 1979). In the last several years the repeat structure of the O-polysaccharide of many of these biovars has been chemically defined (Barton-Willis *et al.*, 1987; Das *et al.*, 1994; Gross *et al.*, 1988; Gbozdyak *et al.*, 1989; Knirel *et al.*, 1986,

1988b,c,d, 1993; Shashkov *et al.*, 1990, 1991; Smith *et al.*, 1985; Vinogradov *et al.*, 1991a,b,c). The general features of the O-polysaccharides from these biovars include an usually high level of either D- or L-rhamnose and a much more homogeneous composition compared with the O-polysaccharides from serogroups of *P. aeruginosa*. The O-polymers from *P. syringae* biovars are composed of both linear and branched repeat units of three to five monosaccharide components. Other monosaccharide units in the repeat unit for different strains include D-fucose, D-*N*-acetyl-4-fucosamine, D-*N*-acetyl-3-fucosamine, D-*N*-acetylquinivosamine, and *N*-acetylglucosamine.

2.2. Lipopolysaccharide Heterogeneity

The LPS synthesized by strains of *Pseudomonas*, like that of many other gram-negative species, consists of a heterogeneous set of molecules of different sizes and some heterogeneity in chemical composition. Structural microheterogeneity in several regions of the LPS from a given *P. aeruginosa* strain has been demonstrated (Kropinski *et al.*, 1985; Wilkinson, 1983). This includes the presence of nonstoichiometric levels of specific groups in the core region mentioned above including variations in the amount or presence of alanine, phosphate, and ethanolamine. LPS is also heterogeneous in size because of the variable number of O-antigen repeat units from one molecule to the next (Kropinski *et al.*, 1985; Rivera *et al.*, 1988a). When separated by sodium dodecyl sulfate polyacrylamide gel electrophoresis (SDS-PAGE), LPS isolates present a ladderlike pattern of bands representing molecules of different sizes ranging from the fastest moving molecules containing only core + lipid A to molecules with as many as 80 O-repeat units (Knirel, 1990; Rivera *et al.*, 1988a). Gel exclusion chromatographic analysis of LPS resolves three or four distinct size classes. A large percentage (>90%) of the molecules is composed of core-lipid A lacking an O-polysaccharide (Chester and Meadow, 1975; Kropinski *et al.*, 1985; Rivera *et al.*, 1988a; Wilkinson and Galbraith, 1975). The other size classes are sets of molecules with a limited distribution of O-polymer lengths.

Additional structural heterogeneity exists in the O-antigen repeats. Most common is the nonstoichiometric O-acetylation of sugars in the O3 and O13 subgroups, amidation of carboxyl groups of uronic acids in the O6 serogroup, and only partial epimerization of diamino-dioxyuronic acids in the O2 serogroup (Knirel, 1990). It is not known if the variations in modification are within or between O-polymers.

Recent studies (Rivera and McGroarty, 1989; Rivera *et al.*, 1988a) indicate that *P. aeruginosa* produces two chemically distinct forms of LPS known as A-band and B-band LPS. The B-band LPS is the O-antigen-containing LPS and determines the O-specificity of the bacterium, whereas the A-band LPS contains shorter chains of predominantly neutral polysaccharide and lacks phosphate but contains stoichiometric amounts of sulfate. The neutral polysaccharide of the A-band is composed mainly of D-rhamnose with lesser amounts of 3-*O*-methyl-rhamnose, ribose, glucose, and mannose (Arsenault *et al.*, 1991). The polysaccharide is a D-rhamnan polysaccharide comprised of the repeat structure: → 3)D-Rha(α1 → 3)D-Rha(α1 → 2)D-Rha(α1 → (Kocharova *et al.*, 1988; Yokota *et al.*, 1987). This second population of molecules with a distinct sugar repeat forms a second ladder on SDS-PAGE which can be resolved when the mixture of LPS molecules is separated on a gel filtration column and the fractions are analyzed by SDS-PAGE (Rivera *et al.*, 1988a). Interestingly, the A-band sugar repeat structure is identical to that reported for the O-polysaccharide chain in the LPS of *P. syringae* pv *morsprunorum* C28 (Smith *et al.*, 1985). The presence of D-rhamnose in this molecule is in contrast to the presence of only L-rhamnose in some of the O-antigens (Knirel, 1990). This D-rhamnan polymer appears to be present on many serotypes of *P. aeruginosa* and is detected with specific monoclonal antibodies (Kocharova *et al.*, 1988; Sawada *et al.*, 1985a). This common polysaccharide LPS has been isolated from a rough mutant of *P. aeruginosa* strain AK1401, and it was shown that the rhamnan polymer is digested by a rhamnase borne by a *P. syringae* pv *morsprunorum* typing phage A7 (Rivera *et al.*, 1992). It was shown that A-band core-lipid A isolates contain only low levels of KDO, heptose, and amino sugars. Monoclonal antibodies specific to the inner and outer core regions of B-band LPS indicate that the inner core of the A-band molecules is similar to B-band LPS but that the outer core regions contain different epitopes and structures (Rivera and McGroarty, 1989). This rhamnan polymer is also present on nontypable clinical isolates of *P. aeruginosa* from patients with cystic fibrosis (Lam *et al.*, 1989; Sawada *et al.*, 1985a). Thus, A-band LPS is considered present on many or most strains of *P. aeruginosa*. It has been suggested that the rhamnan polymer is attached to the same molecules as the O side chains (Hatano *et al.*, 1993). However, the O-polysaccharide-containing LPS and D-rhamnan LPS have been separated and shown to be on distinct populations of core-lipid A molecules (Rivera *et al.*, 1988a, 1992; Rivera and McGroarty, 1989). Thus, it appears that many strains of *P. aeruginosa* can produce two chemically and antigenically distinct LPS molecules. Because the D-rhamnan attached to the A-band LPS is significantly shorter than the O-polymer on the B-band species (Rivera *et al.*, 1988a), the common antigen is usually masked on the surface of the cell and cannot be detected by immunological techniques (Hatano *et al.*, 1995; McGroarty and Rivera, 1990).

2.3. Biosynthesis and Genetic Regulation of LPS

Little is known about the biosynthetic or regulatory genes controlling the production of LPS by *P. aeruginosa* or any other pseudomonad species. If the synthesis is similar to that of enteric bacteria, the $(KDO)_2$-lipid A, core, and the O-polysaccharide are synthesized separately on the inner membrane (Raetz, 1990). The initial steps synthesis of the KDO_2-lipid A for *Escherichia coli* are acylation of UDP-glucosamine at the 2 and 3 positions with 3-hydroxymyristyl-ACP as the acyl donor. Then two diacyl glucosamine moieties are combined via a β, $1 \rightarrow 4$ linkage, and the $4'C$ position is phosphorylated forming a product termed Lipid IV_a. Two KDO residues are attached to the $6'$ position using CMP-KDO as the precursor. The KDO_2-lipid IV_a is further acylated with fatty acids on the hydroxy groups of the hydroxymyristyl moieties attached to the nonreducing glucosamine. Only a few mutants in the genes involved in lipid A synthesis have not been isolated or characterized, presumably because of the importance of lipid A for cell viability (Raetz, 1987).

In contrast, genes involved in the synthesis of the core oligosaccharide of *E. coli* have been studied in detail. Cells lacking all but the two inner KDO moieties are viable, and mutants in the enzymes involved in core synthesis, termed rough or R mutants, have been studied in detail (Mäkelä and Stocker, 1984). The genes involved in core synthesis are termed the *rfa* genes and, depending on where in the biosynthetic process the mutation lies, the resulting core structure may be more or less complete, giving a variety of chemotypes termed Ra through Re, respectively. Mutations in the synthesis of the O-specific units will also be rough or R mutants, but these mutants always synthesize a complete core (Ra type). The enzymes involved in the synthesis of the core of *E. coli* include five transferases, and the *rfa* genes are in a main cluster at 81 min on the genetic map.

For *E. coli* the O-repeat units are synthesized by enzymes coded by genes in the *rfb* gene cluster. The assembly of the O-repeat unit takes place by sequential transfer of the monosaccharide residues to their appropriate acceptors on the undecaprenol carrier (Mäkelä and Stocker, 1984). This assembly requires a specific sugar transferase. Then the completed repeat unit on the carrier is transferred to the growing polysaccharide chain, still attached to the carrier lipid, via the action of a polymerase coded by the *rfc* gene. Mutants lacking a functional polymerase have single O-units in their LPS and are termed SR phenotypes.

The attachment of the O-polysaccharide to the complete core is a complex reaction in *E. coli* (Mäkelä and Stocker, 1984), and this ligase has not been studied in detail (Raetz, 1990).

Only a few rough mutants of *P. aeruginosa* have been isolated and studied in detail. Kropinski and co-workers have reported isolating and characterizing several *rfa*-like mutants of O5 serotype strains which synthesize incomplete core structures and have a rough phenotype. These include strains equivalent to the *Salmonella* Rc chemotype, e.g., strain AK1012 (Jarrell and Kropinski, 1977), and to Rd1 chemotype, e.g., strain AK1282 (Jarrell and Kropinski, 1981). In addition, a mutant O5 serotype strain, which is missing the O-polysaccharide polymerase and which expresses an SR phenotype, strain AK1401, has been isolated and characterized (Berry and Kropinski, 1986). Rowe and Meadow (1983) also isolated a rough mutant of O3 strains which lacked rhamnose and glucose. More recently, the composition of core-deficient mutants derived from O3, O5, and O6 serotypes have been described (Dasgupta *et al.*, 1994). Some of these mutants produced LPS deficient in D-glucose and/or L-rhamnose. Mutants which produce a complete core also produced A-band LPS with a ladder of increasing numbers of D-rhamnan repeats. However, mutants with a deficient core, in some cases, do not produce either A-band or B-band ladders in gels, indicating that the cores for A-band and B-band LPS share biosynthetic pathways. It has also been shown that *algC*, the gene that codes for phosphoglucomutase, is required to synthesize a complete lipopolysaccharide core (Coyne *et al.*, 1994). Presumably this enzyme is involved in producing glucose substrates for core synthesis. *AlgC* mutants also do not synthesize A-band polysaccharide (Coyne *et al.*, 1994), confirming the similarity of core structures for A-band and B-band LPS molecules. The enzyme phosphoglucomutase is also required for synthesizing O side chains in O5 serotypes (Goldberg *et al.*, 1993). The AlgC protein is also a critical enzyme involved in the synthesis of alginate by this organism. Complex genetic control of the alginate biosynthetic genes, including *algC*, may couple the synthesis of LPS and alginate. This is discussed in detail later in the section on alginate synthesis.

The *rfb* regions for both B-band (Evans *et al.*, 1994; Goldberg *et al.*, 1992; Lightfoot and Lam, 1993) and A-band (Lightfoot and Lam, 1991, 1993) LPS have been cloned, and it has been shown that one of the A-band gene products is involved in converting GDP-D-mannose to GDP-rhamnose (Lightfoot and Lam, 1991). The genes involved in synthesizing A-band and B-band LPS are on two separate gene clusters on the *P. aeruginosa* genome. The A-band LPS genes are located at approximately 12 min and the B-band genes near 37 min (Lightfoot and Lam,

1993). Recently the *rfbA* gene involved in B-band LPS biosynthesis was isolated, cloned, sequenced, and expressed (Goldberg *et al.*, 1992).

The pattern of LPS heterogeneity is affected by growth conditions. For instance, during chronic lung infections in cystic fibrosis patients, the serotype of *P. aeruginosa* appears to change. The cells become nontypable and lose the ability to synthesize the population of LPS with long O-polymers (Hancock *et al.*, 1983; Penketh *et al.*, 1983). These strains are polyagglutinable, react with antibodies from different serotypes, and are thought to exhibit a common antigen on the cell surface (Fomsgaard *et al.*, 1988). These changes in LPS patterns in nontypable CF isolates of *P. aeruginosa* also give the cells low-level resistance to aminoglycoside antibiotics (Bryan *et al.*, 1984). The serotype changes during the course of CF infection have been followed. In many cases, the original strain was initially typable but lost the serotype specificity during prolonged infection (Lam *et al.*, 1989). Further, the initial isolates produced both A-band, i.e., common antigen LPS, and the O-serotype specific B-band molecules. During the course of infections, however, the cells lose the B-band O-polysaccharide but retain A-band antigen on the cell surface. Thus, the A-band LPS may comprise the polyagglutinable common antigen in these strains (Lam *et al.*, 1989).

Specific environmental signals alter LPS heterogeneity in a more reversible manner. It has been shown that changes in the carbon source changes LPS composition and the cell's resistance to polymyxin (Gilleland and Conrad, 1980). Growth temperature also affects the size heterogeneity of LPS. Cells grown at lower temperatures produce higher amounts of long-chain O-polymer containing B-band LPS than cells grown at high temperatures (Kropinski *et al.*, 1987; McGroarty and Rivera, 1990). Interestingly the heterogeneity of A-band LPS is unaffected by growth temperature (McGroarty and Rivera, 1990). Stress conditions other than high temperature also decrease the amount of very long chain O-polysaccharides bound to B-band LPS. These stresses include high salt concentrations, low pH, and high osmotic strength in the medium. Growth limiting levels of phosphate also dramatically decrease the amount of O-polymer-containing B-bands and increase the amount of A-band LPS (McGroarty and Rivera, 1990). Under conditions in which high amounts of long chain O-polymers were synthesized, the cells were agglutinated by serotype-specific monoclonal antibodies but not by monoclonals specific to the A-band D-rhamnan. Under stress conditions, however, where long chain B-band molecules were diminished, A-band LPS could be detected. Presumably the longer O-polymers mask the shorter A-band molecules from detection by antibodies.

2.4. Lipopolysaccharide Physical Properties

Lipopolysaccharides form very tight complexes on the surface of the outer membrane, helping to prevent deleterious materials, such as detergents and antibiotics, from penetrating the cell. Part of the tight interactions are caused by strong ionic interactions. Lipopolysaccharides are anionic due to the high levels of phosphate groups and the presence of acidic sugars (e.g., KDO). On the surface of the cell, LPS is bound with cations, mainly the metal cations, Ca^{2+} and Mg^{2+} (Coughlin et al., 1983a). The physical properties of LPS are dramatically affected by the type of cations bound (Coughlin et al., 1983b). It has been shown that different polyamines and polycationic antibiotics have differential affinities for the LPS from P. aeruginosa dependent on the charge of the cation (Peterson et al., 1985). Cationic antibiotics, such as the aminoglycosides and polymyxins, disrupt LPS aggregates by ionically binding to the core region, displacing metal cations and rigidifying the LPS head group (Day, 1980; Loh et al., 1984; Peterson et al., 1985). This disruption of LPS interaction by aminoglycoside antibiotics may explain the mechanism for transport of these antibiotics across the outer membrane, termed the self-promoted uptake pathway (Hancock, 1984). It has been shown that the aminoglycoside, amikacin, disrupts the cell envelope of P. aeruginosa, presumably by displacing metal cations that stabilize the LPS in the outer membrane (Walker and Beveridge, 1988). Polymyxin also binds to multiple sites in the core and lipid A regions (Moore et al., 1986) and disrupts LPS structure (Peterson et al., 1985). Mutants of P. aeruginosa supersusceptible to aminoglycoside antibiotics, in some cases, are altered in their LPS structure and have an increased affinity of polycations (Rivera et al., 1988b). The ionic groups critical for tight interactions between LPS molecules are located in the core and lipid A. LPS-defective mutants which lack part of their core structure have increased susceptibility to dyes, detergents, and antibiotics (Kropinski et al., 1978). The susceptibility to these agents increases with the level of core defect. The roughest mutants exhibit the greatest susceptibility. The lack of ionic interactions in the core of these strains presumably decreases the strength of interactions between the molecules. In a study of mutant strains lacking either the A-band or B-band polysaccharide, strains which contained only the B-band polymer also bound the aminoglycoside, gentamicin, with higher affinity than in strains with A-band LPS (Kadurugamuwa et al., 1993). Strains with B-band LPS were also more prone to gentamicin killing, but cells containing both polymers bound gentamicin with the highest affinity and were the most susceptible to antibiotic attack. Presumably, the interactions between these two types of LPS are critical in forming a tight barrier and in binding ions within the complex.

2.5. Biological Activity of Lipopolysaccharide

The immunogenicity of the polysaccharide of LPS is one of its critical biological activities, but this is discussed separately in the next section. Besides eliciting an immune response, LPS has other specific biological activities. The polysaccharide portion is a virulence factor, and its presence increases the pathogenicity of the cell compared with rough mutant strains (Cryz *et al.*, 1984; Kropinski and Chadwick, 1975). The presence of the long O-polymer apparently protects the cells from opsonization and phagocytosis by polymorphonuclear leukocytes during infection, making smooth cells more resistant to serum killing than rough mutants (Engels *et al.*, 1985). Adjuvant and interferon-inducing activities of LPS from *P. aeruginosa* have been attributed to the lipid A portion of the molecule alone. The ability to inhibit tumor cell growth, however, may require both the lipid A and core structures (Cho *et al.*, 1979). Recently it has been shown that the antitumor activity of polysaccharides and lipopolysaccharides results from the induction of TNF-α in monocytes by these polymers (Otterlie *et al.*, 1993). During infection, LPS also serves as an adhesin helping bind the bacteria to cell surfaces within the infected host. Recently, it has been shown that LPS from *P. aeruginosa* binds to the glycolipid asialo GM1, and this binding may serve as the main interaction between bacteria and host (Gupta *et al.*, 1994).

2.6. Immunology of LPS

During the course of infection *Pseudomonas* induces the production of antibodies to the O-polymer and, to less of a degree, to the core region and lipid A (Gaston *et al.*, 1986; Kronborg *et al.*, 1992; Lam *et al.*, 1983). The different O-polymers produced by different strains of this organism define the serotype groupings of *P. aeruginosa*, as described above. In addition, it has been shown that serum samples from cystic fibrosis patients infected with *P. aeruginosa* contain antibodies to the common antigen, A-band polysaccharide (Lam *et al.*, 1989). The immunogenicity may vary for different LPS molecules (Knirel, 1990). The production of antibodies directed against O-antigen has a protective role during experimental *Pseudomonas* infection (Cryz *et al.*, 1985; Knirel, 1990; Pennington *et al.*, 1986). Immunization with rough LPS lacking an O-polymer does not afford general protection to infection

(Cryz *et al.*, 1985). Monoclonal and polyclonal antibodies against specific serotype O-antigens are becoming increasingly important in identifying, diagnosing, and treating *P. aeruginosa* infections (Fomsgaard *et al.*, 1989; Knirel, 1990; Lam *et al.*, 1987a,b; Lang *et al.*, 1989; Sawada *et al.*, 1984, 1985b, 1987). These antibodies have also been used in defining structures in the O-polysaccharide (Lam *et al.*, 1992) and outer core region (De Kevit and Lam, 1994) of LPS. It is reported that monoclonal antibodies to the outer core of LPS have protective activity against infection (Terashima *et al.*, 1991) whereas others have indicated that anticore antibodies are not opsonic and do not provide protection from infection (Sadoff *et al.*, 1985). Antibodies to the D-rhamnan of the B-band LPS reportedly do not have protective activity against infection (Makarenko *et al.*, 1993).

3. CAPSULAR POLYSACCHARIDES AND SLIME

3.1. Introduction

Pseudomonas aeruginosa produces an extracellular slime layer (Haynes, 1951) which, in liquid medium, causes the medium to become viscous after a few days. This extracellular slime or capsular material can be released from the cell surface by homogenizing cells, sedimenting the intact cells, and recovering the slime layer from the supernatant solution by ethanol precipitation (Brown *et al.*, 1969). When grown on agar plates, these cells produce "slimy" or mucoid colonies (Phillips, 1969). Early studies on the composition of slime were contradictory. One group indicated that the slime contains a gluconate (Haynes, 1951) whereas others could not detect uronic acids but indicated that the material was composed largely of mannan (Eagon, 1956, 1962). The group of Warren and Gray (1954, 1955) reported that one of the polysaccharides produced by *P. aeruginosa* is hyaluronic acid. Careful analysis of crude capsular isolates indicated that it is predominantly polysaccharide (Brown *et al.*, 1969; Liu *et al.*, 1961; Sadoff, 1974) but also contains nucleic acid, proteins (Brown *et al.*, 1969; Pier *et al.*, 1978b), LPS (Pier *et al.*, 1978a), and a toxic glycolipoprotein (Arsenis and Dimitracopoulos, 1986; Bartell *et al.*, 1970; Dimitracopoulos and Bartell, 1979; Koepp *et al.*, 1981; Sensakovic and Bartell, 1977). From this mixture, two polysaccharides of particular interest have been purified and characterized. High molecular weight polysaccharides which have immuno cross-reactivity with LPS have been isolated and carefully characterized (Pier and Bennet; 1986; Pier, 1982, 1983; Pier *et al.*, 1978, 1981, 1983). A second extracellular polysaccharide, alginate, produced mainly by mucoid strains isolated from sputum cultures from patients with cystic fibrosis (Doggett *et al.*, 1971) has been studied in great detail. This alginate has been characterized as similar to the alginic acid produced by brown seaweed (Linker and Jones, 1966b) and by the soil bacterium *Acotobacter vinelandii* (Gorin and Spencer, 1966). Mucoid strains which produce large amounts of alginate seem to lose this ability in subculturing on nutrient plates. The change is detected by changes in colony morphology and by the isolation of nonmucoid revertant colonies (Gorvan, 1975; Lam *et al.*, 1980; Zierdt and Williams, 1975). In the next sections I outline the structure and properties of these two extracellular polysaccharides.

3.2. O-Capsular Polysaccharide-Structure

High molecular weight polysaccharides have been purified from culture supernatant solutions of serotypes IT-I, IT-2 (Pier and Bennet, 1986; Pier, 1983; Pier *et al.*, 1978a, 1983) and IT-3 (Pier *et al.*, 1984) using selective precipitation of nucleic acids, ethanol precipitation of the polysaccharides, heating the resuspended polysaccharides under acidic conditions, and extracting the heated samples with chloroform and with phenol (Pier, 1983). These high molecular weight polysaccharides have the same serological specificity and antigenic reactivity as the LPS serotype determinants (Pier, 1982, 1983; Pier *et al.*, 1978b, 1981). This polysaccharide is termed O-capsule in this review. Analysis of the composition of IT-1 and IT-2 O-capsules indicate that they are serologically identical and chemically similar to the O-polysaccharide chain from purified LPS (Pier and Bennet, 1986; Pier, 1978a, 1983). Differences reported in the composition between O-capsule and lipopolysaccharide O-polymer are as follows: purified O-capsular polysaccharide from IT1 and IT2 strains contains mannose, arabinose, xyulose, and galactose, sugars not in the O-polysaccharide chains of the LPS from these strains (Pier, 1978a, 1983). Furthermore, the O-capsule lacks lipid A (Cross *et al.*, 1983) and LPS core sugars KDO, glucosamine, heptose, and galactosamine, sugars which are detected in the O-polysaccharide from LPS isolates (Pier, 1978b, 1983). It has been shown that the high mannose detected in early studies of the high molecular weight polysaccharides was from a polymannose contaminate in the nutrient medium which copurified with the O-capsule (Pier, 1983; Pier *et al.*, 1983). In addition, careful analysis of the IT2 O-capsular polysaccharide indicated that it was composed of *N*-acetyl fucosamine and glucose in a molar ratio of 2:1, identical to the structure of the O-polysaccharide of LPS from this serotype (Pier and Bennet, 1986). Thus the high amounts of arabinose, galactose, and xylose reported in earlier studies of the O-capsule may have resulted from contamination by other polysaccharides (Pier *et al.*, 1984; Pier, 1983). The sugar composition and ^{13}C NMR analysis also indicated that the structure of O-capsular polysaccharide from an IT3 serotype strain is identical to the O-side chain of the LPS from this immunotype (Pier *et al.*, 1984). Thus, the main difference between O-capsule and LPS-O polysaccharide is size. The O-capsular polysaccharides are reported to be $> 10^5$ (Pier *et al.*, 1978b, 1981, 1984) whereas the O-polysaccharide from LPS of IT3 immunotype has a molecular weight of approximately 3×10^4 (Pier *et al.*, 1978a).

3.3. O-Capsule Function

The O-capsule of *P. aeruginosa* is similar to a class of polysaccharides described for strains of *E. coli* (Goldman *et al.*, 1982; Ørskov *et al.*, 1977). O-capsular polysaccharides from *E. coli* strains have been described which are immunochemically identical to the O-polysaccharide of LPS, lack lipid A and the sugars and phosphate of the core oligosaccharide, have a similarly high ($> 10^5$) molecular weight, and co-purify with LPS preparations. The O-capsule from several serotypes of *E. coli* has been detected in commercial LPS isolates of several serotypes (Peterson and McGroarty, 1985). *E. coli* strains which produce an O-antigen, high molecular weight polysaccharide are not agglutinated by O-serum unless heated (Goldman *et al.*, 1982), suggesting that this capsular polysaccharide masks the antigenic determinants on the surface of the cell from the action of O-specific serum. The heating step appears to release O-capsular polysaccharide from the *E. coli* cell surface and to allow agglutination by O-antiserum (Goldman *et al.*, 1982). It is not known how the O-capsule is associated with the cell surface of *P. aeruginosa*. For *E. coli* it was suggested that the O-capsule associates with the LPS O side chains on the cell surface through secondary and tertiary interactions (Goldman *et al.*, 1982). This would explain the copurification of the two macromolecules in *E. coli* (Peterson and McGroarty, 1985). O-capsule from IT1 serotype of *P. aeruginosa* co-purifies with LPS but is released by allowing the samples to stand at room temperature for three days (Pier *et al.*, 1978b). The studies of the O-capsule from *E. coli* suggest that this capsular polysaccharide protects the cell by preventing deleterious antiserum interaction with the cell surface (Goldman *et al.*, 1982).

3.4. O-Capsule as an Antigen

Because the O-capsular polysaccharides of *P. aeruginosa* are nontoxic and nonpyrogenic (Pier *et al.*, 1978b, 1981) and therefore likely lack lipid A, they have been tested as vaccines to protect against infection.

Pier and co-workers have immunized mice (Markham and Pier, 1983; Pier *et al.*, 1978b, 1981, 1984) rabbits (Pier, 1983, Pier *et al.*, 1984) and humans (Pier, 1982; Pier and Bennet, 1986; Pier and Thomas, 1983) with O-capsular polysaccharide. It has been shown that the material is safe (Pier, 1982), induces an antibody which is immunologically identical to the antibody against LPS O-polysaccharide, and induces a titer nearly as high as that induced by purified LPS (Pier, 1983). Further, the antibody titers to O-polysaccharide following immunization of healthy humans with O-capsule were similar to levels in serum from patients who survived acute infections (Pier and Bennet, 1986). O-capsule also protects mice against challenge from *P. aeruginosa* (Markham and Pier, 1983; Pier *et al.*, 1978a), and serum from immunized humans mediates opsonophagocytosis killing of *P. aeruginosa* cells by human leukocytes (Pier, 1982). All of these studies suggest that the O-capsular polysaccharide has strong vaccine potential, although as in the case with LPS, a number of different antigens/antibodies would have to be included to protect against various *P. aeruginosa* O-serotypes.

3.5. Alginate—Introduction

Alginate was first isolated from brown algae and characterized by Stanford (1883). That the major component of this polysaccharide is uronic acid was determined many years later by Schoeffel and Link (1933). They incorrectly identified β-D-mannuronate as the only component whereas Fisher and Dorfel (1955) determined that there were variable amounts of a second uronic acid, α-L-guluronate. It has only recently been shown that certain microorganisms synthesize this same polysaccharide as capsular material. The two organisms first identified as producers of alginate were *Azotobacter vinlandii* (Gorin and Spencer, 1966) and *Pseudomonas aeruginosa* (Linker and Jones, 1966a). Additional pseudomonads which produce alginate include *P. fluorescens*, *P. mendocina*, *P. putida* (Govan *et al.*, 1981) and *P. syringae* pv. *glycinea* (Fett *et al.*, 1986).

3.6. Alginate—Structure and Physical Properties

Alginates are composed of a family of related acidic polysaccharides with common structural features: they are unbranched $(1 \rightarrow 4)$ linked polysaccharides consisting of the two uronic acids, β-D-mannuronic acid and its C5-epimer α-L-guluronate. The proportion and distribution of these two sugars in the polymer vary depending on the source of the alginate, and differences in these proportions significantly influence the physical properties of the polysaccharide. The ratios of the two sugars can vary with growth conditions for alginates from brown seaweed (Stockton *et al.*, 1980) and from *A. vinlandii* (Larsen and Haug, 1971), but for isolates from *P. aeruginosa* the ratio of mannuronate:guluronate is relatively constant regardless of growth conditions (Gacesa and Russell, 1990). For most strains of *P. aeruginosa* this ratio is never less than one (Russell and Gacesa, 1988). Bacterial alginates are substituted by O-acetyl groups (Sherbrock–Cox *et al.*, 1984). Initial studies indicated that the O-acetyl groups are found exclusively on the D-mannuronate residues (Davidson *et al.*, 1977), and later analysis showed that some of the D-mannuronate groups are 2,3-di-O-acetylated (Skjak–Braek *et al.*, 1986). The function of the O-acetylation is not clear but the presence of O-acetyl groups may alter the susceptibility of alginates to degradation by free radicals and may alter their immunogenicity.

The physical properties of alginates are highly dependent on the chemistry of the monosaccharide components, their arrangements in the polymer, and the degree of O-acetylation. Because of the spatial arrangement of the sugar rings of D-mannuronate and L-guluronate, the two monosaccharide components have different energetically favored chair configurations. β-D-Mannuronate adopts the $^4C_1$ configuration whereas α-L-guluronate normally assumes the $^1C_4$ form. The differences in the conformations significantly influence the glycosidic linkage that they can make. When the two uronic acids are combined, different three-dimensional structures form. When the subunits are arranged in blocks of poly-β-D-mannuronate or poly-α-L guluronate, the polymannuronate regions form ribbonlike structures because of diequatorial linkages. In contrast, in blocks of polyguluronate, the subunits are linked diaxially, and the ribbons are buckled. In this highly folded or buckled conformation, adjacent ribbons can combine and align together so that the uronic acid moieties can complex divalent cations. Such aligned complexes appear as a cross section through an egg box and are often called "egg-box structures" (Rees, 1972). The binding of divalent ions, and especially calcium ions, in such conformations is very strong because of ionic interactions with the carboxyl moieties and also as a consequence of hydroxyl oxygens participating in the chelation complex (Gacesa and Russell, 1990). Thus, polyguluronate blocks preferentially bind Ca^{2+} whereas polymannuronate and mixed polymers show no preference between Ca^{2+} and Mg^{2+} (Smidsrød, 1974). ^1H NMR analysis of the alginates from *P. aeruginosa* indicate that these isolates completely lack polyguluronate blocks (Sherbrock-Cox *et al.*, 1984; Skjak–Braek *et al.*, 1986). Chemical analysis of alginates from different pathovars of *P. syringae* showed that none contain more than 28% guluronate (Fett *et al.*, 1986), but trace amounts of polyguluronate have been detected (Osmond *et al.*, 1986). The evidence suggests that most alginates from *Pseudmomonas* species contain only minimal or no poly-G block structure.

The viscosity of alginate in solution is a function of its high molecular weight and anionic charge (Donnan and Rose, 1950). Alginates from *P. aeruginosa* are reported to be relatively homogeneous in size compared to those from seaweed, although the isolates decrease in molecular weight with storage and/or during purification, probably because of the presence of extracellular alginate lyase (Russell and Gacesa, 1988). The best estimates of the size of *P. aeruginosa* alginates range from 28 Kda to 1,500 Kda although greater than 50% of the molecules are between 277 Kda and 885 Kda (Russell and Gacesa, 1989). The solubility of these polymers depends on pH and interactions with cations. As expected, alginates precipitate at pH's below their pK_a, i.e., below pH 3.4–3.7 (Haug and Larsen, 1961) although the exact pH at which the polymers become insoluble somewhat depends on the monomer composition and O-acetyl content (Delben *et al.*, 1982). In the presence of divalent or polyvalent cations, regardless of the composition, alginates form gels (Russell and Gacesa, 1988). Alginates begin to precipitate at about 3 mM Ca^{2+} (Haug and Larsen, 1971), the approximate concentration of this cation in body fluids. This may have important implications for materials produced by *P. aeruginosa* during clinical infections. Because alginate from this bacterium is predominantly in random block structures of the two monosaccharides, this polymer binds large quantities of water (Rees, 1972). This high level of water binding makes the polymers less rigid or brittle, and they occupy a very large volume. However, the presence of O-acetyl groups dramatically alters the water-binding capacity of calcium alginates. The O-acetylation dramatically increases the overall volume of the gel and decreases the rigidity (Skajak-Braek *et al.*, 1989). It is thought that the acetyl groups disrupt binding of calcium and thus weaken the organized structure of the calcium-induced gel.

3.7. Alginate—Biosynthesis

The pathway for the biosynthesis of alginate in *P. aeruginosa* is similar to that characterized for *A. vinelandii* (Pidar and Bucke, 1975) and for the brown algae *Fucus gardneri* (Lin and Hassid, 1966) and is depicted in Figure 2 (DeVault *et al.*, 1988; May and Chakrabarty, 1994; Narbad *et al.*, 1990). The starting point of the pathway is considered to be fructose 6-phosphate (F6P), which is isomerized to mannose 6-phosphate (M6P) by an important enzyme, phosphomannose isomerase (PMI). This is a critical first committed step in the synthesis of alginate and strongly

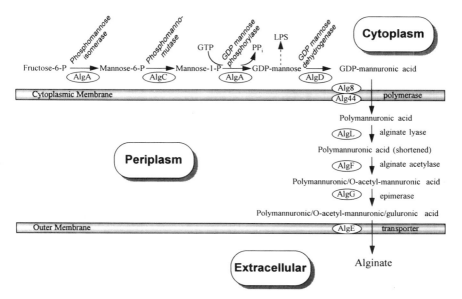

Figure 2. Proposed biosynthetic pathway of alginate in *Pseudomonas aeruginosa*. The likely intracellular site for each of the reactions is indicated, and the genes coding for the enzymes are noted.

favors the synthesis of M6P. In the second reaction in the pathway, the phosphate is transferred from the 6-carbon to the 1-carbon, producing mannose 1-phosphate (M1P) through the action of phosphomannomutase (PMM). It has also been shown recently that this second enzyme is required for synthesizing of lipopolysaccharide core (Coyne *et al.*, 1994) and perhaps O side chains (Goldberg *et al.*, 1993). The next enzyme, GDP-mannose pyrophosphorylase (GMP), catalyzes the formation of GDP-mannose from M1P and GTP. Interestingly, the same protein that catalyzes the first step in the reaction, PMI, is a bifunctional enzyme and contains the catalytic activity of GMP (Darzins *et al.*, 1986). As discussed previously, the product of this reaction, GDP-D-mannose, is also a precursor for LPS core sugars for both A-band and B-band lipopolysaccharides (Lightfoot and Lam, 1993). Thus, at this step in the pathway there is competition for intermediates between LPS and alginate biosynthesis (see Fig. 2). The subsequent step in alginate synthesis, the oxidation of GDP-mannose to GDP-mannuronic acid, is catalyzed by GDP-mannose dehydrogenase (GMD). This is a highly favored reaction and is the step that commits carbon flow into alginate (Roychoudhury *et al.*, 1989; Tatnell *et al.*, 1994). Thus, when this enzyme is maximally active in *P. aeruginosa* cells, the energetics favor carbon flow into the synthesis of alginate at the expense of the LPS pathway. This may explain why highly mucoid strains of *P. aeruginosa* often lack O-polysaccharide (Hancock *et al.*, 1983; Penketh *et al.*, 1983).

The product of the dehydrogenase, GDP-mannuronic acid, is the immediate precursor of the polymerase, but the details of the polymerization reaction and the remainder of the pathway for the synthesis and secretion of alginate are not well characterized. The structural genes for components of the polymerase have been sequenced (*alg*8 and *alg*44, see below), and the results suggest that the components are part of a membrane complex (Maharaj *et al.*, 1993). The subsequent enzymes involved in polymer modification and secretion, alginate lyase, epimerase, and acetyl transferase, are not required for polymer synthesis (Boyd *et al.*, 1993; Chitnis and Ohman, 1990; Franklin and Ohman, 1993). Evidence suggests that the alginate lyase is located in the periplasm along with the epimerase and acetyl transferase. This lyase may function to control the length of the alginate polymer (May and Chakrabarty, 1994), and there is evidence that alginate lyase is important for cell detachment and cell sloughing (Boyd and Chakrabarty, 1994). It has been reported that the epimerase, which converts D-mannuronate to L-guluronate, works on the polymannuronate polymer (Chitnis and Ohman, 1990; Franklin *et al.*, 1994). The action of the epimerase affects the viscous properties of the alginate because of the effect of the levels of L-guluronate on the physical properties of the polymer. The O-acetylation of mannuronate residues is catalyzed by alginate acetylase, and it has been suggested that the acetylated acyl carrier protein involved in fatty acid biosynthesis may be the donor of acetyl groups for O-acetylation (May and Chakrabarty, 1994). And finally, secretion of alginate from the periplasm appears to involve an outer membrane protein, the AlgE protein. Little is known about the mechanism of action of the AlgE protein other than it forms a nonporinlike channel which is affected by GDP-mannuronic acid (Chu *et al.*, 1991; Grabert *et al.*, 1990; Rehm *et al.*, 1994).

3.8. Regulation of Alginate Synthesis

Most of the structural genes for alginate biosynthetic enzymes have been identified, mapped, and sequenced (DeVault *et al.*, 1988; May and Chakrabarty, 1994). Figure 3 indicates the names and loci for these biosynthetic genes. The alginate gene *alg60* is known to be required for alginate synthesis, and although its function is unknown, it has been suggested that it is part of the polymerase complex (May and Chakrabarty, 1994). All of the biosynthetic genes except *algC* map in the 34 min

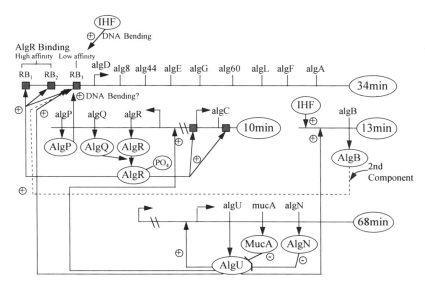

Figure 3. Genetic control of the biosynthetic enzymes involved in alginate synthesis. The biosynthetic genes located on the operan at 34 min and the alg C gene at 10 min on the *P. aeruginosa* chromosome are controlled by regulatory genes found at 10, 13, and 68 min. Activation of transcription is noted by a ⊕ symbol, and inhibition is indicated by ⊖. The direction of transcription is indicated by the symbol ⌐⟶.

region on the *P. aeruginosa* chromosome (see Fig. 3) and function as an operonic structure (Chitnis and Ohman, 1993). The *algC* gene maps at 10 min, close to a cluster of regulatory genes. Environmental factors, such as growth temperature (Leitão *et al.*, 1992), dissolved oxygen (Leitão and Sá–Correia, 1993), growth phase (Leitão and Sá–Correia, 1995), cell attachment (Davies *et al.*, 1993), nitrogen limitation, high osmolarity and ethanol (DeVault *et al.*, 1989, 1990) regulate the synthesis of alginate by activating the biosynthetic genes. Infection of the lungs of CF patients also often induces a genotypic transition of *P. aeruginosa* to a mucoid form. This form makes copious amounts of alginate which clog the airways and provides a protective coating for the infecting cells (May and Chakrabarty, 1994). Some of the environmental triggers that induce alginate production are mirrored in the CF lung. The defective chloride channel in CF lungs results in accumulation of thick mucus and a salty environment. However, mucoid isolates of *P. aeruginosa* from CF patients are locked in a state of constitutive expression of alginate.

The activities of alginate biosynthetic enzymes are very low even in mucoid strains and are greatly reduced in nonmucoid strains (Deretic *et al.*, 1987; Padgett and Phibbs, 1986; Piggott *et al.*, 1981, 1982). The activation of these genes and thus of alginate synthesis by environmental triggers (Deretic *et al.*, 1987, 1994; Fujiwara *et al.*, 1993; Zielinski *et al.*, 1991, 1992) and by conversion to the mucoid state (May and Chakrabarty, 1994) occurs by activation of the *algC* and *algD* promoters (see Fig. 3). The *algD* promoter controls the gene cluster at 34 min whereas *algC* is regulated by its own promoter. Activation of both of these promoters is mediated by three regulatory proteins AlgR, AlgQ, and AlgP (May and Chakrabarty, 1994). AlgR binds to at least three regions upstream of the *algD* gene and at least one site upstream and one site within the *algC* gene. Binding of AlgR to these sites activates the *algD* operon and *algC* gene (Deretic *et al.*, 1994; Fujiwara *et al.*, 1993). AlgR is part of a two-component signal transduction system. The AlgQ protein, which undergoes autophosphorylation, can also phosphorylate AlgR, and the phosphorylated form of AlgR has a significantly greater affinity for the promoter regions (Fujiwara *et al.*, 1993; Zielinski *et al.*, 1992). The presence of Alg R binding sites far upstream from the *algD* promoter suggests that mechanisms of DNA looping or bending are involved in AlgR binding and activation. Elements that are known to be involved in regulation of alginate biosynthesis and may be involved in DNA bending at the promoter include AlgP, a histone-like protein (Deretic *et al.*, 1992; Kato *et al.*, 1990), and the histonelike IHF (integration host factor, Wozniak, 1994; Wozniak and Ohman, 1993), both of which enhance the expression of the biosynthetic operon. Note that the regulatory genes, *algR*, *algQ* and *algP* are located close to the *algC* gene at 10 min on the chromosome but are transcribed in opposite directions (May and Chakrabarty, 1994).

AlgB is another regulatory gene essential for high levels of alginate synthesis (Goldberg and Ohman, 1987). This protein, like AlgR, is also part of a two-component signal transduction system and is required to activate the *algD* promoter (Wozniak and Ohman, 1991). The *algB* gene is located at 13 min on the chromosome (see Fig. 3) and, like the *algD* promoter, is activated by the DNA binding-bending IHF protein (Wozniak and Ohman, 1993).

An additional regulatory gene, *algU*, codes for an alternate sigma factor similar to σ^E of *E. coli* (DeVries and Ohman, 1994; Martin *et al.*, 1993a, 1994; Wozniak and Ohman, 1991, 1994; Yu *et al.*, 1995). It has been shown that this sigma factor is required for *algD*, *algB*, and *algR* transcription (see Fig. 3, Wozniak and Ohman, 1994). The AlgU protein also activates its own synthesis (DeVries and Ohman, 1994) by binding just proximal to the *algU* coding region (Martin *et al.*, 1994). The *algU* gene, located at 68 min on the chromosome, is linked to two other regulatory genes, *mucA and algN* (DeVries and Ohman, 1994; Schurr *et al.*, 1994), and these three genes may form an operon (DeVries and Ohman, 1994). There is evidence that mutations in either the *mucA* (Martin *et al.*, 1993c) or the *algN* (Martin *et al.*, 1993b) result in conversion to the mucoid state and that such mutations are the cause for mucoidy in CF isolates. Normally the MucA and AlgN proteins counteract AlgU activation and thus suppress mucoidy in the wild-type state. It is thought that MucA and perhaps AlgN indirectly control the AlgU by inhibiting its activity on promoters, such as the *algD* promoter. It has been proposed that AlgU/MucA or AlgU/AlgN may work as sigma/antisigma systems controlling AlgU in response to environmental stresses (Deretic *et al.*, 1994).

3.9. Immunology of Alginate

Alginate is immunogenic when injected in experimental animals (Bryan *et al.*, 1983), and cystic fibrosis patients infected with mucoid strains of *P. aeruginosa* produce circulating antibodies against alginate (Bryan *et al.*, 1983; Høiby and Axelsen, 1973; Schiotz, 1982). In spite of inducing high titers of antialginate antibodies, the bacteria are not cleared from the lungs of CF patients, and the antibodies fail to mediate opsonic killing of the bacteria in vitro (Ames *et al.*, 1985; Pier *et al.*, 1987). The inability to elicit opsonic antibodies may be a trait unique to humans. Vaccination of animals with large molecular sized polymers of alginate induces opsonic antibody responses (Pier *et al.*, 1994). Nonmucoid strains are usually responsible for primary infections in CF patients (Pier *et al.*, 1986) but become mucoid during chronic infection (Høiby, 1974a; Huang and Doggett, 1979). One reason why the mucoid bacteria are not cleared from the lung may be that the extracellular capsule impairs opsonization (Baltimore and Mitchell, 1980). Høiby has speculated that alginate may be inhibiting the opsonizing effect of antibody to *P. aeruginosa* (Høiby, 1974b). This inhibition may just be passive, forming a physical barrier. However, addition of extracellular material from *P. aeruginosa* in in vitro experiments decreases the mobility and phagocytic ability of neutrophils (Laharrague *et al.*, 1984; Schwarzmann and Boring, 1971). More recently it has been shown that alginate inhibits chemotaxis of polymorphonuclear leukocytes (Stiver *et al.*, 1988) and reduces nonopsonic phagocytosis by such leukocytes (Bodey *et al.*, 1987). Inhibition of lymphocyte and neutrophil functions by alginate may require that the polysaccharide be intact and of high molecular weight.

Degradation of alginate by alginase or by physicochemical treatment partially reverses its inhibitory properties (Bayer *et al.*, 1992; Mai *et al.*, 1993a). Thus, in CF patients, mutations to mucoidy result in a selective advantage for cells of *P. aeruginosa.* The production of large amounts of the glycocalyx may produce a passive barrier to antibodies and also actively inhibit activity of cells in the immune response.

3.10. Alginate Function

As noted in the previous section, alginate has an immunoprotective function, providing a barrier to antibody molecules and actively inhibiting functions of cells in the immune system. Alginate may also be protective due to its activity as a free radical scavenger, protecting the bacterium against free radical attack by phagocytic cells (Learn *et al.*, 1987; Simpson *et al.*, 1993). In addition, the glycocalyx functions to help adhere the bacteria to cell and tissue surfaces in the host (Baker, 1990; Costerton and Marrie, 1983). A large amount of evidence indicates the importance of capsular material in attaching bacteria to cells. Also, mucoid *P. aeruginosa* strains adhere better than nonmucoid cells to normal epithelium (Ramphal and Pyle, 1983), and alginase, antibiotics, and antibodies to alginate inhibit this binding (Mai *et al.*, 1993b; Ramphal and Pier, 1985). This attachment may be an important early step in colonization especially in the lower respiratory tract in patients with impaired pulmonary clearance (Baker, 1990). It is likely that the alginate binds to lectins and mucin glycoproteins on the tissue cell surfaces (Ceri *et al.*, 1986; Smedley *et al.*, 1986), and this binding may vary depending on the mannuronate/guluronate ratios of the alginate (McArthur and Ceri, 1983). This interaction is also very sensitive to Ca^{2+} concentrations, which may vary significantly in CF patients (Russell and Gacesa, 1988). In clinical situations, the bacterial cell's ability to adhere to and colonize in-dwelling prosthetic devices is a crucial factor in the pathogenesis of infection. Interference with alginate-facilitated adherence may be a way of controlling or preventing infections.

The alginate gel may form a protective barrier for certain classes of antibiotics. Because the gel is permeable, molecules with molecular weights up to 10^6 can permeate to the cell surface (Russell and Gacesa, 1988). Thus, it does not provide a diffusion barrier to nutrients. But because of its high negative charge, it can act as a cation exchanger and thus may allow trapping of nutrient molecules (Costerton *et al.*, 1990). The biofilm bacteria are more resistant to antibacterial agents because the alginate prevents penetration of antibiotics (Costerton *et al.*, 1990; Evans *et al.*, 1991; Faber *et al.*, 1990). Bacteria in a biofilm are also protected from physical forces, such as heating and drying, and, to some extent, they are protected from bacteriophages (Costerton *et al.*, 1990).

REFERENCES

Alms, T. H., and Bass, J. A., 1965, Mouse protective antigen from *Pseudomonas aeruginosa*, *Tex. Rep. Biol. Med.* **23:** 140.

Ames, P., DesJardins, D., and Pier, G. B., 1985, Opsonophagocytic killing activity of rabbit antibody to *Pseudomonas aeruginosa* mucoid exopolysaccharide, *Infect. Immun.* **49:** 281–285.

Arsenault, T. L., Hughes, D. W., MacLean, D. B., Szarek, W. A., Kropinski, A. M. B., and Lam, J. S., 1991, Structural studies on the polysaccharide portion of "A-band" lipopolysaccharide from a mutant (AK1401) of *Pseudomonas aeruginosa* strain PAO1, *Can. J. Chem.* **69:** 1273–1280.

Arsenis, G., and Dimitracopoulos, G., 1986, Chemical composition of the extracellular slime glycolipoprotein of *Pseudomonas aeruginosa* and its relation to gentamicin resistance, *J. Med. Microbiol.* **21:** 199–202.

Baker, N. R., 1990, Adherence and the role of alginate, in: *Pseudomonas Infection and Alginates. Biochemistry, Genetics and Pathology* (P. Gacesa and N. J. Russell, eds.), Chapman and Hall, New York, pp. 95–108.

Baltimore, R. S., and Mitchell, M., 1980, Immunologic investigations of mucoid strains of *Pseudomonas aeruginosa*: Comparison of susceptibility to opsonic antibody in mucoid and nonmucoid strains, *J. Infect. Dis.* **141:** 238–247.

Bartell, P. F., Orr, T. E., and Chudio, B., 1970, Purification and chemical composition of the protective slime antigens of *Pseudomonas aeruginosa*, *Infect. Immun.* **2:** 543–548.

Barton-Willis, P. A., Wang, M. C., Staskawicz, B., and Keen, N. T., 1987, Structural studies on the O-chain polysaccharides of lipopolysaccharides from *Pseudomonas syringae* pv *glycinae*, *Physiol. Mol. Plant Pathol.* **30:** 187–197.

Bayer, A. S., Park, S., Ramos, M. C., Nast, C. C., Eftekhar, F., and Schiller, N. L., 1992, Effects of alginase on the natural history and antibiotic therapy of experimental endocarditis caused by mucoid *Pseudomonas aeruginosa*, *Infect. Immun.* **60:** 3979–3985.

Berry, D., and Kropinski, A. M., 1986, Effect of lipopolysaccharide mutations and temperature on plasmid transformation efficiency in *Pseudomonas aeruginosa*, *Can. J. Microbiol.* **32:** 436–438.

Bhat, R., Marx, A., Galanos, C., and Conrad, R. S., 1990, Structural studies of lipid A from *Pseudomonas aeruginosa* PAO1: Occurrence of 4-amino-4-deoxyarabinose, *J. Bacteriol.* **172:** 6631–6636.

Bodey, G. P., Bolivar, R., Fainstein, V., and Jadeja, L., 1987, Infections caused by *Pseudomonas aeruginosa*, *Pediatr. Res.* **22:** 429–431.

Boyd, A., and Chakrabarty, A. M., 1994, Role of alginate lyase in cell detachment of *Pseudomonas aeruginosa*, *Appl. Environ. Microbiol.* **60:** 2355–2359.

Boyd, A., Ghosh, M., May, T. B., Shinabarger, D., Keogh, R., and Chakrabarty, A. M., 1993, Sequence of the *algL* gene of *Pseudomonas aeruginosa* and purification of its alginate lyase product, *Gene* **131:** 1–8.

Brown, M. R. W., and Richards, R. M. E., 1964, Effect of polysorbate (Tween) 80 on the resistance of *Pseudomonas aeruginosa* to chemical inactivation, *J. Pharm. Pharmacol.* London **16:** Supplement, 51T–55T.

Brown, M. R. W., Foster, J. H. S., and Clamp, J. R., 1969, Composition of *Pseudomonas aeruginosa* slime, *Biochem. J.* **112:** 521–525.

Bryan, L. E., Kureishi, A., and Rabin, H. R., 1983, Detection of antibodies to *Pseudomonas aeruginosa* alginate extracellular polysaccharide in animal and cystic fibrosis patients by ELISA, *J. Clin. Microbiol.* **18:** 276–282.

Bryan, L. E., O'Hara, K., and Wong, S., 1984, Lipopolysaccharide changes in impermeability-type aminoglycoside resistance in *Pseudomonas aeruginosa*, *Antimicrob. Agents Chemother.* **26:** 250–255.

Ceri, H., McArthur, H. A. I., and Whitfield, C., 1986, Association of alginate from *Pseudomonas aeruginosa* with two forms of heparin-binding lectin from rat lung, *Infect. Immun.* **51:** 1–5.

Chester, I. R., and Meadow, P. M., 1975, Heterogeneity of the lipopolysaccharide from *Pseudomonas aeruginosa*, *Eur. J. Biochem.* **58:** 273–282.

Chitnis, C. E., and Ohman, D. E., 1990, Cloning of *Pseudomonas aeruginosa algG*, which controls alginate structure, *J. Bacteriol.* **172:** 2894–2900.

Chitnis, C. E., and Ohman, D. E., 1993, Genetic analysis of the alginate biosynthetic cluster of *Pseudomonas aeruginosa* shows evidence of an operonic structure, *Mol. Microbiol.* **8:** 583–590.

Cho, Y., Tanamoto, K.-I., Oh, Y., and Homma, J. Y., 1979, Difference of chemical structures of *Pseudomonas aeruginosa* lipopolysaccharide essential for adjuvanticity and anti-tumor and interferon-inducing activities, *FEBS Lett.* **105:** 120–122.

Chu, L., May, T. B., Chakrabarty, A. M., and Misra, T. K., 1991, Nucleotide sequence and expression of the *algE* gene involved in alginate biosynthesis by *Pseudomonas aeruginosa*, *Gene* **107:** 1–10.

Costerton, J. W., and Marrie, T. J., 1983, The role of the bacterial glycocalyx in resistance to antimicrobial agents, in: *Medical Microbiology. Vol 3. Role of the Envelope in the Survival of Bacterial in Infection* (C. S. F. Easmon *et al.*, eds.), Academic Press, New York, pp. 63–85.

Costerton, J. W., Brown, M. R. W., Lam, J., Lam, K., and Cochrane, D. M. G., 1990, The microcolony mode of growth *in vivo*—an ecological perspective, in: *Pseudomonas Infection and Alginates. Biochemistry, Genetics and Pathology* (p. Gacesa and N. J. Russell, eds.), Chapman and Hall, New York, pp. 76–94.

Coughlin, R. T., Tonsager, S., and McGroarty, E. J., 1983a, Quantitation of metal cations bound to membranes and extracted lipopolysaccharide of *Escherichia coli*, *Biochemistry* **22:** 2002–2007.

Coughlin, R. T., Haug, A., and McGroarty, E. J., 1983b, Physical properties of defined lipopolysaccharide salts, *Biochemistry* **22:** 2007–2013.

Coyne, M. J., Jr., Russell, K. S., Coyle, C. L., and Goldberg, J. B., 1994, The *Pseudomonas aeruginosa algC* gene encodes phosphoglucomutase required for the synthesis of a complete lipopolysaccharide core, *J. Bacteriol.* **176:** 3500–3507.

Cross, A., Allen, J. R., Burke, J., Ducel, G., Harris, A., John, J., Johnson, D., Lew, M., MacMillan, M., Meers, P., Skalova, R., Wenzel, R., and Tenney, J., 1983, Nosocomial infections due to *Pseudomonas aeruginosa*: Review of recent trends, *Rev. Infect. Dis.* **5** (Suppl.): S837–S845.

Cryz, S. J., Jr., Pitt, T. L., Fürer, E., and Germanier, R., 1984, Role of lipopolysaccharide in virulence of *Pseudomonas aeruginosa*, *Infect. Immun.* **44:** 508–513.

Cryz, S. J., Jr., Meadow, P. M., Fürer, E., and Germanier, R., 1985, Protection against fatal *Pseudomonas aeruginosa* sepsis by immunization with smooth and rough lipopolysaccharides, *Eur. J. Clin. Microbiol.* **4:** 180–185.

Darzins, A., Frantz, B., Vanags, R. I., and Chakrabarty, A. M., 1986, Nucleotide sequence analysis of the phosphomannose isomerase gene (*pmi*) of *Pseudomonas aeruginosa* and comparison with the corresponding *Escherichia coli manA* gene, *Gene* **42**: 293–302.

Das, S., Ramm, M., Kochanowski, H., and Basu, S., 1994, Structural studies of the side chain of outer membrane lipopolysaccharide from *Pseudomonas syringae* pv coriandricola W-43, *J. Bacteriol.* **176**: 6550–6557.

Dasgupta, T., and Lam, J. S., 1995, Identification of *rfbA*, involved in B-band lipopolysaccharide biosynthesis in *Pseudomonas aeruginosa* serotype O5, *Infect. Immun.* **63**: 1674–1680.

Dasgupta, T., DeKievit, T. R., Masoud, H., Altman, E., Richards, J. C., Sadovskaya, I., Speert, D. P., and Lam, J. S., 1994, Characterization of lipopolysaccharide-deficient mutants of *Pseudomonas aeruginosa* derived from serotypes O3, O5, and O6, *Infect. Immun.* **962**: 809–817.

Davidson, I. W., Sutherland, I. W., and Lawson, C. J., 1977, Localization of O-acetyl groups in bacterial alginate, *J. Gen. Microbiol.* **98**: 603–606.

Davies, D. G., Chakrabarty, A. M., and Geesey, G. G., 1993, Exopolysaccharide production in biofilms: Substratum activation of alginate gene expression by *Pseudomonas aeruginosa*, *Appl. Environ. Microbiol.* **59**: 1181–1186.

Day, D. F., 1980, Gentamicin-lipopolysaccharide interactions in *Pseudomonas aeruginosa*, *Curr. Microbiol.* **4**: 277–281.

De Kevit, T. R., and Lam, J. S., 1994, Monoclonal antibodies that distinguish inner core, outer core, and lipid A regions of *Pseudomonas aeruginosa* lipopolysaccharide. *J. Bacteriol.* **176**: 7129–7139.

Delben, F., Cesaro, A., Paoletti, S., and Crescenzi, V., 1982, Monomer composition and acetyl content as main determinants of the ionization behavior of alginates, *Carbohyd. Res.* **100**: C46–C50.

Deretic, V., Gill, J. F., and Chakrabarty, A. M., 1987, Gene *algD* coding for GDP-mannose dehydrogenase is transcriptionally activated in mucoid *Pseudomonas aeruginosa*, *J. Bacteriol.* **169**: 351–358.

Deretic, V., Mohr, C. D., and Martin, D. W., 1991, Mucoid *Pseudomonas aeruginosa* in cystic fibrosis: Signal transduction and histone-like elements in the regulation of bacterial virulence, *Mol. Microbiol.* **5**: 1577–1583.

Deretic, V., Shurr, M. J., Boucher, J. C., and Martin, D. W., 1994, Conversion of *Pseudomonas aeruginosa* to mucoidy in cystic fibrosis: Environmental stress and regulation of bacterial virulence by alternative sigma factors, *J. Bacteriol.* **176**: 2773–2780.

DeVault, J. D., Zielinski, N. A., Berry, A., and Chakrabarty, A. M., 1988, Biochemistry, genetics and regulation of alginate synthesis by *Pseudomonas aeruginosa*, in: *Genetics and Molecular Biology of Industrial Microorganisms* (C. L. Hershberger, S. W. Queener, and G. Hegeman, eds.), American Society for Microbiology, Washington, D.C., pp. 200–214.

DeVault, J. D., Berry, A., Misra, T. K., and Chakrabarty, A. M., 1989, Environmental sensory signals and microbial pathogenesis: *Pseudomonas aeruginosa* infection in cystic fibrosis, *Bio/Technology* **7**: 352–357.

DeVault, J. D., Kimbara, K., and Chakrabarty, A. M., 1990, Pulmonary dehydration and infection in cystic fibrosis: Evidence that ethanol activates alginate gene expression and induction of mucoidy in *Pseudomonas aeruginosa*, *Mol. Microbiol.* **4**: 737–745.

DeVries, C. A., and Ohman, D. E., 1994, Mucoid-to-nonmucoid conversion in alginate-producing *Pseudomonas aeruginosa* often results from spontaneous mutations in *altT*, encoding a putative sigma factor, and shows evidence for autoregulation, *J. Bacteriol.* **176**: 6677–6687.

Dimitracopoulos, G., and Bartell, P. F., 1979, Phage-related surface modifications of *Pseu-*

domonas aeruginosa: Effects on the biological activity of viable cells, *Infect. Immun.* **23:** 87–93.

Doggett, R. G., Harrison, G. M., and Carter, R. E., Jr., 1971, Mucoid *Pseudomonas aeruginosa* in patients with chronic illness, *Lancet* **i:** 236–237.

Donnan, F. G., and Rose, R. C., 1950, Osmotic pressure, molecular weight, and viscoscity of sodium alginate, *Can. J. Res.* **28(B):** 105–113.

Drewry, D. T., Symes, K. S., Gray, G. W., and Wilkinson, S. G., 1975, Studies of polysaccharide fractions from the lipopolysaccharide of *Pseudomonas aeruginosa* NCTC 1999, *Biochem. J.* **149:** 93–106.

Eagon, R. G., 1956, Studies on polysaccharide formation by *Pseudomonas fluorescens, Can. J. Microbiol.* **2:** 673–676.

Eagon, R. G., 1962, Composition of an extracellular slime produced by *Pseudomonas aeruginosa, Can. J. Microbiol.* **8:** 585–586.

Engels, W., Endert, J., Kamps, M. A. F., and van Boven, C. P. A., 1985, Role of lipopolysaccharide in opsonization and phagocytosis of *Pseudomonas aeruginosa, Infect. Immun.* **49:** 182–189.

Evans, D. J., Alison, D. G., Brown, M. R. W., and Gilbert, P., 1991, Susceptibility of *Pseudomonas aeruginosa* and *Escherichia coli* biofilms toward ciprofloxacin: Effect of specific growth rates, *J. Antimicrob. Chemother.* **27:** 177–184.

Evans, D. J., Pier, G. B., Coyne, M. J., Jr., and Goldberg, J. B., 1994, The *rfb* locus from *Pseudomonas aeruginosa* strain PA103 promotes the expression of O antigen by both LPS-rough and LPS-smooth isolates from cystic fibrosis patients, *Mol. Microbiol.* **13:** 427–434.

Fahy, P. C., and Lloyd, A. B., 1983, The fluorescence pseudomonads, in: *Plant Bacterial Diseases: A Diagnostic Guide* (P. C. Fahy and G. J. Persley, eds.), Academic Press, New York, pp. 144–188.

Farber, B. F., Kaplan, M. H., and Clogston, A. G., 1990, *Staphlococcus epidermidis* extracted slime inhibits the antimicrobial action of glycopeptide antibiotics, *J. Infect. Dis.* **161:** 37–40.

Fett, W. F., Osman, S. F., Fishman, M. L., and Siebles III, T. S., 1986, Alginate production by plant-pathogenic pseudomonads, *Appl. Environ. Microbiol.* **52:** 466–473.

Fisher, F. G., and Dorfel, H., 1955, Die Polyuronsauren der Braunakgen (Kohlenhydrate der Algen I), *Z. Physiol. Chem. Hoppe-Seylers* **302:** 186–203.

Fisher, M. W., Devlin, H. B., and Gnabaski, F. J., 1969, New immunotype schema for *Pseudomonas aeruginosa* based on protective antigens, *J. Bacteriol.* **98:** 835–836.

Fomsgaard, A., Conrad, R. S., Galanos, C., Shand, G. H., and Høiby, N., 1988, Comparative immunochemistry of lipopolysaccharides from typable and polyagglutinable *Pseudomonas aeruginosa* strains isolated from patients with cystic fibrosis, *J. Clin. Microbiol.* **26:** 821–826.

Fomsgaard, A., Dinesen, B., Shand, G. H., Pressler, T., and Høiby, N., 1989, Anti-lipopolysaccharide antibodies and differential diagnosis of chronic *Pseudomonas aeruginosa* lung infection in cystic fibrosis, *J. Clin. Microbiol.* **27:** 1222–1229.

Franklin, M. J., and Ohman, D. E., 1993, Identification of *algF* in the alginate biosynthetic gene cluster of *Pseudomonas aeruginosa* which is required for alginate acetylation, *J. Bacteriol.* **175:** 5057–5065.

Franklin, M. J., Chitnis, C. E., Gacesa, P., Sonesson, A., White, D. C., and Ohman, D. E., 1994, *Pseudomonas aeruginosa* AlgG is a polymer level alginate C5-mannuronan epimerase, *J. Bacteriol.* **176:** 1821–1830.

Fujiwara, S., Zielinski, N. A., and Chakrabarty, A. M., 1993, Enhancer-like activity of

AlgR1-binding site in alginate gene activation: Positional, orientational, and sequence specificity, *J. Bacteriol.* **175:** 5452–5459.

Gacesa, P., and Russell, N. J., 1990, The structure and properties of alginate, in: *Pseudomonas Infection and Alginates. Biochemistry, Genetics and Pathology* (P. Gacesa and N. R. Russell, eds.), Chapman and Hall, New York, pp. 29–49.

Gaston, M. A., Vale, T. A., Wright, B., Cox, P., and Pitt, T. L., 1986, Monoclonal antibodies to the surface antigens of *Pseudomonas aeruginosa. FEMS Microbiol. Lett.* **37:** 357–361.

Gilleland, H. E., and Conrad, R. S., 1980, Effects of carbon sources on chemical composition of cell envelopes of *Pseudomonas aeruginosa* in association with polymyxin resistance, *Antimicrob. Agents Chemother.* **17:** 623–628.

Goldberg, J. B., and Ohman, D. E., 1987, Construction and characterization of *Pseudomonas aeruginosa algB* mutants: Role of *algB* in high-level production of alginate, *J. Bacteriol.* **169:** 1593–1602.

Goldberg, J. B., Hatano, K., Meluleni, G. M., and Pier, G. B., 1992, Cloning and surface expression of *Pseudomonas aeruginosa* O antigen in *Escherichia coli, Proc. Natl. Acad. Sci. USA* **89:** 10716–10720.

Goldberg, J. B., Hatano, K., and Pier, G. B., 1993, Synthesis of lipopolysaccharide O side chains by *Pseudomonas aeruginosa* PAO1 requires the enzyme phosphomannomutase, *J. Bacteriol.* **175:** 1605–1611.

Goldman, R. C., White, D., Ørskov, F., Ørskov, I., Rick, P. D., Lewis, M. S., Bhattacharjee, A. K., and Leive, L., 1982, A surface polysaccharide of *Escherichia coli* 0111 contains O-antigen and inhibits agglutination of cells by O-antiserum, *J. Bacteriol.* **151:** 1210–1221.

Gorin, P. A. T., and Spencer, J. F. T., 1966, Exocellular alginic acid from *Azotobacter vinelandii, Can. J. Chem.* **44:** 993–998.

Govan, J. R. W., 1975, Mucoid strains of *Pseudomonas aeruginosa*: The influence of culture medium on the stability of mucus production, *J. Med. Microbiol.* **8:** 513–522.

Govan, J. R. W., Fyfe, J. A. M., and Jarman, T. R., 1981, Isolation of alginate-producing mutants of *Pseudomonas fluorescens, Pseudomonas putida*, and *Pseudomonas mendocina, J. Gen. Microbiol.* **125:** 217–220.

Grabert, E., Wingender, J., and Winkler, U. K., 1990, An outer membrane protein characteristic of mucoid strains of *Pseudomonas aeruginosa, FEMS Microbiol. Lett.* **68:** 83–88.

Gross, M., Mayer, H., Widemann, C., and Rudolph, K., 1988, Comparative analysis of the lipopolysaccharides of a rough and a smooth strain of *Pseudomonas syringae* pv *phaseolicola, Arch. Microbiol.* **149:** 372–376.

Gupta, S. K., Berk, R. S., Masinick, S., and Hazlett, L. D., 1994, Pili and lipopolysaccharide of *Pseudomonas aeruginosa* bind to the glycolipid asialo GM1, *Infect. Immun.* **62:** 4572–4579.

Gvozdyak, R. I., Yakovleva, L. M., Gubanova, N. Ya., Zakharova, I., Ya., and Zdorovenko, G. M., 1989, Lipopolysaccharides from pathovars of *Pseudomonas syringae* pv. *atrofaciens*, in: *Proceedings of the 7th International Conference on Plant Pathogenic Bacteria* (Z. Klemen, ed.), Akadémiai Kiadó, Budapest, Hungary, pp. 131–136.

Habs, I., 1957, Untersuchungen über die O-antigene von *Pseudomonas aeruginosa, Z. Hyg. Infectionskr.* **144:** 218–228.

Hancock, R. E. W., 1984, Alterations in outer membrane permeability, *Ann. Rev. Microbiol.* **38:** 237–264.

Hancock, R. E. W., Mutharia, L. M., Chan, L., Darveau, R. P., Speert, D. P., and Pier, G. B., 1983, *Pseudomonas aeruginosa* isolates from patients with cystic fibrosis: A class of serum-sensitive, nontypable strains deficient in lipopolysaccharide O side chains, *Infect. Immun.* **42:** 170–177.

Hatano, K., Goldberg, J. B., and Pier, G. B., 1993, *Pseudomonas aeruginosa* lipopolysaccharide: Evidence that the O side chains and common antigen are on the same molecule, *J. Bacteriol.* **175:** 5117–5128.

Hatano, T., Goldberg, J. B., and Pier, G. B., 1995, Biologic activities of antibodies to the neutral-polysaccharide component of the *Pseudomonas aeruginosa* lipopolysaccharide are blocked by O side chains and mucoid exopolysaccharide (alginate), *Infect. Immun.* **63:** 21–26.

Haug, A., and Larsen, B., 1961, Separation of uronic acids by paper electrophoresis, *Acta Chem. Scan.* **15:** 1395–1396.

Haug, A., and Larsen, B., 1971, Biosynthesis of alginate: Part II. Polymannuronic acid C5-epimerase from *Azotobacter vinelandii* (Lipman), *Carbohyd. Res.* **17:** 297–308.

Haynes, W. C., 1951, *Pseudomonas aeruginosa*—its characterization and identification, *J. Gen. Microbiol.* **5:** 939–950.

Høiby, N., 1974a, Epidemiological investigations of the respiratory tract bacteriology in patients with cystic fibrosis, *Acta Pathol. Microbiol. Scand.* **B 82:** 541–550.

Høiby, N., 1974b, *Pseudomonas aeruginosa* infection in cystic fibrosis: Relationship between mucoid strains of *Pseudomonas aeruginosa* and the humoral immune response, *Acta Pathol. Microbiol. Scand.* **B 82:** 551–558.

Høiby, N., and Axelsen, N. H., 1973, Identification and quantitation of precipitins against *Pseudomonas aeruginosa* in patients with cystic fibrosis by means of crossed immunoelectrophoresis with intermediate gel, *Acta Pathol. Microbiol. Scand.* **B 81:** 298–308.

Homma, J. Y., 1976, A new antigenic schema and live-cell slide-agglutination procedure for the infrasubspecific serologic classification of *Pseudomonas aeruginosa*, *Jpn J. Exp. Med.* **46:** 329–336.

Homma, J. Y., 1982, Designation of the thirteen O-group antigens of *Pseudomonas aeruginosa*; an amendment for the tentative proposals in 1976, *Jap. J. Exp. Med.* **52:** 317–320.

Horton, D., Rodemeyer, G., and Haskell, T., 1977, Analytical characterization of lipopolysaccharide antigens from seven strains of *Pseudomonas aeruginosa*, *Carbohyd. Res.* **55:** 35–47.

Huang, N. N., and Doggett, R. G., 1979, Antibiotic therapy of *Pseudomonas aeruginosa*, in: *Pseudomonas aeruginosa: Clinical Manifestations of Infection and Current Therapy* (R. G. Doggett, ed.), Academic Press, New York, pp. 411–444.

Jarrell, K. F., and Kropinski, A. M., 1977, The chemical composition of the lipopolysaccharide from *Pseudomonas aeruginosa* PAO and a spontaneously derived rough mutant, *Microbios* **19:** 103–116.

Jarrell, K. F., and Kropinski, A. M., 1981, *Pseudomonas aeruginosa* bacteriophage φPLS27 lipopolysaccharide interactions, *J. Virol.* **40:** 411–420.

Kadurugamuwa, J. L., Lam, J. S., and Beveridge, T. J., 1993, Interaction of gentamicin with the A band and B band lipopolysaccharides of *Pseudomonas aeruginosa* and its possible lethal effects, *Antimicrob. Agents Chemother.* **37:** 715–721.

Karunaratne, D. N., Richards, J. C., and Hancock, R. E., 1992, Characterization of lipid A from *Pseudomonas aeruginosa* O-antigenic B band lipopolysaccharide by 1D and 2D NMR and mass spectral analysis, *Arch. Biochem. Biophys.* **299:** 368–376.

Kato, J., Misra, T. K., and Chakrabarty, A. M., 1990, Alg R3, a protein resembling eukaryotic histone H1, regulates alginate synthesis in *Pseudomonas aeruginosa*, *Proc. Natl. Acad. Sci. USA* **87:** 2887–2891.

Knirel, Y. A., 1990, Polysaccharide antigens of *Pseudomonas aeruginosa*, *Crit. Rev. Microbiol.* **17:** 217–304.

Knirel, Y. A., Zdorovenko, G. M., Dashunin, V. M., Yakoleva, L. M., Shashkov, A. S., Zakharova, I. Y., Gvozdiak, R. I., and Kochetkov, N. K., 1986, Antigenic polysaccharide of bacteria. 15. Structure of the repeating unit of O-specific polysaccharide chain of *Pseudomonas wieringae* lipopolysaccharide, *Bioorg. Khim.* **12:** 1253–1262.

Knirel, Y. A., Vinogradov, E. V., Kocharova, N. A., Paramonov, N. A., Kochetkov, N. K., Dmitriev, B. A., Stanislavsky, E. S., and Lanyi, B., 1988a, The structure of O-specific polysaccharides and serological classification of *Pseudomonas aeruginosa*, *Acta Microbiol. Hung.* **35:** 3–24.

Knirel, Y. A., Zdorovenko, g. M., Shashkov, A. S., Gubanova, N. I., Yakovleva, L. M., and Gvozdiak, R. I., 1988b, Antigenic polysaccharide of bacteria. 27. Structure of the O-specific polysaccharide chain of lipopolysaccharides from *Pseudomonas syringae* pv *atrofaciens* 2399, *phaseolicola* 120a, and *Pseudomonas holci* 8299, belonging to serogroup VI, *Bioorg. Khim.* **14:** 92–99.

Knirel, Y. A., Zdorovenko, G. M., Shashkov, A. S., Mamyan, S. S., Gubanova, N. I., and Solyanik, L. P., 1988c, Antigenic polysaccharide of bacteria. 30. Structure of the polysaccharide chain of the *Pseudomonas syringae* pv. *syringae* 281 (serogroup I) lipopolysaccharide, *Bioorg. Khim.* **14:** 166–171.

Knirel, Y. A., Zdorovenko, G. M., Yakovleva, L. M., Shashkov, A. S., Solyanik, L. P., and Zakharova, I. Y., 1988d, Antigenic polysaccharide of bacteria. 28. Structure of the polysaccharide chain of *Pseudomonas syringae* pv *atrofaciens* K-1025 and *Pseudomonas holci* 90a (serogroup II), *Bioorg. Khim.* **14:** 166–171.

Knirel, Y. A., Shashkov, A. S., Paramonov, N. A., Zdorovenko, G. M., Solyanic, L. P., and Yakovleva, L. M., 1993, Somatic antigens of pseudomonads: Structure of the O-specific polysaccharide of *Pseudomonas syringae* pv *tomato* 140(R), *Carbohydr. Res.* **243:** 199–204.

Kocharova, N. A., Knirel, Y. A., Kochetokov, N. K., and Stanislavsky, E. S., 1988, Characterization of a D-rhamnan derived from preparations of *Pseudomonas aeruginosa* lipopolysaccharide, *Bioorg. Khim.* **14:** 701–703.

Koepp, L. H., Orr, T., and Bartell, P. F., 1981, Polysaccharide of the slime glycolipoprotein of *Pseudomonas aeruginosa*, *Infect. Immun.* **33:** 788–794.

Kronborg, G., Fomsgaard, A., Galanos, C., Freudenberg, M. A., and Høiby, N., 1992, Antibody responses to lipid A, core, and O sugars of *Pseudomonas aeruginosa* lipopolysaccharide in chronically infected cystic fibrosis patients, *J. Clin. Microbiol.* **30:** 1848–1855.

Kropinski, A. M. B., and J. S. Chadwick, J. S., 1975, The pathogenicity of rough strains of *Pseudomonas aeruginosa* for *Galleria mellonella*, *Can J. Microbiol.* **21:** 2084–2088.

Kropinski, A. M. B., Chan, L., and Milazzo, F. H., 1978, Susceptibility of lipopolysaccharide-defective mutants of *Pseudomonas aeruginosa* strain PAO to dyes, detergents, and antibiotics, *Antimicrob. Agents Chemother.* **13:** 494–499.

Kropinski, A. M., Jewell, B., Kuzio, J., Milazzo, F., and Berry, D., 1985, Structure and functions of *Pseudomonas aeruginosa* lipopolysaccharide, *Antibiot. Chemother.* **36:** 58–73.

Kropinski, A. M., Lewis, V., and Berry, D., 1987, The effect of growth temperature on the lipids, outer membrane proteins, and lipopolysaccharides of *Pseudomonas aeruginosa* PAO, *J. Bacteriol.* **169:** 1960–1966.

Laharrague, P. F., Corberand, J. X., Fillola, G., Gleizes, B. J., Fontamilles, A. M., and Gyrard, E., 1984, In vitro effect of the slime of *Pseudomonas aeruginosa* on the function of human polymorphonuclear neutrophils, *Infect. Immun.* **44:** 760–762.

Lam, J., Chan, R., Lam, K., and Costerton, J. W., 1980, Production of mucoid micro colonies by *Pseudomonas aeruginosa* within infected lungs in cystic fibrosis, *Infect. Immun.* **28:** 546–556.

Lam, J. S., Mutharia, L. M., Hancock, R. E. W., Høiby, N., Lam, K., Baek, L., and Costeron, J. W., 1983, Immunogenicity of *Pseudomonas aeruginosa* outer membrane antigens examined by crossed immunoelectrophoresis, *Infect. Immun.* **42:** 88–98.

Lam, J. S., MacDonald, L. A., and Lam, M. Y. C., 1987a, Production of monoclonal antibodies against serotype strains of *Pseudomonas aeruginosa*, *Infect. Immun.* **55:** 2854–2856.

Lam, J. S., MacDonald, L. A., Lam, M. Y. C., Duscheshe, L. G. M., and Southam, G. G., 1987b, Production and characterization of monoclonal antibodies against serotype strains of *Pseudomonas aeruginosa*, *Infect. Immun.* **55:** 1051–1057.

Lam, M. Y. C., McGroarty, E. J., Kropinski, A. M., MacDonald, L. A., Pedersen, S. S., Høiby, N., and Lam, J. S., 1989, Occurrence of a common lipopolysaccharide antigen in standard and clinical strains of *Pseudomonas aeruginosa*, *J. Clin. Microbiol.* **27:** 962–967.

Lam, J. S., Handelsman, M. Y. C., Chivers, T. R., and MacDonald, L. A., 1992, Monoclonal antibodies as probes to examine serotype-specific and cross-reactive epitopes of lipopolysaccharides from serotypes O2, O5, and O16 of *Pseudomonas aeruginosa*, *J. Bacteriol.* **174:** 2178–2184.

Lang, A. B., Fürer, E., Larrick, J. W., and Cryz, S. J., Jr., 1989, Isolation and characterization of a human monoclonal antibody that recognizes epitopes shared by *Pseudomonas aeruginosa* immunotype 1, 3, 4, and 6 lipopolysaccharides, *Infect. Immun.* **57:** 3851–3855.

Lanyi, B., 1966/67, Serological properties of *Pseudomonas aeruginosa*, *Acta Microbiol. Acad. Sci. Hung.* **13:** 295–318.

Larsen, B., and Haug, A., 1971, Biosynthesis of alginate I. Composition and structure of alginate produced by *Actobacter vinelandii*, *Carbohyd. Res.* **17:** 287–296.

Learn, D. B., Brestel, E. P., and Seetharama, S., 1987, Hypochlorite scavenging by *Pseudomonas aeruginosa* alginate, *Infect. Immun.* **55:** 1813–1818.

Leitão, J. H., and Sá-Correia, I., 1993, Oxygen-dependent alginate synthesis and enzymes in *Pseudomonas aeruginosa*, *J. Gen. Microbiol.* **139:** 441–445.

Leitão, J. H., and Sá-Correia, I., 1995, Growth-phase-dependent alginate synthesis, activity of biosynthetic enzymes, and transcription of alginate genes in *Pseudomonas aeruginosa*, *Arch. Microbiol.* **163:** 217–222.

Leitão, J. H., Fialho, A. M., and Sá-Correia, I., 1992, Effects of growth temperature on alginate synthesis and enzymes in *Pseudomonas aeruginosa* variants, *J. Gen. Microbiol.* **138:** 605–610.

Lightfoot, J., and Lam, J. S., 1991, Molecular cloning of genes involved with expression of A-band lipopolysaccharide, an antigenically conserved form, in *Pseudomonas aeruginosa*, *J. Bacteriol.* **173:** 5624–5630.

Lightfoot, J., and Lam, J. S., 1993, Chromosomal mapping, expression, and synthesis of lipopolysaccharide in *Pseudomonas aeruginosa*: A role for guanosine diphospho(GDP)-D-mannose, *Mol. Microbiol.* **8:** 771–782.

Lin, T. Y., and Hassid, W. Z., 1966, Pathway of alginic acid synthesis in the marine brown algae *Fucus gardneri* silva, *J. Biol. Chem.* **241:** 5284–5297.

Linker, A., and Jones, R. S., 1966a, A new polysaccharide resembling alginic acid isolated from pseudomonads, *J. Biol. Chem.* **241:** 3845–3851.

Linker, A., and Jones, R. S., 1966b, A polysaccharide resembling alginic acid from a *Pseudomonas* microorganism, *Nature* (London) **204:** 187–188.

Liu, P. V., and Wang, S. P., 1990, Three new major somatic antigens of *Pseudomonas aeruginosa*, *J. Clin. Microbiol.* **28:** 922–925.

Liu, P. V., Abe, Y., and Bates, J. C., 1961, The roles of various fractions of *Pseudomonas aeruginosa* in its pathogenesis, *J. Infect. Dis.* **108**: 218–228.

Liu, P. V., Matsumoto, H., Kusama, H., and Bergan, T., 1983, Survey of heat-stable major somatic antigens of *Pseudomonas aeruginosa*, *Int. J. Syst. Bacteriol.* **33**: 256–264.

Loh, B., Grant, C., and Hancock, R. E. W., 1984, Use of the fluorescent probe 1-*N*-phenolnaphthylamine to study the interactions of amminoglycoside antibiotics with the outer membrane of *Pseudomonas aeruginosa*, *Antimicrob. Agents Chemother.* **26**: 546–551.

Lüderitz, O., Freudenberg, M. A., Galanos, C., Lehmann, V., Rietschel, E. T., and Shaw, D. H., 1982, Lipopolysaccharides of gram-negative bacteria, *Curr. Top. Membr. Transp.* **17**: 79–151.

Maharaj, R., May, T. B., Wang, S.-K., and Chakrabarty, A. M., 1993, Sequence of the *alg8* and *alg44* genes involved in the synthesis of alginate by *Pseudomonas aeruginosa*, *Gene* **136**: 267–269.

Mai, G. T., Seow, W. K., Pier, G. B., McCormack, J. G., and Thong, Y. H., 1993a, Suppression of lymphocyte and neutrophil functions by *Pseudomonas aeruginosa* mucoid exopolysaccharide (alginate): Reversal by physicochemical, alginase, and specific monoclonal antibody treatments, *Infect. Immun.* **61**: 559–564.

Mai, G. T., McCormack, J. G., Seow, W. K., Pier, G. B., Jackson, L. A., and Thong, Y. H., 1993b, Inhibition of adherence of mucoid *Pseudomonas aeruginosa* by alginase, specific monoclonal antibodies, and antibiotics, *Infect. Immun.* **61**: 4338–4343.

Makarenko, T. A., Kocharova, N. A., Edvabnaya, L. S., Tsvetkov, Y. E., Kholodkova, E. V., Knirel, Y. A., Backinowsky, L. V., Kochetkov, N. K., and Stanislavsky, E. S., 1993, Immunological studies of an artificial antigen with specificity of a common polysaccharide antigen of *Pseudomonas aeruginosa*, *FEMS Immun. Med. Microbiol.* **7**: 251–256.

Mäkelä, P. H., and Stocker, B. A. D., 1984, Genetics of lipopolysaccharide. 1984, in: *Handbook of Endotoxin. Vol 1: Chemistry of Endotoxins* (E. T. Reietchel, ed.), Elsevier/North-Holland Biomedical, Amsterdam, pp. 59–137.

Markham, R. B., and Pier, G. B., 1983, Immunologic basis for mouse protection provided by high-molecular-weight polysaccharide from immunotype 1 *Pseudomonas aeruginosa*, *Rev. Infect. Dis.* **5**: S957–S962.

Martin, D. W., Holloway, B. W., and Deretic, V., 1993a, Characterization of a locus determining the mucoid status of *Pseudomonas aeruginosa*: AlgU shows sequence similarities with a *Bacillus* sigma factor, *J. Bacteriol.* **175**: 1153–1164.

Martin, D. W., Shurr, M. J., Mudd, M. H., and Deretic, V., 1993b, Differentiation of *Pseudomonas aeruginosa* into the alginate-producing form: Inactivation of *mucB* causes conversion to mucoidy, *Mol. Microbiol.* **9**: 497–506.

Martin, D. W., Shurr, M. J., Mudd, M. H., Govan, J. R. W., Holloway, B. W., and Deretic, V., 1993c, Mechanism of conversion to mucoidy in *Pseudomonas aeruginosa* infecting cystic fibrosis patients, *Proc. Natl. Acad. Sci. USA* **90**: 8377–8381.

Martin, D. W., Schurr, M. J., Yu, H., and Deretic, V., 1994, Analysis of promoters controlled by the putative sigma factor AlgU regulating the conversion to mucoidy in *Pseudomonas aeruginosa*: Relationship to σ^E and stress response, *J. Bacteriol.* **176**: 6688–6696.

Masoud, H., Altman, E., Richards, J. C., and Lam, J. S., 1994, A general strategy for structural analysis of the oligosaccharide region of lipooligosaccharides. Structure of the oligosaccharide component of *Pseudomonas aeruginosa* IATS serotype O6 mutant R5 rough-type lipopolysaccharide, *Biochemistry* **33**: 10568–10578.

Masoud, H., Sadovskaya, I., De Kievit, T., Altman, E., Richards, J. C., and Lam, J. S., 1995. Structural elucidation of the lipopolysaccharide core region of the O-chain-deficient mutant strain A28 from *Pseudomonas aeruginosa* serotype O6 (international antigenic typing scheme), *J. Bacteriol.* **177:** 6718–6726.

May, T. B., and Chakrabarty, A. M., 1994, *Pseudomonas aeruginosa*: Genes and enzymes of alginate synthesis, *Trends Microbiol.* **2:** 151–157.

McArthur, H. A. I., and Ceri, H., 1983, Interaction of a rat lung lectin with exopolysaccharides of *Pseudomonas aeruginosa, Infect. Immun.* **42:** 574–578.

McGroarty, E. J., and Rivera, M., 1990, Growth-dependent alterations in production of serotype-specific and common antigen lipopolysaccharides in *Pseudomonas aeruginosa* PAO1, *Infect. Immun.* **58:** 1030–1037.

Moore, R. A., Bates, N. C., and Hancock, R. E. W., 1986, Interactions of polycationic antibiotics with *Pseudomonas aeruginosa* lipopolysaccharide and lipid A studied by using dansyl-polymyxin, *Antimicrob. Agents Chemother.* **29:** 496–500.

Narbad, A., Gacesa, P., and Russell, N. J., 1990, Biosynthesis of alginate, in: *Pseudomonas Infection and Alginates. Biochemistry, Genetics and Pathogenicity* (P. Gacesa and N. R. Russell, eds.), Chapman and Hall, London, pp. 181–205.

Ørskov, I., Ørskov, F., Jann, B., and Jann, K., 1977, Serology, chemistry, and genetics of O and K antigens of *Escherichia coli, Bacteriol. Rev.* **41:** 667–710.

Osmond, S. F., Fett, W. F., and Fishman, M. L., 1986, Exopolysaccharides of the phytopathogen *Pseudomonas syringae* pv. *glycinea, J. Bacteriol.* **166:** 66–71.

Otterlie, M., Sundan, A., Skjåk–Bræk, G. Ryan, L., Smidsrød, O., and Espevik, T., 1993, Similar mechanisms of action of defined polysaccharides and lipopolysaccharides: Characterization of binding and tumor necrosis factor alpha induction, *Infect. Immun.* **61:** 1917–1925.

Padgett, P. J., and Phibbs, P. V., Jr., 1986, Phosphomannomutase activity in wild-type and alginate-producing strains of *Pseudomonas aeruginosa, Curr. Microbiol.* **14:** 187–192.

Pastushenko, L. T., and Simonovich, I. D., 1979, Serochemical groups of phytopathogenic bacteria of *Pseudomonas* genus, *Mikrobiol. Zh.* **41:** 330–339.

Penketh, A., Pitt, T., Roberts, D., Hodson, M. E., and Batten, J. C., 1983, The relationship of phenotype changes in *Pseudomonas aeruginosa* to the clinical condition of patients with cystic fibrosis, *Am. Rev. Respir. Dis.* **127:** 605–608.

Pennington, J. E., Small, G. J., Lostrom, M. E., and Pier, G. B., 1986, Polyclonal and monoclonal antibody therapy for experimental *Pseudomonas aeruginosa* pneumonia, *Infect. Immun.* **54:** 239–244.

Peterson, A. A., and McGroarty, E. J., 1985, High-molecular-weight components in lipopolysaccharides of *Salmonella typhimurium, Salmonella minnesota*, and *Escherichia coli, J. Bacteriol.* **162:** 738–745.

Peterson, A. A., Hancock, R. E. W., and McGroarty, E. J., 1985, Binding of polycationic antibiotics and polyamines to lipopolysaccharides of *Pseudomonas aeruginos, J. Bacteriol.* **164:** 1256–1261.

Phillips, I., 1969, Identification of *Pseudomonas aeruginosa* in the clinical laboratory, *J. Med. Microbiol.* **2:** 9–16.

Pidar, D. F., and Bucke, C., 1975, The biosynthesis of alginic acid by *Azotobacter vinlandii, Biochem. J.* **152:** 617–622.

Pier, G. B., 1982, Safety and immunogenicity of high molecular weight polysaccharide vaccine from immunotype I *Pseudomonas aeruginosa, J. Clin. Invest.* **69:** 303–308.

Pier, G. B., 1983, Immunochemistry of *Pseudomonas aeruginosa* lipopolysaccharides and high-molecular-weight polysaccharides, *Rev. Infect. Dis.* **5:** S950–S956.

Pier, G. B., and Thomas, D. M., 1983, Characterization of the human immune response to polysaccharide vaccine from *Pseudomonas aeruginosa, J. Infect. Dis.* **148:** 206–213.

Pier, G. B., and Bennet, S. E., 1986, Structural analysis and immunogenicity of *Pseu-*

domonas aeruginosa immunotype 2 high molecular weight polysaccharide, *J. Clin. Invest.* **77:** 491–495.

Pier, G. B., Sidberry, H. F., and Sadoff, J. C., 1978a, Protective immunity induced in mice by immunization with high-molecular-weight polysaccharide from *Pseudomonas aeruginosa, Infect. Immun.* **22:** 919–925.

Pier, G. B., Sidberry, H. F., Zolyomi, S., and Sadoff, J. C., 1978b, Isolation and characterization of a high molecular weight polysaccharide from the slime of *Pseudomonas aeruginosa, Infect. Immun.* **22:** 908–918.

Pier, G. B., Sidberry, H. F., and Sadoff, J. C., 1981, High molecular weight polysaccharide antigen from *Pseudomonas aeruginosa* immunotype 2, *Infect. Immun.* **34:** 461–468.

Pier, G. B., Cohen, M., and Jennings, H., 1983, Further purification and characterization of high molecular weight polysaccharide from *Pseudomonas aeruginosa, Infect. Immun.* **42:** 936–941.

Pier, G. B., Pollack, M., and Cohen, M., 1984, Immunochemical characterization of high-molecular-weight polysaccharide from Fisher immunotype 3 *Pseudomonas aeruginosa, Infect. Immun.* **45:** 309–313.

Pier, G. B., DesJardins, D., Aguilar, T., Barnard, M., and Speert, D. P., 1986, Polysaccharide surface antigens expressed by non-mucoid isolates of *Pseudomonas aeruginosa* from cystic fibrosis patients, *J. Clin. Microbiol.* **24:** 189–196.

Pier, G. B., Saunders, J. M., Ames, P., Edwards, M. S., Auerbach, H., Goldfarb, J., Speert, D. P., and Hurwitch, S., 1987, Opsonophagocytic killing antibody to *Pseudomonas aeruginosa* mucoid exopolysaccharide in older, non-colonized cystic fibrosis patients, *N. Engl. J. Med.* **317:** 793–798.

Pier, G. B., DesJardin, D., Grout, M., Garner, C., Bennett, S. E., Pekoe, G., Fuller, S. A., Thornton, M. O., Harkonen, W. S., and Miller, H. C., 1994, Human immune response to *Pseudomonas aeruginosa* mucoid exopolysaccharide (alginate) vaccine, *Infect. Immun.* **62:** 3972–3979.

Piggott, N. H., Sutherland, I. W., and Jarman, T. R., 1981, Enzymes involved in the biosynthesis of alginate by *Pseudomonas aeruginosa, Eur. J. Appl. Microbiol. Biotechnol.* **13:** 179–183.

Piggott, N. H., Sutherland, I. W., and Jarman, T. R., 1982, Alginate synthesis by mucoid strains of *Pseudomonas aeruginosa* PAO, *Eur. J. Appl. Microbiol. Biotechnol.* **16:** 131–135.

Raetz, C. R. H., 1987, Structure and biosynthesis of lipid A, in: *Excherichia coli and Salmonella typhimurium. Cellular and Molecular Biology* (F. C. Neidhardt, ed.), ASM Publications, Washington, D.C., pp. 498–503.

Raetz, C. R. H., 1990, Biochemistry of endotoxins, *Ann. Rev. Biochem.* **59:** 129–170.

Ramphal, R., and Pyle, M., 1983, Adherence of mucoid and non-mucoid *Pseudomonas aeruginosa* to acid-injured tracheal cells, *Infect. Immun.* **41:** 345–351.

Ramphal, R., and Pier, G., 1985, Role of *Pseudomonas aeruginosa* mucoid exopolysaccharide in adherence to tracheal cells, *Infect. Immun.* **47:** 1–4.

Rees, D. A., 1972, Shapley polysaccharides, *Biochem. J.* **126:** 257–273.

Rehm, B. H. A., Boheim, G., Tommassen, J., and Winkler, U. K., 1994, Overexpression of *algE* in *Escherichia coli*: Subcellular localization, purification, and ion channel properties, *J. Bacteriol.* **176:** 5639–5647.

Rivera, M., and McGroarty, E. J., 1989, Analysis of a common antigen lipopolysaccharide from *Pseudomonas aeruginosa, J. Bacteriol.* **171:** 2244–2248.

Rivera, M., Bryan, L. E., Hancock, R. E. W., and McGroarty, E. J., 1988a, Heterogeneity of lipopolysaccharides from *Pseudomonas aeruginosa*: Analysis of lipopolysaccharide chain length, *J. Bacteriol.* **170:** 512–521.

Rivera, M., Hancock, R. E. W., Sawyer, J. G., Haug, A., and McGroarty, E. J., 1988b,

Enhanced binding of polycationic antibiotics to lipopolysaccharide from an aminoglycoside-supersusceptible, *tolA* mutant strain of *Pseudomonas aeruginosa*, *Antimicrob. Agents Chemother.* **32:** 649–655.

Rivera, M., Chivers, T. R., Lam, J. S., and McGroarty, E. J., 1992, Common antigen lipopolysaccharide from *Pseudomonas aeruginosa* AK1401 as a receptor for bacteriophage A7, *J. Bacteriol.* **174:** 2407–2411.

Rowe, P. S. N., and Meadow, P. M., 1983, Structure of the core oligosaccharide from the lipopolysaccharide of *Pseudomonas aeruginosa* PAC1R and its defective mutants, *Eur. J. Biochem.* **132:** 329–337.

Roychoudhury, S., May, T. B., Gill, J. F., Sigh, S. K., Feingold, D. S., and Chakrabarty, A. M., 1989, Purification and characterization of guanosine diphospho-D-mannose dehydrogenase: A key enzyme in the biosynthesis of alginate by *Pseudomonas aeruginosa*, *J. Biol. Chem.* **264:** 9380–9385.

Russell, N. J., and Gacesa, P., 1988, Chemical structure and physical properties of alginate, *Mol. Asp. Med.* **10:** 21–31.

Russell, N. J., and Gacesa, P., 1989, Physicochemical properties of alginate from mucoid strains of *P. aeruginosa* isolated from CF patients, *Antibiot. Chemother.* **42:** 62–66.

Sadoff, J. C., 1974, Cell-wall structures of *Pseudomonas aeruginosa* with immunological significance: A brief review, *J. Infect. Dis.* **Suppl. 130:** S61–S64.

Sadoff, J. C., Wright, D. C., Futrovsky, S., Sidberry, H., Collins, H., and Kaufmann, B., 1985, Characterization of mouse monoclonal antibodies against *Pseudomonas aeruginosa*, *Antibiot. Chemother.* **36:** 134–146.

Sawada, S., Suzuki, M., Kawamura, T., Fujinaga, S., Masuho, Y., and Tomibe, K., 1984, Protection against infection with *Pseudomonas aeruginosa* by passive transfer of monoclonal antibodies to lipopolysaccharides and outer membrane proteins, *J. Infect. Dis.* **150:** 570–576.

Sawada, S., Kawamura, T., Masuho, Y., and Tomibe, K., 1985a, A new common polysaccharide antigen of strains of *Pseudomonas aeruginosa* detected with a monoclonal antibody, *J. Infect. Dis.* **151:** 1290–1299.

Sawada, S., Kawamura, T., Masuho, Y., and Tomibe, K., 1985b, Characterization of a human monoclonal antibody to lipopolysaccharides of *Pseudomonas aeruginosa* serotype 5: A possible candidate as an immunotherapeutic agent for infections with *P. aeruginosa*, *J. Infect. Dis.* **152:** 965–970.

Sawada, S., Kawamura, T., and Masuho, Y., 1987, Immunoprotective human monoclonal antibodies against five major serotypes of *Pseudomonas aeruginosa*, *J. Gen. Microbiol.* **133:** 3581–3590.

Schiotz, P. O., 1982, Systemic and mucosal immunity and non-specific defense mechanisms in cystic fibrosis patients, *Acta Paediat. Scand.* **Suppl. 301:** 55–62.

Schoeffel, E., and Link, K. P., 1933, Isolation of α- and β-D-mannuronic acid, *J. Biol. Chem.* **100:** 397–405.

Schurr, M. J., Martin, D. W., Mudd, M. H., and Deretic, V., 1994, Gene cluster controlling conversion to alginate-overproducing phenotype in *Pseudomonas aeruginosa*: Functional analysis in a heterologous host and role in the instability of mucoidy, *J. Bacteriol.* **176:** 3375–3382.

Schwarzmann, S., and Boring III, J. R., 1971, Antiphagocytic effect of slime from a mucoid strain of *Pseudomonas aeruginosa*, *Infect. Immun.* **3:** 762–767.

Sensakovic, J. W., and Bartell, P. F., 1977, Glycolipoprotein from *Pseudomonas aeruginosa* as a protective antigen against *P. aeruginosa* infection in mice, *Infect. Immun.* **18:** 304–309.

Shashkov, A. S., Zdorovenko, G. M., Daeva, E. D., Yakovleva, L. P., Solyanik, L. P.,

Gvozdyak, R. I., Knirel, Y. A., and Kochetkov, N. K., 1990, Antigenic polysaccharide of bacteria. 37. Structure of the polysaccharide chain of the lipopolysaccharide chain of the lipopolysaccharide of *Pseudomonas syringae* pv *tabaci* (serotroup VII), *Bioorg. Khim.* **16:** 90–97.

Shashkov, A. S., Vinogradov, E. V., Daeva, E. D., Knirel, Y. A., Zdorovenko, G. M., Gubanova, N. Y., Yakovleva, L. M., and Zakhrova, I. Y., 1991, Somatic antigens of pseudomonads: Structure of O-specific polysaccharide chain of *Pseudomonas syringae* pv *lachrymans* 7591 (serogroup IX) lipopolysaccharide, *Carbohydr. Res.* **212:** 301–305.

Sherbrock-Cox, V., Russell, N. J., and Gacesa, P., 1984, The purification and chemical characterization of the alginates present in extracellular material produced by mucoid strains of *Pseudomonas aeruginosa*, *Carbohyd. Res.* **135:** 147–154.

Simpson, J. A., Smith, S. E., and Dean, R. T., 1993, Alginate may accumulate in cystic fibrosis lung because the enzymatic and free radical capacities of phagocytic cells are inadequate for its degradation, *Biochem. Mol. Biol. Int.* **30:** 1021–1034.

Skjak-Braek, G., Grasdalen, H., and Larsen, B., 1986, Monomer sequence and acetylation pattern in some bacterial alginates, *Carbohydr. Res.* **154:** 239–250.

Skjak-Braek, G., Zanetti, F., and Paoletti, S., 1989, Effect of acetylation on some solution and gelling properties of alginates, *Carbohyd. Res.* **185:** 131–138.

Smedley, Y. M., James, S. L., Hodges, N. A., and Marriott, C., 1986, The effect of calcium on mucus/alginate mixtures, in: *Abstracts of the 14th Annual Meeting of the European Workshop Group for Cystic Fibrosis*, p. 36.

Smidsrød, O., 1974, Molecular basis for some physical properties of alginates in the gel state, *Frad. Disc. Chem. Soc.* **57:** 263–274.

Smith, A. R. W., Zamze, S. E., Munro, S. M., Carter, K. J., and Hignett, R. C., 1985, Structure of the side chain of lipopolysaccharide from *Pseudomonas syringae* pv *morsprunorum* C28, *Eur. J. Biochem.* **149:** 73–78.

Stanford, E. C. C., 1883, On algin: A new substance obtained from some of the commoner species of marine algae, *Chem. News* **47:** 254–257.

Stiver, H. G., Zachidniak, W., and Speert, D. P., 1988, Inhibition of polymorphonuclear leukocyte chemotaxis by the mucoid exopolysaccharide of *Pseudomonas aeruginosa*, *Clin. Invest. med.* **11:** 247–252.

Stockton, B. L., Evans, V., Morris, E. R., and Rees, D. A., 1980, Circular dichroism analysis of the block structure of alginates from *Alaria esculenta*, *Int. J. Biol. Macromol.* **2:** 176–178.

Tatnell, P. J., Russell, N. J., and Gacesa, P., 1994, GDP-mannose dehydrogenase is the key regulatory enzyme in alginate biosynthesis in *Pseudomonas aeruginosa*: Evidence from metabolite studies, *Microbiology* **140:** 1745–1754.

Terashima, M., Uezumi, I., Tomio, T., Kato, M., Irie, K., Okuda, T., Yokota, S–I., and Noguchi, H., 1991, A protective human monoclonal antibody directed to the outer core region of *Pseudomonas aeruginosa* lipopolysaccharide, *Infect. Immun.* **59:** 1–6.

Thomassen, M. J., Demko, C. A., and Doershuk, C. F., 1987, Cystic fibrosis: A review of pulmonary infections and interventions, *Pediatr. Pulmonol.* **3:** 334–351.

Vinogradov, E. V., Daeva, E. D., Shashkov, A. S., Knirel, Y. A., Zdorovenko, G. M., Yakovleva, L. M., Gubanova, N. Y., and Solyanik, L. P., 1991a, Somatic antigens of pseudomonads: Structure of the O-specific polysaccharide chain of *Pseudomonas gladiola* pv *allucola* 8494 (serogroup X) lipopolysaccharide, *Carbohydr. Res.* **212:** 313–320.

Vinogradov, E. V., Shashkov, A. S., Knirel, Y. A., Zdorovenko, G. M., Solyanik, L. P., Gubanova, N. Y., and Yakovleva, L. M., 1991b, Somatic antigens of pseudomonads: Structure of the O-specific polysaccharide chain of *Pseudomonas syringae* pv *tabaci* 225 (serotroup VIII) lipopolysaccharide, *Carbohydr. Res.* **212:** 307–311.

Vinogradov, E. V., Shashkov, A. S., Knirel, Y. A., Zdorovenko, G. M., Solyanik, L. P., and Gvozdyak, R. I., 1991c, Somatic antigens of pseudomonads: Structure of the O-specific polysaccharide chain of *Pseudomonas syringae* pv *syringae* (cerasi) 435 lipopolysaccharide, *Carbohydr. Res.* **212:** 295–299.

Walker, S. G., and Beveridge, T. J., 1988, Amikacin disrupts the cell envelope of *Pseudomonas aeruginosa* ATCC 9027, *Can. J. Microbiol.* **34:** 12–18.

Warren, G. H., and Gray, J., 1954, The depolymerization of bacterial polysaccharides by hyaluronidase preparations, *J. Bacteriol.* **67:** 167–170.

Warren, G. H., and Gray, J., 1955, Studies on the properties of a polysaccharide constituent produced by *Pseudomonas aeruginosa*, *J. Bacteriol.* **70:** 152–157.

Wilkinson, S. G., 1983, Composition and structure of lipopolysaccharides from *Pseudomonas aeruginosa*, *Rev. Infect. Dis.* **5:** S941–S949.

Wilkinson, S. G., and Galbraith, L., 1975, Studies of lipopolysaccharides from *Pseudomonas aeruginosa*, *Eur. J. Biochem.* **52:** 331–343.

Wozniak, D. J., 1994, Integration host factor and sequences downstream of the *Pseudomonas aeruginosa algD* transcription start site are required for expression, *J. Bacteriol.* **176:** 5068–5076.

Wozniak, D. J., and Ohman, D. E., 1991, *Pseudomonas aeruginosa* AlgB, a two-component response regulator of the NtrC family, is required for *algD* transcription, *J. Bacteriol.* **173:** 1406–1413.

Wozniak, D. J., and Ohman, D. E., 1993, Involvement of the alginate *algT* gene and integration host factor in the regulation of the *Pseudomonas aeruginosa algB* gene, *J. Bacteriol.* **175:** 4145–4153.

Wozniak, D. J., and Ohman, D. E., 1994, Transcriptional analysis of the *Pseudomonas aeruginosa* genes *algR*, *algB*, and *algD* which reveals a hierarchy of alginate gene expression which is modulated by *algT*, *J. Bacteriol.* **176:** 6007–6014.

Yokota, S-I., Kaya, S., Sawada, S., Kawamura, T., Araki, Y., and Ito, E., 1987, Characterization of a polysaccharide component of lipopolysaccharide from *Pseudomonas aeruginosa* IID 1008 (ATCC 27584) as D-rhamnan, *Eur. J. Biochem.* **167:** 203–209.

Yokota, S., Ochi, H., Ohtsuka, H., Kato, M., and Noguchi, H., 1989, Heterogeneity of the L-rhamnose residue in the outer core of *Pseudomonas aeruginosa* lipopolysaccharide, characterized by using human monoclonal antibodies, *Infect. Immun.* **57:** 1691–1696.

Yokota, S., Terashima, M., Chiba, J., and Noguchi, H., 1992, Variable cross-reactivity of *Pseudomonas aeruginosa* lipopolysaccharide-core-specific monoclonal antibodies and its possible relationship with serotype, *J. Gen. Microbiol.* **138:** 289–296.

Young, J. M., Dye, D. W., Bradbury, J. F., Panagopoulos, C. G., and Robbs, C. F., 1978, A proposed nomenclature and classification for plant pathogenic bacteria, *N.Z. J. Agric. Res.* **21:** 153–177.

Yu, H., Schurr, M. J., and Deretic, V., 1995, Functional equivalence of *Echerichia coli* σ^E and *Pseudomonas aeruginosa* AlgU: *E. coli rpoE* restores mucoidy and reduces sensitivity to reactive oxygen intermediates in *algU* mutants of *P. aeruginosa*, *J. Bacteriol.* **177:** 3259–3268.

Zielinski, N. A., Chakrabarty, A. M., and Berry, A., 1991, Characterization and regulation of the *Pseudomonas aeruginosa algC* gene encoding phosphomannomutase, *J. Biol. Chem.* **266:** 9754–9763.

Zielinski, N. A., Maharaj, R., Roychoudhury, S., Danganan, C. E., Hendrickson, W., and Chakrabarty, A. M., 1992, Alginate synthesis in *Pseudomonas aeruginosa*: Environmental regulation of the *alg C* promoter, *J. Bacteriol.* **174:** 7680–7688.

Zierdt, C. H., and Williams, R. L., 1975, Serotyping of *Pseudomonas aeruginosa* isolates from patients with cystic fibrosis of the pancreas, *J. Clin. Microbiol.* **1:** 521–526.

Lipids of *Pseudomonas* 4

HOLLY C. PINKART and DAVID C. WHITE

1. INTRODUCTION

Lipids are generally defined as fatty acids, alcohols, hydrocarbons, and compounds containing these substances which are soluble in organic solvents. The lipids most commonly found in bacteria are phospholipids, glycolipids, ornithine amide lipids, fatty acids, and lipopolysaccharides. Phospholipids generally constitute ~40% of the cytoplasmic membrane of bacteria and up to 25% of the outer membrane (mainly localized in the inner leaflet). A generalized structure for a *Pseudomonas* membrane is shown in Figure 1. It has been found that the predominant phospholipid in both the inner and outer membranes in most *Pseudomonas* species is phosphatidylethanolamine (Wilkinson, 1988). Ornithine amide lipids are localized in the outer membrane. Lipopolysaccharides are located in the outer leaflet of the outer membrane of gram-negative bacteria. Glycolipids are generally found as storage lipids located in intracellular inclusions but can also be found in the membranes of *P. diminuta* and *P. vesicularis* and gram-positive bacteria (Wilkinson, 1988). Carotenoids and hydrocarbons may be found in the cytoplasmic membrane.

2. LIPIDS OF THE GENUS PSEUDOMONAS

2.1. Membrane Lipids

2.1.1. Phospholipids

Phospholipids usually constitute 90% of the cellular lipids of fluorescent pseudomonads. The primary phospholipids in *Pseudomonas* are

HOLLY C. PINKART • U.S. Environmental Protection Agency, Microbial Ecology Branch, Gulf Breeze, Florida 32514.   DAVID C. WHITE • Department of Microbiology and Center for Environmental Biotechnology, University of Tennesssee, Knoxville, Tennessee 37932.

Pseudomonas, edited by Montie. Plenum Press, New York, 1998.

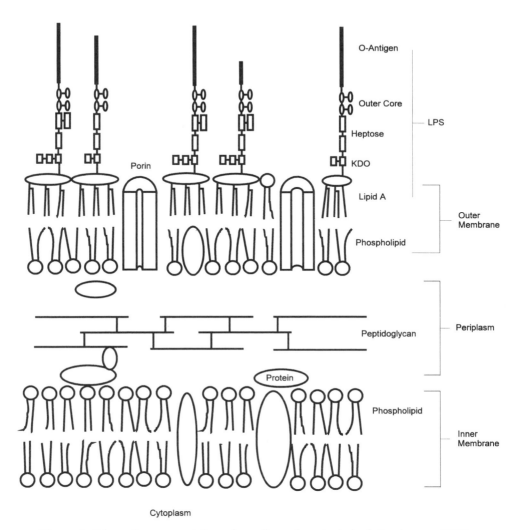

Figure 1. Schematic representation of a cell envelope typical of *Pseudomonas*. KDO, 2-keto-3-deoxyoctulosonate; LPS, lipopolysaccharide.

phosphatidylethanolamine, phosphatidylglycerol, and diphosphatidyl-glycerol (cardiolipin). Phosphatidylcholine, lysophosphatidylethano-lamine, methyl-substituted and dimethyl-substituted phosphatidyle-thanolamine, and glucosyl-substituted phosphatidylglycerol have been identified in *Pseudomonas* (Table I). Phosphatidylserine is detected in

Table I. Phospholipids in *Pseudomonas* and Their Structures[a]

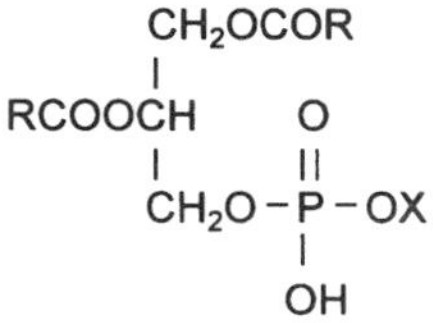

$$CH_2OCOR$$
$$|$$
$$RCOOCH \quad O$$
$$| \quad ||$$
$$CH_2O-P-OX$$
$$|$$
$$OH$$

Lipid	—X
Phosphatidic acid	—H
Phosphatidylglycerol	—CH$_2$CH(OH)CH$_2$OH
Diphosphatidylglycerol (cardiolipin)	—CH$_2$CH(OH)CH$_2$OOPO$_3$H-sn-1,2-diacylglycerol
Phosphatidylserine	—CH$_2$CH(NH$_2$)CO$_2$H
Phosphatidylcholine (lecithin)	—CH$_2$CH$_2$N$^+$(CH$_3$)$_3$
Phosphatidyl-*N*-monomethyleth-anolamine	—CH$_2$CH$_2$NHCH$_3$
Phosphatidyl-*N,N*-dimethylethanol-amine	—CH$_2$CH$_2$N(CH$_3$)$_2$

[a]R = fatty acyl groups.

most *Pseudomonas* strains in trace amounts but has not been shown to be a major phospholipid (Table II). The inner cytoplasmic membrane and the inner leaflet of the outer membrane are comprised of these phospholipids. Table II lists the phospholipids currently known in *Pseudomonas* species. *P. diminuta* and *P. vesicularis* (recently reclassified as *Brevundimonas*) are unusual in their lack of phosphatidylethanolamine and diphosphatidylglycerol, and both contain 6-*O*-phosphatidylglucosyldiacylglycerol (Fig. 2). The halophile *P. halosaccharolytica* contains glucosyl-substituted phospholipids, as shown in Figure 3 (Wilkinson, 1988).

2.1.2. Ornithine Amide Lipids

Acylornithines (ornithine amide lipids) are quite common in some members of the *Pseudomonas* genus. They have been isolated most commonly from the fluorescent pseudomonads and from *P. diminuta* and *P. vesicularis* (Segers *et al.*, 1994). For the strains studied, these lipids serve functions similar to those of phospholipids. In phosphate-limited cultures, *P. fluorescens* produces ornithine lipids as the sole polar lipid (Minnikin and Abdolrahimzadeh, 1974). A generalized structure for

Table II. Phospholipids in the Genus *Pseudomonas*

Strain	PE	PG	DPG	PC	PS	LysoPE	glucosaminyl PG	6-O-phosphati-dylglucosyl-diacylglycerol	Phosphatidyl-N-methyl ethanolamine	Phosphatidyl-N,N-dimethyl-ethanolamine
Pseudomonas aeruginosa[a]	+	+	+							
Pseudomonas alcaligenes[a]	+	+	+							
Pseudomonas aureofaciens[a]	+	+	+	+		+				
Pseudomonas carboxydovorans[a]	+	+	+	+						
Pseudomonas carophylli[b,h]	+	+	+							
Pseudomonas cepacia[a,h]	+	+								
Pseudomonas coronafaciens[c]	+	+	+							
Pseudomonas diazotrophicus[d]	+			+						
Pseudomonas diminuta[c,h]		+							+	+
Pseudomonas fluorescens[a]	+	+	+							
Pseudomonas gladioli[b,h]	+	+	+							
Pseudomonas gardneri[e]	+	+	+							
Pseudomonas halophila[f]	+	+	+					+		
Pseudomonas halosaccharolytica[a]	+	+	+				+			
Pseudomonas mildenbergii[a]	+	+	+							
Pseudomonas pickettii[b,h]	+	+	+							
Pseudomonas pseudomallei[g,h]	+	+								
Pseudomonas putida[a]	+	+	+							
Pseudomonas rubescens[c]	+	+								
Pseudomonas savastanoi[c]	+	+	+							
Pseudomonas solanacearum[a,h]	+	+	+							
Pseudomonas stutzeri[c]	+	+	+			+				
Pseudomonas syncyanea[c]	+	+	+							
Pseudomonas syringae[c]	+	+	+							
Pseudomonas vesicularis[a,i]		+						+		

[a]Wilkinson, 1988; [b]Galbraith and Wilkinson, 1991; [c]Wilkinson et al., 1973; [d]Taylor et al., 1993; [e]Bouzar et al., 1994; [f]Franzmann et al., 1990; [g]Yabucchi et al., 1992; [h]Now classified as *Burkholderia* (Yabuuchi et al., 1994); [i]Now classified as *Brevundimonas* (Segers et al., 1994).

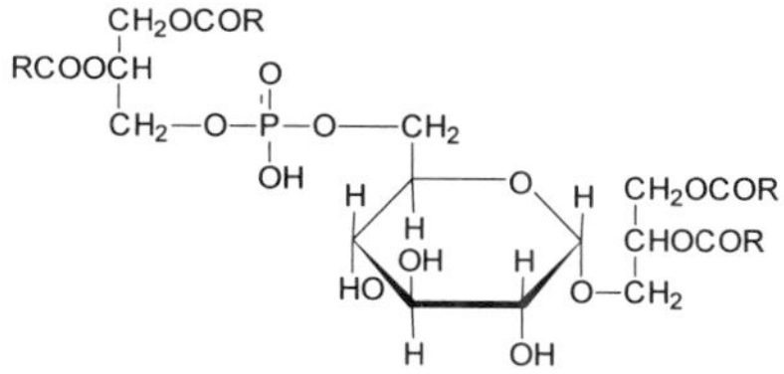

Figure 2. Structure of 6-*O*-phosphatidyl-glucosyldiacylglycerol found commonly in *P. diminuta* and *P. vesicularis*. R = fatty acyl groups.

acylornithine lipids is shown in Figure 4. The definitive mechanism for tproduction of amino lipids is still under investigation, and their true function is still unknown.

2.2. Fatty Acids

2.2.1. Non-Hydroxy Fatty Acids

The fatty acids found in *Pseudomonas* are shown in Tables III and IV. High concentrations of free fatty acids are not found in bacteria because they would lyse the membrane. Detection of free fatty acids in an extract usually means that some hydrolysis has occurred. However, reports of fatty acid methyl esters and ethyl esters produced by *P. fluorescensw* and *P. fragi* have been documented. The production of these compounds as associated with decomposition of refrigerated beef (Edwards *et al.*, 1987). Non-hydroxy fatty acids are generally found in the inner and outer membranes, covalently linked to phospholipids via ester and amide bonds. It has been shown that phospholipid inner leaflet of the outer membrane contains mainly saturated fatty acids, whereas the cytoplasmic membrane contains more of a distribution of saturated and unsaturated fatty acids. Almost all *Pseudomonas* species contain the saturated fatty acids hexadecanoic acid and octadecanoic acid. Branced-chain fatty acids are not common in *Pseudomonas* but can be found in some species (Table III). These fatty acids may occur as saturated or unsaturated forms. Unsaturated fatty acids are common in *Pseudomonas* and commonly constitute 25–40% of the total fatty acids (although these

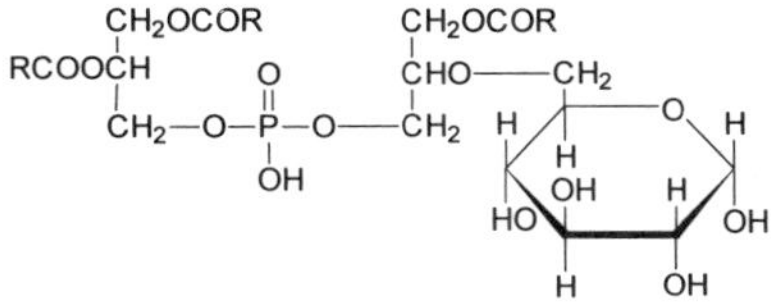

Figure 3. Structure of 1-*O*-glucosylphosphatidylglycerol found in *P. halosaccharolytica*. R = fatty acyl groups.

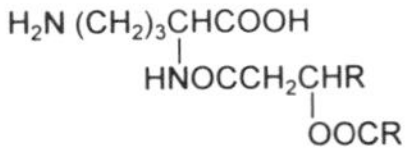

H₂N (CH₂)₃CHCOOH
HNOCCH₂CHR
OOCR

Figure 4. Generalized structure of an ornithine amide lipid. R = fatty acyl groups.

percentages may vary depending on growth conditions). Many species can make monounsaturated forms of hexadecanoic acid and octadecanoic acid. Monounsaturated fatty acids in *Pseudomonas* have the unsaturation predominantly in the cis configuration. Localization seven carbons from the alkyl (ω) end of the molecule suggests biosynthesis via the "anaerobic bacterial pathway" (Wilkinson, 1988). These bacteria also can make the fatty acids cyclopropyl 17:0 and cyclopropyl 19:0 (Table IV). Because of this variety of fatty acid structures, fatty acid profiles can be used in both clinical and agricultural applications to classify strains (Denny *et al.*, 1988; Franzmann and Tindall, 1990; Galbraith and Wilkinson, 1991; Janse, 1991; Rosello-Mora *et al.*, 1994; Stead, 1992; Yabuuchi *et al.*, 1992). These fatty acids are most useful for differentiating species rather than major groups of *Pseudomonas* if the isolates are grown under carefully specified conditions.

2.2.2. Hydroxy Fatty Acids

Almost all gram-negative bacteria contain Lipid A, which constitutes the inner portion of LPS (Fig. 1). Many hydroxy fatty acids are localized to Lipid A. Lipid A usually also contains amide and ester-linked non-hydroxy fatty acids in small amounts (primarily hexadecanoic acid). Hydroxy fatty acids are commonly used in combination with non-hydroxy fatty acid profiles to classify *Pseudomonas*. The hydroxy fatty acids are most useful for differentiating *Pseudomonas* into groups, whereas the non-hydroxy fatty acids are most useful for further differentiation into species (Stead, 1992). Table V shows the distribution of these types of fatty acids among *Pseudomonas*. Hydroxy fatty acids range in chain length from ten to eighteen carbons and may be saturated or unsaturated. The saturated forms are more common than the unsaturated forms, and branching is uncommon in *Pseudomonas* species, although it has been detected in *P. rubescens*. While the hydroxy fatty acids and non-hydroxy fatty acids demonstrate great utility in classifying *Pseudomonas* groups and species, it must be kept in mind that lipid profiles can vary greatly with environmental conditions (see section, "Lipid Alteration in Response to Environmental Conditions"). Great care must be taken with respect to growth conditions when classifying organisms by using lipid profiles.

Recently a distinct group of gram-negative motile rods, previously

Table III. Hydroxy Fatty Acids of Pseudomonas Species

Strain	Hydroxy fatty acids												
	3OH10	3OH11	2OH12	3OH12	3OH12:1	3OH13	2OH14	3OH14	2OH16	2OH16:1	3OH16	2OH18:1	3OH18
Pseudomonas aeruginosa[a]	+		+	+									
Pseudomonas alcaligenes[a]	+	+		+									
Pseudomonas andropogonis[a]									+	+		+	
Pseudomonas aureofaciens[a]	+		+	+	+								
Pseudomonas carboxydovorans[b]				+									+
Pseudomonas carophylli[c,p]								+	+	+	+	+	
Pseudomonas cepacia[a,p]	+		+					+	+	+	+	+	
Pseudomonas cichorii[a]				−				+	+	+	+	+	
Pseudomonas cocovenenans[d]									+	+		+	
Pseudomonas coronafaciens[d]	+		+	+									
Pseudomonas corrugata[a]	+		+	+	+	+					+		
Pseudomonas diazotrophicus[e]								+					
Pseudomonas diminutaf			+	+				+					
Pseudomonas fluorescens[a,q]	+		+	+									+

(continued)

Table III. (*Continued*)

Strain	3OH10	3OH11	2OH12	3OH12	3OH12:1	3OH13	2OH14	3OH14	2OH16	2OH16:1	3OH16	2OH18:1	3OH18
Pseudomonas gladioli[c,p]	+								+	+	+	+	
Pseudomonas gardneri[g]		+											
Pseudomonas halophila[h]					+	+							
Pseudomonas halosaccharolytica[i]			+										
Pseudomonas indigofera[d]	+												
Pseudomonas marginalis[a]	+			+									
Pseudomonas marginata[j]	−			+			+						
Pseudomonas mendocina[d]	+			+				+	+		+	+	
Pseudomonas mildenbergii[k]	+			+									
Pseudomonas oleovorans[l]								+					
Pseudomonas pickettii[c,p]				+					+	+		+	
Pseudomonas pseudoalcaligenes[a]	+			+									
Pseudomonas pseudomallei[m,p]								+	+	+	+	+	
Pseudomonas putida[a]	+		+	+									
Pseudomonas rubescens[f,r]		+	+	+		+		+		+			
Pseudomonas rubrisubalbicans[a]	+		+	+			+						+

Pseudomonas saccharo- phila[d]				+			
Pseudomonas savastanoi[d]	+	+					
Pseudomonas solana- cearum[n,p]			+		+		+
Pseudomonas stutzeri[o]	+		+		+		
Pseudomonas syncyanea[f]	+	+					
Pseudomonas syringae[n]	+	+					
Pseudomonas vesicu- laris[i,q]		+					
Pseudomonas viridiflava[d]	+	+		+	+	+	
Pseudomonas woodsii[d]	+	+		+	+	+	+

[a]Stead, 1992.
[b]Mayer et al., 1989.
[c]Galbraith and Wilkinson, 1991.
[d]MIDI.
[e]Taylor et al., 1993.
[f]Wilkinson et al., 1973.
[g]Bouzar et al., 1994.
[h]Franzmann and Tindall, 1990.
[i]Wilkinson, 1988.
[j]Dees et al., 1983.
[k]Roussel and Asselineau, 1980.
[l]de Smet et al., 1983.
[m]Yabuuchi et al., 1992.
[n]Janse et al., 1991.
[o]Rosello—Mora et al., 1994.
[p]Now classified as *Burkholderia* (Yabuuchi et al., 1994).
[q]Now classified as *Brevundimonas* (Segers et al., 1994).
[r]Contains small amounts of br3OH13 and br3OH 15.

Table IV. Saturated and Branched Fatty Acids of *Pseudomonas* Species

Strain	Saturated fatty acids							Terminally branched saturated fatty acids									
	10:0	12:0	14:0	15:0	16:0	17:0	18:0	i11:0	i13:0	i14:0	i15:0	i15:1	a15:0	i16:0	i17:0	i17:1	a17:0
Pseudomonas aeruginosa[a]		+	+		+		+										
Pseudomonas alcaligenes[a]	+	+	+	+	+	+	+										
Pseudomonas andropogonis[a]			+		+		+								+		
Pseudomonas aureofaciens[a]		+	+		+		+										
Pseudomonas carboxydovorans[b]			+				+										
Pseudomonas caryophylli[c-p]				+	+	+	+										
Pseudomonas cepacia[a,p]			+	+	+	+	+										
Pseudomonas cichorii[a]		+			+		+										
Pseudomonas cocovenenans[d]			+		+		+										
Pseudomonas coronafaciens[d]		+	+		+		+										
Pseudomonas corrugata[a]	+	+			+		+										
Pseudomonas diazotrophicus[e]					+		+										
Pseudomonas diminuta[f,q]			+	+	+	+	+										
Pseudomonas fluorescens[a]		+	+		+		+										
Pseudomonas gladioli[c,p]			+	+	+	+	+										
Pseudomonas gardneris[g]			+					+			+		+	+	+		
Pseudomonas halophila[h]		+	+				+				+		+	+	+		
Pseudomonas halosac-charolytica[i]					+		+										
Pseudomonas indigofera[d]	+	+	+	+	+		+										
Pseudomonas marginalis[a]		+	+		+		+										
Pseudomonas marginata[j]			+	+		+	+										
Pseudomonas mendocina[d]		+	+		+	+	+					+					
Pseudomonas mildenbergii[k]	+	+	+		+		+										
Pseudomonas oleovorans[l]		+	+		+												+

Pseudomonas pickettii[c,p]		+	+	+	+				+
Pseudomonas pseudoalcaligenes[a]	+	+	+	+	+			+	
Pseudomonas pseudomallei[m,p]	+	+	+	+	+				
Pseudomonas putida[a]	+	+	+		+				
Pseudomonas rubescens[f,q]	+	+	+	+	+	+		+	+
Pseudomonas rubrisubalbicans[a]	+	+	+	+	+	+			
Pseudomonas saccharophila[d]		+	+	+		+			
Pseudomonas savastanoi[d]	+	+	+	+	+				
Pseudomonas solanacearum[n,p]	+	+	+	+	+	+			
Pseudomonas stutzeri[o]	+	+	+	+	+			+	
Pseudomonas syncyanea[f]	+	+	+	+	+				+
Pseudomonas syringae[n]	+	+	+	+	+				
Pseudomonas vesicularis[i,q]		+	+	+					
Pseudomonas viridiflava[d]	+	+	+	+					
Pseudomonas woodsii[d]	+	+	+	+					

[a] Stead, 1992.
[b] Mayer et al., 1989.
[c] Galbraith and Wilkinson, 1991.
[d] MIDI.
[e] Taylor et al., 1993.
[f] Wilkinson et al., 1973.
[g] Bouzar et al., 1994.
[h] Franzmann and Tindall, 1990.
[i] Wilkinson, 1988.
[j] Dees et al., 1983.
[k] Roussel and Asselineau, 1980.
[l] de Smet et al., 1983.
[m] Yabuuchi et al., 1992.
[n] Janse et al., 1991.
[o] Rosello–Mora et al., 1994.
[p] Now classified as Burkholderia (Yabuuchi et al., 1994).
[q] Now classified as Brevundimonas (Segers et al., 1994).

Table V. Unsaturated and Cyclopropyl Fatty Acids of *Pseudomonas* Species

Strain	Unsaturated fatty acids									Cyclopropyl FA	
	15:1	16:1ω7cis	16:1ω7trans	16:1ω9cis	17:1ω8cis	17:1ω8trans	18:1ω7cis	18:1ω7trans	18:1ω9cis	cyclo 17:0	cyclo 19:0
Pseudomonas aeruginosa[a]		+					+			+	+
Pseudomonas alcaligenes[a]		+	+		+					+	
Pseudomonas andropogonis[a]		+					+			+	+
Pseudomonas aureofaciens[a]		+					+			+	+
Pseudomonas carboxydovorans[b]							+				
Pseudomonas carophylli[c,p]		+					+			+	+
Pseudomonas cepacia[a,p]		+					+			+	+
Pseudomonas cichorii[a]		+					+			+	+
Pseudomonas cocovenenans[d]		+								+	
Pseudomonas coronafaciens[d]		+									+

Species									
Pseudomonas corrugata[a]	+					+		+	+
Pseudomonas diazotrophicus[e]						+		+	+
Pseudomonas diminuta[f,q]	+			+	+				+
Pseudomonas fluorescens[a]	+	+				+		+	+
Pseudomonas gladioli[c,p]	+		+			+		+	+
Pseudomonas gardneri[o]	+			+					
Pseudomonas halophilas[g,r]	+		+			+	+		
Pseudomonas halosaccharolytica[h]	+					+		+	+
Pseudomonas indigofera[d]	+								
Pseudomonas marginalis[a]	+							+	
Pseudomonas marginata[i]	+			+		+		+	+
Pseudomonas mendocina[d]	+			+					
Pseudomonas mildenbergii[j]	+					+		+	+

(continued)

Table V. (*Continued*)

Strain	15:1	16:1ω7cis	16:1ω7trans	16:1ω9cis	17:1ω8cis	17:1ω8trans	18:1ω7cis	18:1ω7trans	18:1ω9cis	cyclo 17:0	cyclo 19:0
		Unsaturated fatty acids								Cyclopropyl FA	
Pseudomonas oleovorans[k]		+					+			+	+
Pseudomonas pickettii[c,p]		+					+			+	+
Pseudomonas pseudoalcaligenes[a]		+	+		+		+			+	+
Pseudomonas pseudomallei[i,p]		+					+			+	+
Pseudomonas putida[a]		+		+			+	+		+	+
Pseudomonas rubescens[f,x]	+	+			+		+				
Pseudomonas rubrisubalbicans[a]		+					+				
Pseudomonas saccharophila[d]		+									
Pseudomonas savastanoi[d]		+					+			+	+
Pseudomonas solanacearum[m,p]	+	+					+			+	+

Species						
Pseudomonas stutzeri[n]	+				+	+
Pseudomonas syncyanea[f]	+	+			+	+
Pseudomonas syringae[m]	+			+	+	+
Pseudomonas vesicularis[h,q]	+				+	+
Pseudomonas viridiflava[d]	+				+	+
Pseudomonas woodsii[d]	+		+		+	+

[a]Stead, 1992.
[b]Mayer et al., 1989.
[c]Galbraith and Wilkinson, 1991.
[d]MIDI.
[e]Taylor et al., 1993.
[f]Wilkinson et al., 1973.
[g]Franzmann and Tindall, 1990.
[h]Wilkinson, 1988.
[i]Dees et al., 1983.
[j]Roussel and Asselineau, 1980.
[k]de Smet et al., 1983.
[l]Yabuuchi et al., 1992.
[m]Janse et al., 1991.
[n]Rosello–Mora et al., 1994.
[o]Bouzar et al., 1994.
[p]Now classified as *Burkholderia* (Yabuuchi et al., 1994).
[q]Now classified as *Brevundimonas* (Segers et al., 1994).
[r]Contains 10ME18.

called *Pseudomonas*, has been reclassified as *Sphingomonas*. The genus *Sphingomonas* forms a phylogenetically tight group in the α-4 subclass of the *Proteobacteria* based on 16S rRNA sequence homology (Takeuchi *et al.*, 1994). Some *Sphingomonas* are nonmotile and nonfermentative, but all contain a class of unusual 'signature' components: 18–21-carbon straight-chain, saturated, monounsaturated, and cyclopropane-containing dihydrosphingosines in a ceramide glycolipid containing uronic acid and amide-linked 2-hydroxy straight-chain saturated fatty acids. In addition, they conatin a long-chain respiratory benzoquinone with a side chain of 10 isoprenoid units (ubiquinone Q-10) (Yabuuchi *et al.*, 1990). *Sphingomonas spp.* do not contain detectable ester or amid-linked, 3-OH fatty acids and lack the lipopolysaccharide components or structures characteristic of gram-negative bacteria. These former pseudomonads have important roles in biotechnology including the ability to degrade a diverse range of environmental pollutants (White *et al.*, 1996).

2.3. Storage Lipids

2.3.1. Polyhydroxyalkoanates (PHA)

PHA is a storage lipid common to many *Pseudomonas* species. *Burkholderia* (*Pseudomonas*) *cepacia*, *B. pickettii*, *B. pseudomallei*, *B. carophyllii*, *B. gladioli*, *B. solanacearum*, and *P. saccharophila* all produce and accumulate poly-β-hydroxybutyrate (PHB) as a carbon storage polymer, a characteristic which as been used to distinguish these organisms from other pseudomonads (Figure 5). Although some fluorescent pseudomonads cannot make PHB, they can accumulate other forms of PHA, most notably poly-β-hydroxydecanoate, which has been studied in *P. putida and poly-β-hydroxyoctanoate in P. oleovorans. In all cases, these polyhydroxyalkoanates accumulate in intracellular inclusions. Biosynthesis of these polymer depends on a combination of nutrient limitation (especially nitrogen limitation) and carbon source excess (Anderson and Dawes, 1990). PHA accumulation is greatest in fluorescent pseudomonads when they grow on decanoate, but accumulation also accurs when they frow on other fatty acids (hydroxy and non-hydroxy), glucose, fructose, and glycerol. P. oleovorans* accumulates poly-β-hydroxyoctanoate when gown on C_8–C_{12} alkanes, 1-octene, and C_8–C_{10} alcohols (de Smet *et al.*, 1983). Several investigators have shown that PHA monomer length depends on the growth source. The monomer is generally the carbon length of the growth substrate, or one or two C_2 units shorter following beta oxidation (Huijberts *et al.*, 1992). Interestingly, the concentration of unsaturated monomers in PHA depends on growth temperature, simi-

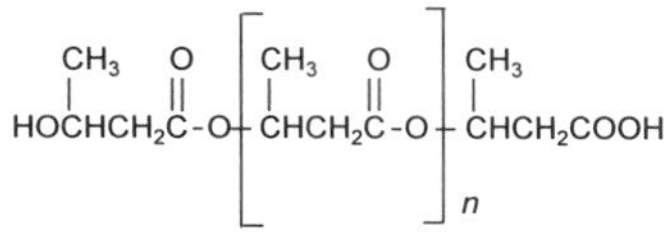

$$\text{HOCHCH}_2\overset{\overset{\displaystyle CH_3}{|}}{C}\text{-O}\left[\overset{\overset{\displaystyle CH_3}{|}}{C}\text{HCH}_2\overset{\overset{\displaystyle O}{||}}{C}\text{-O}\right]_n\overset{\overset{\displaystyle CH_3}{|}}{C}\text{HCH}_2\text{COOH}$$

Figure 5. Structure of poly(hydroxybutyrate), n = up to 10,000.

lar to the cell envelope concentration of unsaturated fatty acids (Huijberts *et al.*, 1992).

Production of PHA requires the enzymes 3-ketothiolase, acetoacyl-CoA reductase, and PHA synthase. In *P. putida*, PHA is synthesized from monomers generated from one of three sources. It has been shown that PHA can be produced from de novo fatty acids synthesis, β-oxidation of fatty acids, and fatty acid elongation. The PHA synthase of *P. putida* preferentially incorporates C_8 and C_{10} CoA thioesters, C_6 and CoA thoesters larger tha C_{10} are invorporated less efficiently into PHA (Juijberts *et al.*, 1994). The PHA synthase of *P. oleovorans* accepts CoA thoesters in the range of C_6–C_{14}.

The organization of the genes involved in the biosynthesis of PHA has been most extensively studied in *Alcaligenes eutrophus*. However, investigators have cloned genes of the PHA biosynthetic pathway in *Pseudomonas*, most notably *P. oleovorans* and *P. aeruginosa*. In both of these strains, PHA synthases 1 and 2 have been identified as well as a PHA depolymerase and a protein of unknown function (Steinbuchel *et al.*, 1992). *These proteins have similar molecular weights and amino acid sequence similarities ranging from 53.7–79.6%. In P. aeruginosa* two transcriptional start sites have been identified, one of which is preceded by a σ^{54}-dependent promoter, and the other by a σ^{70}-dependent promoter. An intact RpoN σ factor is required for PHA accumulation from gluconate in *P. aeruginosa*. Theis factor is not required in *P. oleovorans*. It is also notable that *P. oleovorans* can produce only PHA from β-oxidation-derived monomers butnot from luconate, even when complemented with PHA synthases from *P. aeruginosa*. An intact RpoN GS FACTOR IS NOT REQUIRED FOR PHA PRODUCTION FROM GLUCONATE IN *P. putida*, indicating yet another regulatory pathway for this species (Timm and Steinbuchel, 1992).

One final note of interest in the production of PHA by *Pseudomonas* species is the packaging of PHA material. As noted previously, the production of PHB versus longer carbon chain length PHA is usually mutually exclusive. Organisms capable of PHB production do not make longer chain length PHA molecules (and vice versa). However, those that have been engineered to do so through the addition of a plasmid accumulate PHB exclusively in some intracelular grnules, while accumulating the longer carbon chain PHA molecules in other, separate granules (Preusting *et al.*, 1992).

2.4. Exolipids

2.4.1. Rhamnolipid

This glycolipid is produced by *Pseudomonas aeruginosa*. It is of special interest for two reasons. It is one of the few extracellular lipids produced by *Pseudomonas*, and it also possesses surfactant qualities. The surfactant qualities of rhamnolipid cause serious problems in the respiratory tracts of cystic fibrosis patients. Rhamnolipid releases glycoconjugates from tracheal cells (Somerville *et al.*, 1992) and damages tracheal cilia (Hastie *et al.*, 1986), thus increasing airway mucus output and helping to maintain infection. It has also been implicated in aggravating the oxidative burst response of lung macrophages which further damage lung tissue (Kharami *et al.*, 1989). The surfactant qualities of rhamnolipid are helpful in removing hydrocarbons from soil (Van Dyke *et al.*, 1992), and some investigators have shown that biodegradation rates of hydrocarbons increase when rhamnolipid is added to the growth medium (Zhang and Miller, 1992).

P. aeruginosa can synthesize both the mono-and di-rhamnolipids (Rendell *et al.*, 1990) shown in Figure 6. The most common rhamnolipid isolated from *P. aeruginosa* is rhamnosyl-3-hydroxydecanoyl-3-hydroxydecanoate. Rhamnolipids containing the fatty acids 3-hydroxyoctananoate, 3-hydroxydodecanoate, and 3-hydroxydodecenoate substituted

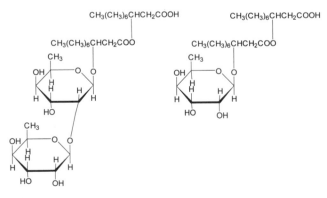

Figure 6. The two forms of rhamnolipid commonly found in *P. aeruginosa*.

at one of the 3-hydroxydecanoate positions have also been identified (Rendell *et al.*, 1990). Rhamnolipid production is greatest during the stationary phase of growth when grown on glycerol, n-alkanes, or glucose, and under some circumstances constitute up to 38% of cell dry weight (de Andres *et al.*, 1991). Production of rhamnolipid also increases under nitrogen limitation and limitation of divalent cations, such as magnesium and iron (Syldak *et al.*, 1985).

A biosynthetic pathway for rhamnolipid has been proposed (Burger *et al.*, 1963) based on studies with the radiolabelled substrates acetate and glycerol. The initial substrate for rhamnolipid biosynthesis is thymidine-diphospho-rhamnose (TDP-rhamnose), a precursor of the O-antigen region of LPS. TDP-rhamnose is the donor molecule for the rhamnose moiety of rhamnolipid. The rhamnose is donated to the β-hydroxydecanoyl-β-hydroxydecanoate molecule via a specific rhamnosyltransferase enzyme. The β-hydroxydecanoyl-β-hydroxydecanoate molecule is the product of a condensation reaction between two β-hydroxydecanoate molecules, although the exact mechanism for this reaction is unknown. The source of the β-hydroxydecanoate can be from one of two sources (Boulton and Ratledge, 1987). The first source is the fatty acid β-oxidation pathway. This is the most likely source if the organism is grown on fatty acids or alkanes. The second source of β-hydroxydecanoate is de novo fatty acid synthesis.

There are four genes in the *rhl* gene cluster (Ochsner and Reiser, 1995). The first two genes, *rhlA* and *rhlB*, code for the two subunits of the rhamnosyltransferase enzyme. *rhlR* codes for a transcriptional activator, and *rhlI* codes for an autoinducer synthetase. The *rhlR* gene is expressed at a low constitutive level from a σ^{70} promoter. The *rhlR* gene product becomes a fully functional transcriptional activator of rhlAB following binding to the autoinducer, *N*-acyl homoserine lactone. The *rhlAB* operon follows a σ^{54} promoter. Autoinducer molecules are produced at a low constitutive level. A high cell density is required to accumulate enough autoinducer locally to activate transcription of the *rhlAB* genes. Autoinducer-mediated activation of transcription in *P. aeruginosa* is a well-studied phenomenon and is linked to expression of elastase (Passador *et al.*, 1993; Pearson *et al.*, 1995), chitinase (Winson *et al.*, 1995), and pyocyanin (Brint and Ohman, 1995; Latifi *et al.*, 1995).

Although much is known about rhamnolipid gentics, regulation, and biosynthesis, there are still unanswered questions about its function. Rhamnolipid increases degradation rate of hydrocarbons, but how the cell processes hydrocarbon-rhamnolipid complexes is still unknown. The enzyme responsible for the condensation of the two β-hydroxydecanoate molecules has not as yet been isolated. Many researchers are now working to optimize production of rhamnolipid as a source of rhamnose (Ochsner *et al.*, 1995) and are studying its use in bioremediatoin to increase bioavailability of hydrocarbons sorbed to soil particles (Finnerty, 1994).

2.4.2. Viscosin

P. fluorescens and *P. viscosa* also make an exolipid which is a peptidolipid called viscosin. This compound is unusual in that it possesses antiviral activity (Kochi, 1951). Little is known at this time about its regulation. Its primary function seems to be to promote the spread of bacteria along surfaces colonized by these plant pathogens (Laycock *et al.*, 1991). The structure of viscosin has been elucidated (Neu, 1990). It contains hydrophobic amino acid moieties (L-leucine, D-serine, L-isoleucine, D-valine, and threonine) and a fatty acid.

3. ALTERATION OF LIPIDS IN RESPONSE TO ENVIRONMENTAL CONDITIONS

3.1. Growth Temperature

Most growth temperature studies have been performed on enteric bacteria, but a few investigators have examined temperature response in *Pseudomonas*. *P. aeruginosa*, a mesophile, grows in temperatures ranging from 15–45 °C. Like the enteric bacteria, it conforms to the rules of homeoviscous adaptation, changing membrane lipids of both the inner and outer membranes (Kropinski *et al.*, 1987). As the temperature increases, the percentage of saturated phospholipid fatty acids (dodecanoic, hexadecanoic and octadecanoic acid) increases whereas the percentage of the unsaturated fatty acids, hexadecenoic and octadecenoic acid, decreases. The fatty acids found in Lipid A of the outer membrane also change as temperature increases. Dodecanoic acid, hexadecanoic acid, and 3-hydroxydodecanoic acid increase whereas 3-hydroxydecanoic and 2-hydroxydodecanoic acid decrease. An increase in the LPS:phospholipid ratio was also noted. This increase in carbon chain length and saturation helps to maintain the proper phase-transition state at increased temperature to maintain normal membrane permeability and to provide an environment suitable for membrane proteins.

Psychrophilic pseudomonads do not conform to the rules of homeoviscous adaptation. When grown in the temperature range of 0–20 °C or

5–30 °C, these *Pseudomonas* species do not show any significant difference in lipid profiles. *Pseudomonas* sp. E-3 showed a small increase in hexadecenoic acid at 5 °C compared with what was observed at 30 °C, but the increase was not significant (Wada *et al.*, 1987). In another experiment, five psychrophilic *Pseudomonas* strains showed small changes in phospholipid composition as temperature decreased from 20 to 0°C (Bhakoo and Herbert, 1980). Small increases were seen in phosphatidylserine and cardiolipin, a decrease in phosphatidylglycerol was noted, and phosphatidylethanolamine content varied at each temperature for each strain. No significant change in fatty acid composition as noted for these strains. It has been hypothesized that because these organisms are generally exposed to low, constant temperature, they have not developed the ability toadapt their membrane lipids as a function of temperature change.

3.2. Oxygen Tension

Oxygen limitation has not been studied extensively in *Pseudomonas* with respect to lipid composition. However, some early studies have shown that oxygen limitation induces the formation of cyclopropyl fatty acids in *P. denitrificans* (Jaques and Hunt, 1980). Further studies have shown that it is an indirect function of oxygen tension. The production of cyclopropyl fatty acids in *P. denitrificans* is directly related to the state of reduction of components of the respiratory system rather than oxygen tension per se (Jacques, 1981) and therefore is associated with starvation response, rather than strictly oxidative stress.

Exposure of *P. aeruginosa* to hyperbaric oxygen tensions (100% O_2) resulted in the formation of giant colonies (Kenward *et al.*, 1980). These cells showed significant increases in readily extractable lipid, free fatty acid and neutral lipids, and a small increase in total phospholipid content. These cells also showed a significant increase in cardiolipin content and a significant decrease in phosphatidylglycerol content. Phosphatidylethanolamine content remained unchanged. These changes are related to adaptation rather than mutation because subculture in normal air restores the normal phenotype.

3.3. Desiccation

Although not much is known about bacterial response to desiccation, some information is available. *P. aureofaciens* lipid profiles are strongly influenced by moisture and nutrient availability (Kieft *et al.*, 1994). The strain studied showed a marked increase in saturated/

unsaturated fatty acid ratios, an increase in trans unsaturated fatty acid to cis unsaturated fatty acid ratios, and an increase in cyclopropyl fatty acids. These changes in fatty acid profiles coincided with entry into the viable but nonculturable state of the starved, desiccated organisms.

3.4. Nutrient Deprivation

Carbon starvation in *Pseudomonas* generally results in the alteration of lipid profiles and content. Prolonged starvation leads to the generation of minicells, which have proportionally more phopholipid. Although the proportions of specific phospholipids do not change greatly in carbon-starved cells, an increase in cardiolipin has been noted for some strains. The bulk of lipid changes are localized to membrane fatty acids (found in phospholipids and ornithine-amide lipids). The ratio of saturated to unsaturated fatty acids incrases. The bulk of the remaining cis unsaturated fatty acids are converted to trans unsaturated fatty acids and to cyclopropyl fatty acids (unpublished data). Although the reasons for these shifts are not completely understood, it is generally believed that the increased saturation of the fatty acids, couple with modifications of the unsaturated fatty acids, create a membrane with a higher phase-transition temperature. It is believed that a membrane with a higher phase-transition temperature creates a more rigid, less permeable cell envelope capable of maintaining envelope integrity during environmental stress, such as starvation. Nitrate and phosphate limitation have different effects on lipid composition. When both of these nutrients are limiting, but carbon is readily available, many *Pseudomonas* species accumulate carbon storage polymer in the form of polyhydroxyalkoanates (PHA) (Anderson and Dawes, 1990). Many strains can accumulate up to 60% or more of their dry weight in nitrogen and/or phosphate-limited conditions. As mentioned previously, most *Pseudomonas* strains accumulate PHA with either short (C_4) or medium carbon chain lengths (C_8–C_{10}). Medium chain PHA monomer length is generally dictated by the carbon source used for growth.

When starved for phosphate, *P. fluorescens* makes ornithine amide lipids virtually to the exclusion of phospholipids, with no apparent adverse affect on cell function (Abdolrazmah and Minnikin, 1974). The same authors also showed that a magnesium-limited chemostat culture of the same organism produces membranes devoid of any ornithine amide lipids. Phosphate starvation in *P. diminuta* decreases its phospholipid content to 0.3% of the membrane lipids, and acidic and neutral glycolipids make up the bulk of the membrane lipid content.

3.5. Solvent Tolerance

Several studies involving membrane alterations of *P. putida* strains resistant to high concentrations of phenol, toluene, ethanol and other organic solvents have been conducted recently (Heipieper *et al.*, 1992; Pinkart *et al.*, 1995; Weber *et al.*, 1994). Most hydrocarbons are toxic to microorganisms because they partition in the membranes, causing swelling and disorganization of the membranes. This disorganization leads to alteration of the cell's permeability, resulting in the leakage of small molecules from the cell and disruption of protonmotive force (Sikkema *et al.*, 1994, 1995). In the past several years, many *P. putida* strains have been discovered that are resistant to the effects of organic solvents. Although the mechanism for resistance is not understood at this time, several changes in membrane composition have been documented in solvent-tolerant and solvent-sensitive strains. One common response seen in cells exposed to toluene, xylene, and phenol is the formation of *trans*-unsaturated fatty acids (Heipieper *et al.*, 1992; Weber *et al.*, 1994). This response is seen in both solvent-tolerant and in solvent-sensitive *P. putida* strains (Pinkart *et al.*, 1995), indicating that it may be a common initial response to membrane damage. The mechanism for formation of trans-unsaturated fatty acids has not been well characterized, but evidence exists for both *de novo* synthesis of *trans*-unsaturated fatty acids (Guckert *et al.*, 1987), and for isomerization of the intact phospholipid (Heipeiper and de Bont, 1994). Some strains also show an increase in cyclopropyl fatty acids when exposed to xylene, which is also a modification of the intact phospholipid. An increase in saturated fatty acids following exposure to organic solvents has been noted in some strains but not in others. Exposure to ethanol decreases saturated fatty acids in *P. putida* S12. Ethanol also increases C_{18} fatty acids relative to C_{16}, whereas toluene causes an opposite reaction. This unusual response could be caused by inhibition of fatty acid biosynthetic enzymes by ethanol. An increase in hydroxy fatty acids has also been seen in the LPS of *P. putida* Idaho following exposure to xylene (Pinkart *et al.*, 1995). This may be a response of the cell to alter outer membrane permeability because strain was shown to have an increase resistance to difloxacin, a hydrophobic antibiotic with the same hydrophobic index as xylene.

3.6. Antibiotic Resistance

Some *Pseudomonas* strains have developed antibiotic resistance through membrane alterations. *P. aeruginosa* strains with increased amounts of KDO and Lipid A were found to be much less permeable to hydrophilic quinolones and had much higher resistance to these antibiotics (Michea-Hamzepour *et al.*, 1991). In an experiment comparing four *P. aeruginosa* strains with respect to lipid content and antibiotic susceptibility, it was found that strains with a higher lipid content were considerably more resistant to ampicillin than strains containing less lipid (Norris *et al.*, 1985).

4. SUMMARY

Although much research has been conducted on the lipid composition of *Pseudomonas*, several areas of study still need attention. No information is yet available about the genetics or regulation of phospholipid or fatty acid biosynthesis in *Pseudomonas*. The enzymology of trans-unsaturated fatty acid formation has not been elucidated. The mechanism for maintenance of membrane integrity in solvent tolerant organisms is still unclear. Studies are currently underway to examine the possibility of using *Pseudomonas* to make copolyesters of PHB and medium chain length PHA (Lee *et al.*, 1995). There is still much to be learned about the function of rhamnolipid in bioavailability. These probelms provide investigators of *Pseudomonas* with several opportunities for study in the future.

REFERENCES

Anderson, A. J., and Dawes, E. A., 1990, Occurence, metabolism, metabolic role, and industrial uses of bacterial polyhydroxyalkoanates, *Microbiol. Rev.* **54:** 450–472.

Bhakoo, M., and Herbert, R. A., 1989, Fatty acid and phospholipid composition of five psychrotrophic *Pseudomonas* species grown at different temperatures. *Arch. Microbiol.* **126:** 51–5.

Bouzar, H., Jones, J. B., Stall, R. E., Hodge, N. C., Minsavage, G. V., Benedict, A. A., and Alverez, A. M., 1994, Physiological, chemical, serological,and pathogenic analysis of a worldwide collection of *Xanthomonas campestris* pv. vesicatoria strains, *Phytopathology* **84:** 663–671.

Boulton, C. A., and Ratledge, C., 1987, Biosynthesis of lipid precursors to surfactant production, in: *Biosurfactants and Biotechnology* (N. Kosaric, W. L. Cairns, and N. C. C. Gray, eds.), M. Dekker, New York, pp. 47–87.

Brint, J. M., and Ohman, D., 1995, Synthesis of multiple exoproducts in *Pseudomonas aeruginosa* is under the control of RhlR and RhlI, another set of regulators in strain PAO1 with homology to the autoinducer-responsive LuxR-LuxI family, *J. Bacteriol.* **177:** 7155–7163.

Burger, M. M., Glaser, L., and Burton, R. M., 1963, The enzymatic synthesis of a rhamnose-containing glycolipid by extracts of *Pseudomonas aeruginosa*, *J. Biol. Chem.* **238:** 2595–2604.

de Andres, C., Espuny, M. J., Robert, M., Mercade, M. E., Manresa, A., and Guinea, J., 1991, Cellular lipid accumulation by *Pseudomonas aeruginosa* 44T1, *Appl. Microbiol. Biotechnol.* **35:** 813–816.

Dees, S. B., Hollis, D. G., Weaver, R. E., and Moss, C. W., 1983, Cellular fatty acid composition of *Pseudomonas marginata* and closely related bacteria, *J. Clin. Microbiol.* **18:** 1073–1078.

Denny, T. P., 1988, Phenotypic diversity in *Pseudomonas syringae* pv. tomato, *J. Gen. Microbiol.* **134:** 1939–1948.

de Smet, M. J., Eggink, G., Witholt, B., Kingma, J., and Wynberg, H., 1983, Characterization of cellular inclusions formed by *Pseudomonas oleovorans* during growth on octaine, *J. Bacteriol.* **154:** 870–878.

de Waard, P., van der Wal, H., Huijberts, G. N. M., and Eggink, G., 1993, Heteronuclear NMR analysis of unsaturated fatty acids in poly(3-hydroxyalkoanates): Study of beta-oxidation in *Pseudomonas putida*. *J. Biol. Chem.* **268:** 315–319.

Edwards, R. A.., Dainty, R. H., and Hibbard, C. M., 1987, Volatile compounds produced by meat pseudomonads and related reference strains during growth on beef stored in air at chill temperatures, *J. Appl. Bacteriol.* **62:** 403–412.

Finnerty, W. R., 1994, Biosurfactants in environmental biotechnology. *Curr. Op. Biotech.* **5:** 291–295.

Franzmann, P. D., and Tindall, B. J., 1990, A chemotaxonomic study of members of the family *Halomonadaceae*, *Syst. Appl. Microbiol.* **13:** 142–147.

Galbraith, L., and Wilkinson, S. G., 1991, Polar lipids and fatty acids of *Pseudomonas carophylli*, *Pseudomonas gladioli*, and *Pseudomonas pickettii*, *J. Gen. Microbiol.* **137:** 197–202.

Guckert, J. B., Ringelberg, D. B., and White, D. C., 1987, Biosynthesis of trans fatty acids from acetate inthe bacterium *Pseudomonas atlantica*, *Can. J. Microbiol.* **33:** 748–754.

Hastie, A. T., Hingley, S. T., Higgins, M. L., Kueppers, F., and Shryok T., 1986, Rhamnolipid from *Pseudomonas aeruginosa* inactivates mammaliam tracheal ciliary axonemes, *Cell. Motil. Cytoskeleton* **6:** 502–509.

Heipieper, H.-J., Deifenbach, R., and Keweloh, H., 1992, Conversion of cis unsaturated ratty acids to trans, a possible mechanism for the protection of phenol-degrading *Pseudomonas* P8 from substrate toxicity, *Appl. Environ. Microbiol.* **58:** 1847–1852.

Heipieper, H.-J., and de Bont, J. A. M., 1994, Adaptation of *Pseudomonas putida* S12 to ethanol and toluene at the level of fatty acid composition of membranes, *Appl. Environ. Microbiol.* **60:** 4440–4444.

Huijberts, G. N. M., DeRijk, T. C., de Waard, P., and Eggink G., 1994, ^{13}C nuclear magnetic resonance studies of *Pseudomonas putida* fatty acid metabolic routes involved in poly(3-hydroxyalkoanate) synthesis, *J. Bacteriol.* **176:** 1661–1666.

Huijberts, G. N. M., Eggink, G., de Waard, P., Huisman, G. W., and Witholt, B., 1992, *Pseudomonas putida* KT2442 cultivated on glucose accumulates poly(3-hydroxyalkoanates) consisting of saturated and unsaturated monomers, *Appl. Environ. Microbiol.* **58:** 536–544.

Jacques, N. A., 1981, Studies on cyclopropane fatty acid synthesis: Correlation between the state of reduction of respiratory components and the accumulation of methylene hexadecanoic acid by *Pseudomonas denitrificans*, *Biochim. Biophys. Acta* **665:** 270–282.

Jacques, N. A., and Hunt, A. L., 1989, Studies on cyclopropane fatty acid synthesis. Effect of carbon source and oxygen tension on cyclopropane fatty acid synthetase activity in *Pseudomonas denitrificans*, *Biochim. Biphys. Acta* **619:** 453–470.

Janse, J. D., 1991, infra and intraspecific classification of *Pseudomonas solanacearum* strains, using whole-cell fatty acid analysis, *Syst. Appl. Microbiol.* **14:** 335–345.

Janse, J. D., 1991, Pathovar discrimination within *Pseudomonas syringae* subsp. savastanoi using whole cell fatty acids and pathogenicity as criteria, *Syst. Appl. Microbiol.* **14:** 79–84.

Karunaratne, D. N., Richards, J. C., and Hancock, R. E. W., 1992, Characterization of Lipid A from *Pseudomonas aeruginosa* O-antigenic B-band lipopolysaccharide by 1D and 2D NMR and mass spectral analysis, *Arch. Biochem. Biophys.* **299:** 268–376.

Kenward, M. A., Alcock, S. R., and Brown, M. R., 1980, Effects of hyperbaric oxygen on the growth and properties of *Pseudomonas aeruginosa*, *Microbios* **28:** 47–60.

Kharami, A., Bibi, Z., Neilson, H., Holby, N., and Doring, G., 1989, Effect of *Pseudomonas aeruginosa* rhamnolipid on human neutrophil and monocyte function, *APMIS* **97:** 1-68–1072.

Kieft, T. L., Ringelberg, D. B., and White D. C., 1994, Changes in ester-linked fatty acid profiles of subsurface bacteria during starvation and dessication in a porous medium, *Appl. Environ. Microbiol.* **60:** 3292–3299.

Kochi, M., Weiss, D. W., Pugh, L. H., and Groupe, V., 1951, Viscosin, a new antibiotic, *Bact. Proc.* 29–30.

Kropinski, A. M. B., Lewis, V., and Berry, D., 1987, Effect of growth temperature on the lipids, outer membrane proteins, and lipopolysaccharides of *Pseudomonas aeruginosa* PAO, *J. Bacteriol.* **169:** 1960–1966.

Latifi, A., Winson, M. D., Foglino, M., Bycroft, B. W., Stewart, G. S., Lazdunski, A., and Williams, P., 1995, Multiple homologues of LuxR and LuxI control expression of virulence determinants and secondary metabolites through quorum sensing in *Pseudomonas aeruginosa* PAO1, *Mol. Microbiol.* **17:** 333–343.

Laycock, M. V., Hildebrand, P. D., Thibault, P., Walter, J. A., and Wright, J. L. C., 1991, Viscosin, a potent peptidolipid biosurfactant and phytopathogenic mediator produced by a pectolytic strain of *Pseudomonas fluorescens*, *J. Agri. Food Chem.* **39:** 483–489.

Lee, E. Y., Jendrossek, D., Schirmer, A., Choi, C. Y., and Steinbuchel, A., 1995, Biosynthesis of copolyesters consisting of 3-hydroxybutyric acid and medium-chain-length 3-hydroxyalkanoic acids from 1,3-butanediol or from 3-hydroxybutyrate by *Pseudomonas* sp. A33, *Appl. Microbiol. Biotechnol.* **42:** 901–909.

Mayer, H., Krauss, J. H., Urbanik-Sypniewska, T., Puvanesarajah, V., Stacey, G., and Auling, G., 1989, Lipid A with 2,3-diamino-2,3-dideoxy-glucose in lipopolysaccharides from slow-growing members of *Rhizobiaceae* and "*Pseudomonas carboxydovarans*," *Arch. Microbiol.* **151:** 111–116.

Michea-Hamzehpour, M., Furet, Y. X., and Pechere, J.-C., 1991, Role of protein D2 and lipopolysaccharide in diffusion of quinolones through the outer membrane of *Pseudomonas aeruginosa*. *Antimicrob. Agents Chemother.* **35**(10): 2091–2097.

Minnikin, D. E., and Abdolrahimzadeh, H., 1974, The replacement of phosphatidylethanolamine and acidic phospholipids by an ornithine-amide lipid and a minor phosphorus-free lipid in *Pseudomonas fluorescens* NCMB 129, *FEBS Lett.* **43:** 257–260.

Monteoliva-Sanchez, M., and Ramos-Cormenzana, A., 1987, Cellular fatty acid composition in moderately halophilic gram-negative rods, *J. Appl. Bacteriol.* **62:** 361–366.

Neu, T. R., Hartner, T., and Poralla, K., 1990, Surface active properties of viscosin: A peptidolipid antibiotic, *Appl. Microbiol. Biotechnol.* **32:** 518–520.

Norris, M. J., Rogers, D. T., and Russell, A. D., 1985, Cell envelope composition and sensitivity of *Proteus Mirabilus*, *Pseudomonas aeruginosa*, and *Serratia marcescens* to polymixin and other antibacterial agents, *Lett. Appl. Microbiol.* **1:** 3–6.

Ochsner, U. A., and Reiser, J., 1995, Autoinducer-mediated regulation of rhamnolipid biosurfactant synthesis *Pseudomonas aeruginosa*, *Proc. Natl. Acad. Sci. USA* **92:** 6424–6428.

Passador, L., Cook, J. M., Gambello, M. J., Rust, L., and Iglewski, B. H., 1993, Expression of *Pseudomonas aeruginosa* virulence genes requires cell-to-cell communication, *Science* **260:** 1127–1130.

Pearson, J. P., Passador, L., Iglewski, B. H., and Greenberg, E. P., 1995, A second N-acylhomoserine lactone signal produced by *Pseudomonas aeruginosa*, *Proc. Natl. Acad. Sci. USA* **92:** 1490–1494.

Pinkart, H. C., Wolfram, J., Rogers, R., and White D. C., 1995, Cell envelope changes in solvent-tolerant and solvent-sensitive *Pseudomonas putida* strains following exposure to *o*-xylene, *Appl. Environ. Microbiol.* **62:** 1129–1132.

Preusting, H., Kingma, J., Huisman, G. W., Steinbuchel, A., and Witholt, B., 1992, Formation of polyester blends by a recombinant strain of *Pseudomonas oleovorans*: Different poly(3-hydroxyalkoantes) are stored in separate granules, *J. Environ. Polym. Degradation* **1:** 11–21.

Rendell, N. B., Taylor, G. W., Somerville, M., Todd, H., Wilson, R., and Cole, P. J., 1990, Characterization of *Pseudomonas* rhamnolipids, *Biochim. Biophys. Acta* **1045:** 189–193.

Rosello-Mora, R. A., Lalucat, J., Dott, W., and Kampfer, P., 1994, Biochemical and chemotaxonomic characterization of *Pseudomonas stutzeri* genomovars, *J. Appl. Bacteriol.* **76:** 226–233.

Roussel, J. and Asselineau, J., 1980, Fatty acid composition of the lipids of *Pseudomonas mildenbergii*: Presence of a fatty acid containing two conjugated double bonds, *Biochim. Biophys. Acta* **619:** 689–692.

Segers, P., Vancanneyt, M., Pot, B., Torck, U., Hoste, B., Dewettinck, D., Falsen, E., Kersters, K., and de Vos, P., 1994, Classification of *Pseudomonas diminuta* (Leifson and High 1954) and *Pseudomonas vesicularis* (Busing, Doll and Freytag 1953) in *Brevundimonas* gen. nov. as *Brevundimonas diminuta* comb. nov. and *Brevundimonas vesicularis* comb. nov., respectively, *Int. J. Syst. Bacteriol.* **44:** 499–510.

Sikkema, J., Weber, F. J., Heipieper, H. J., and de Bont, J. A. M., 1994, Cellular toxicity of lipophilic compounds: Mechanisms, implications, and adaptations, *Biocatalysis* **10:** 113–122.

Sikkema, J., de Bont, J. A. M., and Poolman, B., 1995, Mechanisms of membrane toxicity of hydrocarbons, *Microbiol. Rev.* **59:** 201–222.

Somerville, M., Taylor, G. W., Watson, D., Rendell, N. B., Rutman, A., Todd, H., Davies, J. R., Wilson, R., Cole, P., and Richardson, P. S., 1992, Release of mucus glycoconjugates by *Pseudomonas aeruginosa* rhamnolipid into feline trachea in vivo and human bronchus in vitro, *Am. J. Respir. Cell. Mol. Biol.* **6:** 116–122.

Stead, D. E., 1992, Grouping of plant-pathogenic and some other *Pseudomonas* spp. by using cellular fatty acid profiles, *Int. J. Syst. Bacteriol.* **42:** 281–295.

Steinbuchel, A., Hustede, E., Leibergesell, M., Pieper, U., Timm, A., and Valentin, H., 1992, Molecular basis for biosynthesis and accumulation of polyhydroxyalkanoic acids in bacteria, *FEMS Microbiol. Rev.* **103:** 217–230.

Syldatk, C., Lang, S., Matulovik, U., and Wagner, F., 1985, Production of four interfacial active rhamnolipids from *N*-alkanes or glycerol by resting cells of *Pseudomonas* species DSM 2874, *Z. Naturforsch.* **40:** 61–67.

Takeuchi, M., Sawada, W., Oyaizu, H., and Yolota, A., 1994, Phylogenetic evidence for *Sphingomonas* and *Rhizomonas* as nonphotosynthetic members of the alpha-4 subclass of the *Proteobacteria*, *Int. J. Syst. Bacteriol.* **44:** 308–314.

Taylor, C. J., Carrick, B. J., Galbraith, L., and Wilkinson, S. G., 1993, Polar lipids of *Pseudomonas diazotrophicus*, *FEMS Microbiol. Lett.* **106:** 65–70.

Timm, A., and Steinbuchel, A., 1992, Cloning and molecular analysis of the poly(3-hydroxyalkanoic acid) gene locus of *Pseudomonas aeruginosa* PAO1, *FEBS Eur. J. Biochem.* **209:** 15–30.

Van Dyke, M. W., Couture, P., Brauer, M., Lee, H., and Trevors, J. T., 1993, *Pseudomonas aeruginosa* UG2 rhamnolipid biosurfactants: Structural characterization and their use in removing hydrophobic compounds from soil, *Can. J. Microbiol.* **39:** 1071–1078.

Wada, M., Fukunaga, N., and Sasaki, S., 1987, Effect of growth temperature on phospholipid and fatty acid composition in a phychrotrophic bacterium, *Pseudomonas* sp. strain E-3, *Plant Cell. Physiol.* **28:** 1209–1217.

Weber, F. J., Isken, S., and de Bont, J. A. M., 1994, Cis/trans isomerization of fatty acids as a defense mechanism of *Pseudomonas putida* strains to toxic concentrations of toluene, *Microbiology* **140:** 2013–2017.

White, D. C., Sutton, S. D., and Ringleberg, D. B., 1996, The genus *Sphingomonas*: Physiology and ecology, *Current Opinion in Biotechnology*, July.

Wilkinson, S. G., 1988, Gram-negative bacteria, in: *Microbial Lipids* (C. Ratledge and S. G. Wilkinson eds.), Academic Press, San Diego, Vol. 1, pp. 333–348.

Wilkinson, S. G., Galbraith, L., and Lightfoot, G. A., 1973, Cells walls, lipids, and lipopolysaccharides of *Pseudomonas* species, *Eur. J. Biochem.* **33:** 158–174.

Winson M. K., Camara, M., Latifi, A., Foglino, M., Chabra, S. R., Daykin, M., Bally, M., Chapon, V., Salmond, G. P., and Bycroft, B. W., 1995, Multiple N-acyl-L-homoserine lactone signal molecules regulate production of virulence determinants and secondary metabolites in *Pseudomonas aeruginosa*, *Proc. Natl. Acad. Sci. USA* **92:** 9427–9431.

Yabuuchi, E., Yano, I., Oyaizu, H., Hashimoto, Y., Ezaki, T., and Yamamoto, Y., 1990, Proposals of *Sphingomonas paucimobilis* gen. nov. and comb., nov. *Sphingomonas parapaucimobilis* sp. Nov., *Sphingomonas yanoikuyae* sp. nov., *Sphingomonas adhaesiva* sp. nov., *Sphingomonas capsulata* comb. nov., and two genospecies of the genus *Sphingomonas*, *Microbiol. Immunol.* **34:** 99–119.

Yabuuchi, E., Kosako, Y., Arakawa, M., Hotta, H., and Yano, I., 1992, Identification of Oklahoma isolate as a strain of *Pseudomonas pseudomallei*, *Microbiol. Immunol.* **36:** 1239–1249.

Yabuuchi, E., Kosaka, Y., Oyaizu, H., Yano, I., Hotta, H., Hashimoto, H., Ezaki, T., and Arakawa, M., 1994, Proposal of *Burkholderia* gen. nov. and transfer of seven species of the *Pseudomonas* hoimology group II to the new genus, with the type species *Burkholderia cepacia* (Palleroni and Holmes 1981) comb. nov., *Microbiol. Immunol.* **36:** 1251–1275.

Zhang, Y., and Miller, R. M., 1992, Enhanced octadecane dispersion and biodegradation by a *Pseudomonas* rhamnolipid (biosurfactant), *Appl. Env. Microbiol.* **58:** 3276–3282.

Outer Membrane Proteins 5

ROBERT E. W. HANCOCK and
ELIZABETH A. WOROBEC

1. INTRODUCTION

Pseudomonas aeruginosa is a gram-negative bacterium and as such has an outer membrane. The general functions of the outer membrane include size-dependent exclusion of larger molecules, permitting selective permeability of smaller molecules, uptake of large and small polycations, specific facilitated uptake of certain substrates, export of excreted molecules including proteins, secondary metabolites and siderophores, maintenance of cell shape and growth in low osmolarity medium, binding of phages, bacteriocins, and pili during conjugation, serum resistance, and surface binding of antibodies and complement. Most of these functions have been described for *Pseudomonas aeruginosa* and result from the properties of outer membrane functions. It has been 5 years since the last review of the outer membrane proteins of *Pseudomonas aeruginosa* (*Hancock et al.*, 1990). Because of the vast amount of data available on this topic, we have concentrated on the information published since the previous review, but prior to the release of the *Pseudomonas aeruginosa* genomic sequence (http://www.pseudomonas.com).

2. ROLE IN ANTIBIOTIC SUSCEPTIBILITY

It is now well established that *Pseudomonas aeruginosa* is intrinsically antibiotic-resistant and that the low outer membrane permeability of this organism contributes to this high intrinsic (background) level of resis-

ROBERT E. W. HANCOCK • Department of Microbiology and Immunology, University of British Columbia, Vancouver V6T 1Z4, British Columbia.   ELIZABETH A. WOROBEC • Department of Microbiology, University of Manitoba, Winnepeg R3T 2N2, Manitoba.

Pseudomonas, edited by Montie. Plenum Press, New York, 1998.

tance (reviewed in Hancock *et al.*, 1990). It is clear that reduced outer membrane permeability, by itself, is not sufficient to explain resistance, because antibiotics equilibrate across even the weakly permeable outer membrane of *Pseudomonas aeruginosa* in a few seconds. Thus a secondary mechanism must exist that takes advantage of defective permeability. For β-lactams, inducible chromosomal β-lactamase has been proposed as such a mechanism (Bayer *et al.*, 1987), whereas for other antibiotics, efflux is involved (Li *et al.*, 1994a,b; Poole *et al.*, 1993). There is also evidence that efflux may be involved in intrinsic (basal) resistance of *Pseudomonas aeruginosa* to β-lactams, in that *mexA*, *mexB* or *oprM* interposon mutants are supersusceptible to many β-lactams (Li *et al.*, 1995). However, although efflux pumps have quite loose specificity, they favor hydrophobic (or amphipathic), weakly positively charged compounds, whereas β-lactams are generally anionic or zwitterionic and reasonably hydrophilic. Furthermore, β-lactams act in the periplasm and thus face only one uptake/efflux barrier whereas the usual substrates for efflux pumps must cross two membranous barriers. In addition, there are considerable disparities in the extent to which susceptibility of some β-lactams is affected by *mexA* or *oprM* deletion. Therefore in this case, it might be necessary to exclude other possible secondary effects, e.g., on β-lactamase levels or inducibility, penicillin binding protein levels, or peptidoglycan biosynthesis, before one can definitively conclude that β-lactams are subject to efflux. Nevertheless, it is clear that low outer membrane permeability is an absolute determinant in the intrinsic antibiotic resistance of *P. aeruginosa* because specifically increasing outer membrane permeability by cloning in a large channel porin (Loop 5 deletion mutant of OprD) increases antibiotic susceptibility to many antibiotics by eightfold or more (Huang and Hancock, 1996).

3. IROMPs: FpvA, FptA, PfeA

Pseudomonas aeruginosa, like most bacteria, requires iron for growth. To sequester iron from its environment, this organism produces two iron chelators or siderophores, pyochelin and pyoveridine (Cox and Adams, 1985), which are produced under conditions of iron limitation. Up to seven iron-regulated outer membrane proteins have also been identified (Cornelis *et al.*, 1987), some of which have been classified as receptors for the iron-bound siderophores and hence are involved in the first step in the entry of iron into the cell. In addition to the two aforementioned siderophores, *P. aeruginosa* can utilize siderophores synthesized by other organisms, such as enterobactin (Polle *et al.*, 1990) and

Table I. Outer Membrane Proteins of *Pseudomonas aeruginosa*

Protein name	Other names	Apparent molecular weight[a]	Sequence accession number	Map position	Production conditions	Function	Identified in following fluorescent pseudomonads[d]
FpvA	IROMP	80,000	L10210 U07379	SpeI-J DpnI-B	Iron limitation	Ferripyoverdine uptake	PP, PF, PS, PC, PT
FptA	IROMP	75,000	U03161	SpeI-A DpnI-C	Iron limitation	Ferripyochelin uptake	
PfeA	IROMP	80,000	M98033	—[b]	Iron limitation, enterochelin present	Ferrienterobactin uptake	
OprC	C	70,000	Cloned[c]	–	Anaerobic induction, copper repressible	Copper transport	PST
OprJ	OprK	54,000	Cloned	–	Derepressed by mutation	Efflux	
AlgE	alg 76	54,000	M37181	SpeI-T DpnI-H	Coexpressed with alginate exopolysaccharide	Putative export of alginate	AV, PS, PP
OprN		50,000	—[b]	–	Derepressed by mutation	Efflux	
OprM	OprK	50,000	L23839	SpeI-H DpnI-E	Constitutive low level, can be derepressed	Efflux	
OprP	P	48,000	X53313	SpeI-C DpnI-N	Inducible by low phosphate	Phosphate uptake	PP, PS, PF, PC

(*continued*)

Table I. (*Continued*)

Protein name	Other names	Apparent molecular weight[a]	Sequence accession number	Map position[b]	Production conditions	Function	Identified in following fluorescent pseudomonads[d]
OprO		48,000	M86648	SpeI-C DpnI-N	Inducible by low phosphate, stationary phase	Pyrophosphate uptake	
OprB	D1	46,000	X77131	–	Inducible by glucose	Glucose/sugar uptake	PP, PF
OprD	OprD2, D2, D	45,500	Z14065 X63152	SpeI-F DpnI-A	Low level constitutive	Basic amino and imipenem uptake	
OprE	E1	43,500	D12711	–	Anaerobic induction	Unknown specific substrate	
E	E2	44,000	–		Constitutive	Structural?	
OprF	F	38,000	M94078 M18795	SpeI-L DpnI-C	Constitutive	Major nonspecific porin/structural	PP, PS, PF, PC, PST, AV, PT
OprG	G	25,000	–	–	Low level constitutive. Induced in high iron, high Mg^{2+}	Unknown	
OprH	H1	21,000	M26954	SpeI-M DpnI-A	Low Mg^{2+}, Sr^{2+}, Mn^{2+}, and Ca^{2+}	Gated porin?; polycation/EDTA resistance	PP, PS, PF, PC, PST
OprL	H2	20,500	Z50191	–	Constitutive	Structural	All
OprI	I	8,000	M25761	SpeI-V DpnI-D	Constitutive	Structural	All

[a]Apparent molecular weight on SDS-PAGE. Actual molecular weights can be obtained from the sequences. Apparent molecular weights vary according to the SDS-PAGE system used.
[b]Physical map position not determined.
[c]Gene cloned but accession number not available at press time.
[d]Abbreviations: PP - *Pseudomonas putida*; PS - *P. syringae*; PC - *P. chloraphis*; PT - *P. tolasii*; PST - *P. stutzeri*; AV - *Azobacter vinelandii*; All - all of the above Pseudomonads.

aerobactin (Liu and Shokrani, 1978) produced by enteric bacteria, ferrioxamine B produced by *Streptomyces spp.* (Cornelis *et al.*, 1987), and pyoverdines produced by other fluorescent pseudomonads (Hohnadel and Meyer, 1988).

The induction of high molecular weight, iron-regulated, outer membrane proteins by *P. aeruginosa*, in response to the presence of non-pseudomonad siderophores, occurs with desferral, a derivative of ferrioxamine B, and ferri-enterobactin. The latter enterobactin-mediated protein has been characterized. Poole *et al.* (1990), upon studying enterobactin-mediated iron uptake by a pyoverdine-deficient strain of *P. aeruginosa*, identified a novel 80,000-Da outer membrane protein. Mutant strains incapable of enterobactin-dependent iron uptake were also lacking this protein, suggesting that this protein an enterobactin receptor. The structural gene for this putative enterobactin receptor was later cloned and characterized (Dean and Poole, 1993). The gene product was termed PfeA for *Pseudomonas* ferric enterobactin receptor. The structural gene for this putative enterobactin uptake was expressed in E. coli and was capable of complementing a mutation in the *E. coli fepA* gene which encodes the *E. coli* FepA, enterobactin receptor protein (an 81,000-Da outer membrane protein). The *pfeA* gene codes for an 81,000-Da protein which has more than 60% homology to its *E. coli* counterpart. The most significant homology is in regions involved in ligand binding. In addition, and not surprisingly, the two proteins cross-react. *P. aeruginosa* PfeA protein can replace FepA in *E. coli*, hence reinforcing the similarity between enterobactin-dependent uptake between the two organisms. This suggests the existence of analogous components in *P. aeruginosa*, such as periplasmic and cytoplasmic membrane proteins. The identification of a TonB box at the N-terminus of PfeA and Fur binding sites upstream of the *pfeA* gene also indicates the similarities between the two systems because FepA function depends on interaction with the TonB protein (Lundrigan and Kadner, 1986) and the expression of FepA is under control of the *fur* gene (Earhart, 1987). The extent of the analogy between the two systems is currently under investigation.

The concomitant loss of pyoverdine production and decreased production of a 90,000-Da IROMP in *P. aeruginosa*, led Poole *et al.* (1991) to believe that this protein is a ferri-pyoverdine receptor. In addition, a mutant strain deficient in the production of this protein shows a substantial decrease in pyoverdine-mediated iron transport. Previous to this Cornelis *et al.* (1987) assigned this function to an 80-KDa IROMP. An 85-KDa protein has also been identified as a pyocin S-receptor in a clinical isolate of *P. aeruginosa* (Gensberg *et al.*, 1992). It is likely that these three proteins are the same. Using the above mutant strains, more recently Poole *et al.* (1993) reported cloning the gene for the putative pyoverdine receptor by complementation. The product of the *fpvA* gene is consistent with an 86-kDa mature protein. Sequence analyses revealed homology to highly conserved domains found in TonB-dependent receptors, even though nothing resembling a TonB box was found. FpvA also has regions of homology with the PupA and PupB proteins of *P. putida*. PupA and PupB are IROMPs which function as receptors for ferric pseudobactin, a siderophore which is similar to pyoverdine (Bitter *et al.*, 1991). Some homology was also found with FhuE, the *E. coli* receptor for fungal siderophores coprogen, rhodoturilic acid, and ferrioxamine A.

Two different IROMPs have been described as involved in ferripyochelin uptake by *P. aeruginosa*. Originally, a 14-KDa protein, designated the ferripyochelin-binding protein (FBP) was assigned this role (Sokol and Woods, 1985). However, Heinrichs *et al.* (1991) demonstrated that a high MW protein (75,000 Da) interacts with pyochelin and supports pyochelin-mediated iron transport at nM levels of $FeCl_3$. This group also reported that a strain deficient in this protein transports ferripyochelin at higher ferric chloride concentrations, hence suggesting a putative role of the 14,000-Da protein in a second, lower affinity transport system. They also attributed the failure of earlier studies to observe the 75-Kda protein to the induction of the protein in the late log to stationary phase. The structural gene for the high-affinity ferripyochelin receptor (*fptA*) was cloned and the expression of a 75-Kda protein (Ankenbauer, 1992) substantiated the findings of Heinrichs *et al.* (1991). The mature FptA protein has a molecular mass of 76 KDa, and sequence homology studies revealed considerable homology with FpvA of *P. aeruginosa*, PupA and PupB of *P. putida*, and FhuE of *E. coli* (ankenbauer and Quan, 1994). Like FpvA and PupB, no TonB box was identified. Pyochelin and its receptor are both regulated by the transcriptional activator PchR which in turn is iron-regulated. The *fptA* gene is preceded by a sequence matching the *E. coli* fur-binding site, indicating the importance of iron concentrations in the expression of this receptor. Heinrich and Poole (1993) described a regulator PchR for both pyochelin and FptA biosynthesis.

In their study of the uptake and function of an antipseudomonal cephalosporin antibiotic, Yamano *et al.* (1994) found that the compound was best transported via the ferric iron transport pathway. They also found that the compound was best transported via the ferric iron transport pathway. They also found that mutant strains resistant to this antibiotic lacked a 66-kDa IROMP which, as they predicted, acts as a receptor for the antibiotic. *E. coli* IROMPS Fiu and Cir have been implicated in the uptake of similar antibiotics (Curtis *et al.*, 1988; Nikaido and Rosenberg, 1990), hence suggesting a similarity between the uptake systems and the proteins in question. No further reports have emerged regarding the role of this 66-KDa IROMP.

4. OprC

OprC is a low copy number outer membrane protein which has been shown by Yoshihara and Nakai (1989) to form slightly anion-selective, small diffusion pores. The purified protein runs on SDS-PAGE with an apparent M_r of 70,000. The results, however, from this study are not without controversy because the investigators demonstrated in the same study that the major nonselective porin OprF, of *P. aeruginosa* did not exhibit porin function. In a later study, Nikaido *et al.* (1990) repeated the work of Yoshihara and Nakai and found that in their hands the levels of OprC were so low that they were not able to assess its function accurately. OprC has also been linked to the uptake of anionic antipseudomonal β-lactam antibiotics, such as cefsulodin, piperacillin, and azetreonam (Satake *et al.*, 1990). However, the role of OprC in antibiotic susceptibility was recently reexamined by Yoneyama *et al.* (1995). The *oprC*, *oprD*, and *oprE* genes were disrupted by gene replacement, and the resulting mutant strains were examined for susceptibility to a variety of antiotics. The results demonstrated that OprC is not involved in imipenem permeability nor the permeability of any of a number of antipseudomonal antibiotics, including quinilones, cephalosporins, chloramphenicol, and various penicillin derivatives.

Yamano *et al.* (1993) demonstrated that OprC levels increase fourfold during anaerobic growth and this increase is not caused by a concomitant decrease in iron concentration in the growth medium, nor reduced growth rate in a minimal medium. It was proposed that the increased production of this protein along with OprE aids in the growth of *P. aeruginosa* by enhancing the rate of entry of essential nutrients into the cell during oxygen stress. More recently, Nakae and collaborators sequenced OprC and showed that it is 65% homologous with NosA of *P. stutzeri* (Lee *et al.*, 1989), an outer membrane porin required for producing copper-containing nitrite reductase. Like NosA, it was shown that OprC is made only anaerobically and is repressed by high Cu^{2+} concentrations in the medium. Nitrite reductase is a key element in the anaerobic respiratory process termed denitrification. The predominant single-channel conductance increments of the OprC and NosA porins are virtually identical.

5. OprJ

The overproduction of a 54-KDa outer membrane protein has been reported in several mutant strains of *P. aeruginosa* belonging to the Type2 (*nfxB*) group (Hirai *et al.*, 1987; Legakis *et al.*, 1989; Yoshida *et al.*, 1994). These strains are generally cross-resistant to quinilones, such as norfloxacin, to newly developed cephems like cefpirome, and are hypersensitive to β lactams and aminoglycosides. This protein has only been recently been termed OprJ and its relationship to the other resistance-correlated outer membrane proteins OprK, OprM, and OprN examined (Masuda *et al.*, 1995). In some strains, OprJ is repressed by salicylate in conjunction with an increase in OprN production, suggesting some sort of linked control mechanism, not unlike that seen with OmpC/F in *E. coli*. It seems plausible to imply that OprJ is an efflux protein although no conrete information has been presented.

Purified OprJ retains its SDS-PAGE electrophoretic mobility regardless of reduction or heating. A monoclonal antibody specific for OprJ was used in immunoblot studies (Hosaka *et al.*, 1995) to demonstrate that OprJ is produced by fluoroquinolone-resistant NfxB mutant strains of *P. aeruginosa* but not by other types of fluoroquinolone-resistant strains (NalB, NfxC). This implies that OprJ is a protein novel to NfxB mutant strains.

6. AlgE

Mucoid strains of *Pseudomonas aeruginosa* produce a 54-kD protein termed AlgE (Rehm *et al.*, 1994). The *algE* gene is found as part of the 34 min alginate gene cluster and has been cloned and sequenced (Chu *et al.*, 1991). Purified AlgE forms a large, anion-selective channel. Therefore it has been proposed that it functions in enhancing the excretion of exopolysaccharide precursors out of the cell to a location where they are assembled into the mucoid material of *P. aeruginosa*. The protein has been modeled as an 18-β-strand porin, reminiscent of the Lam B porin (Schirmer *et al.*, 1995).

7. OprN

P. aeruginosa mutant strains possessing cross-resistance to quinilones, imipenem, and chloramphenicol plus a hypersensitivity to cephems, aminoglycosides, and carbenicillin overproduce a 50-kDa outer membrane protein, OprN (Fukuda *et al.*, 1990, 1995). In these strains a decrease in norfloxacin accumulation and in the rate of imipenem uptake was noted, suggesting the presence of an efflux process. Also in these strains, a decrease in OprD was noted, hence the reduced imipenem uptake. Although it has the same apparent molecular weight as OprM, this protein does not change SDS-PAGE mobility after heating and is positively regulated by salicylate (Masuda *et al.*, 1995). In conjunction with salicylate induction, increased resistance to quinolones and carbapenems is linked to the decrease in OprD production and the increased production of OprN. This mirrors the phenotype observed for the OprN-overproducing Type 3 (*nfxC*) mutant strains. OprN levels are also elevated when OprJ overproducing mutant strains are treated with salicylate, suggesting that the two proteins are regulated to obtain a constant copy number in the outer membrane. Masuda *et al.* (1995) have proposed that OprJ, OprM, and OprN act as independent efflux transporters and that OprJ and OprN are cooperatively regulated, much like OmpF and OmpC from the enteric bacteria.

8. OprK

Upon examining a pyoverdine-deficient strain of *P. aeruginosa*, Poole *et al.* discovered the iron-inducible expression of OprK, a 50-kDa outer membrane protein. The same strain is resistant to a number of structurally diverse antimicrobial agents (e.g., ciprofloxacin, nalidixic acid, tetracycline and chloramphenicol). It was originally though that the *oprK* gene belongs to an operon (*mexA-mexB-oprK*) responsible for the efflux of a variety of antibiotics and secondary metabolites (e.g., pyoverdine). However, it was recently found that the *oprK* gene expresses OprM and thus was renamed *oprM* (Li *et al.*, 1995). The OprK protein derives in fact from an *nfxB* mutation derepressing the *oprJ* gene, and thus the protein has now been renamed OprJ (Poole, K., personal communication).

9. OprM

Analysis of nalidixic acid (*nalB* or Type 1) multiple antibiotic-resistant strains of *P. aeruginosa* revealed the overexprssion of a 49-KDa protein termed OprM (Masuda and Ohya, 1995). OprM has been partially characterized (Gotoh *et al.*, 1994b) and found to be heat modifiable. OprM changes molecular mass on SDS-PAGE from 50 kDa to more than 100 kDa upon heating before loading on gels. Strains deficient in the production of OprM are highly susceptible to a number of antibiotics, including quinilones, β-lactams, tetracycline, chloramphenicol, and several penems (Gotoh *et al.*, 1994a). OprM production is not influenced by salicylate, the known synthesis suppressor of some outer membrane proteins (Masuda *et al.*, 1995). Initially evidence indicated that OprM acts much like OprK in a separate, but related, efflux system that Li *et al.* (1994a) call MexC-MexD-OprJ. However it was recently shown that the product of the cloned *oprK* gene is identical to OprM (Li *et al.*, 1995). Li *et al.* (1995) examined the accumulation of a variety of drugs in several mutant strains in the presence and absence of a proton ionophore and found that MexA-MexB-OprM forms an energy-dependent complex capable of pumping various antibacterial agents out of *Pseudomonas aeruginosa*. MexA and MexB resemble cytoplasmic membrane export proteins. A proposed model for efflux has the antibiotics pumped from the cytoplasm to the external face of the cell through a direct channel (Li *et al.*, 1994b). MexB would form a cytoplasmic membrane transporter with the lipoprotein MexA acting as a link between the outer membrane channel, OprM (formerly OprK), and the MexB transporter. Indeed a disruption of any of the three gene makes the strain extremely susceptible to the effluxable antibiotics. Components similar to OprM, MexA, and MexB have also been identified by Morshed *et al.* (1995) using norfloxacin-resistant and *nalB* multiply resistant strains of *P. aeruginosa*. Active efflux has been very well documented in *E. coli* (Thanassi *et al.*, 1995), and the MexA/MexB proteins have homologues in many bacteria (Poole *et al.*, 1993) Although the finding of a similar process in *P. aeruginosa* is not completely unexpected, the concept of efflux as a major player in intrinsic resistance provided a substantial breakthrough in our thinking about how *P. aeruginosa* escapes killing by multiple antibiotics.

10. OprP

OprP was first identified as an outer membrane protein induced upon growth of *P. aeruginosa* in a low phosphate medium (see Hancock *et al.*, 1990; Siehnel *et al.*, 1990 for most citations). A Tn501 mutant lacking OprP was shown to be deficient in the high-affinity, phosphate-starvation-inducible (PST) transport system. Detailed model membrane studies have made this one of the best characterized anion channels in nature. The OprP channel was shown to contain a phosphate-binding site with a K_i of approximately $0.15 \mu M$ (approximately the medium concentration of phosphate at which OprP is optimally induced). Chemical modification studies of OprP indicated that the phosphate-binding site contains a lysine residue. Therefore we have started to change the lysines of OprP systematically to glutamates and glycines (Sukhan and Hancock, 1996). A single lysine, residue-121, changed to either a glutamate or a glycine residue, causes loss of the phosphate-binding site, as judged by model membrane studies.

To determine where lysine-121 exists, a membrane topology model (Fig. 1a) was created by predicting transmembranous amphipathic β-strands (typical of porins; Huang *et al.*, 1995) and constructing linker and epitope insertion mutants at various sites through the protein (utilizing the observation that loop regions interconnecting β-strands can tolerate insertion of extra amino acids, but that the β-strand regions cannot) (Sukhan and Hancock, 1995). Interestingly, in this model, lysine-121 is right in the middle of loop 3 which, in the porins were with defined-three-dimensional structurs, form the narrowest (constriction) region of the porin channel. Indeed lysine-131 of PhoE, which is substituted for glycine in the related *E. coli* porin OmpF (Cowan *et al.*, 1992), is the major contributor to the anion selectivity of PhoE and the cation selectivity of OmpF.

OprP is coregulated with the periplasmic phosphate-binding protein. Biochemical data suggests that these two proteins can interact whereas genetic data suggest they collaborate in phosphate uptake. A mathematical model of phosphate passage through the OprP channel has been developed.

11. OprO

OprO was first identied as an open reading frame upstream of the OprP gene. It shares 76% identical amino acids with OprP (Siehnel *et al.*, 1992). Model membrane studies on the purified OprO porin, after overexpression in *E. coli*, indicate that it is quite similar to OprP in that it forms anion-specific channels with a phosphate-binding site (Hancock *et al.*, 1992). However, the OprO channel is somewhat larger and prefers pyrophosphate over phosphate (cf. OprP for which the situation is reversed).

Utilizing an OprP::Tn501 mutant of *P. aeruginosa*, OprO expression was demonstrated after growth of this mutant under phosphate-starvation conditions into the stationary phase (Siehnel *et al.*, 1992). Like the *oprP* gene, *oprO* contains an inverted repeat element overlapping this

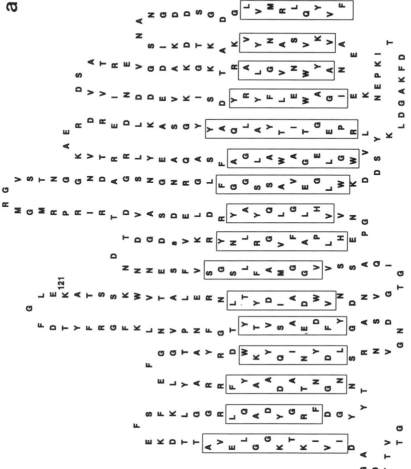

Figure 1. Membrane topology models of selected outer membrane porin proteins of *Pseudomonas aeruginosa*. The boxed regions represent proposed membrane spanning β-strands. The top interconnecting stretches of amino acids represent proposed surface exposed loop regions and the bottom stretches are proposed periplasmic turns. (a) OprP; (b) OprD; (c) OprF.

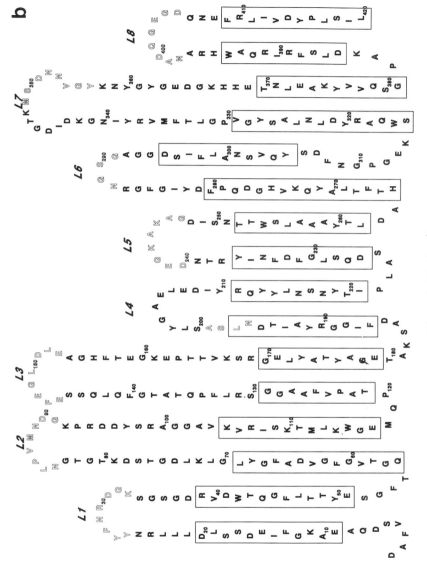

Figure 1. (*Continued*).

C

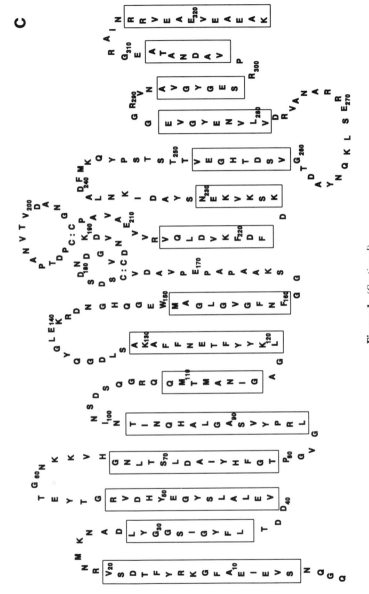

Figure 1. (*Continued*).

Pho box with homology to the cyclic AMP receptor-binding site of the *E. coli rpoH* gene.

12. OprB

OprB was first identified by Hancock and Carey (1980) following the growth of *P. aeruginosa* in a minimal medium supplemented with glucose as the sole carbon source. The protein was identified as a heat-modifiable protein, whose M_r changes from 30,000 to 47,000 upon heating to 100 °C before SDS-PAGE. OprB is coregulated with a periplasmic glucose-binding protein as part of a high-affinity glucose uptake system. Hancock and Carey (1980) originally demonstrated the porin function of OprB by following the efflux of several carbohydrates from phospholipid vesicles reconstituted with purified OprB. Using the liposome swelling assay, Trias *et al.*, (1988) characterized OprB as selective for glucose and xylose. Further analyses of purified OprB from *P. aeruginosa* (Wylie *et al.*, 1993) and *P. putida* (Saravolac *et al.*, 1991) revealed many similarities between the two proteins, including high β-sheet content (40%), as determined by circular dichroism, immunochemical cross-reactivity, amino terminal sequence homology, and a glucose-binding site ($K_s = 380$ mM for *P. aeruginosa* and 110 mM for *P. putida*). The two proteins differ in molecular weight (*P. aeruginosa* OprB is 4500 Da larger) and in ion selectivity. The *P. aeruginosa* OprB is anion-selective, and *P. putida* OprB is cation-selective. The reason for this difference in selectivity is still unknown.

The sequence of the cloned *P. aeruginosa oprB* gene revealed some homology to the maltodextrin-selective LamB porin of *E. coli* (Wylie and Worobec, 1994). Topological modeling by the prediction rules of Paul and Rosenbusch predicted 18 potential membrane-spanning β-strands when placing both N-and C-termini on the periplasmic face of the membrane, similar to that deduced from the crystal structue of LamB (Schirmer *et al.*, 1995). As with LamB, extra membranous loops joining the β-strands are short on the periplasmic face compared with those facing extracellularly. Amino acid comparison between OprB and LamB revealed several conserved and identical residues implicated in sugar binding to LamB. Likewise, OprB has a cluster of five tryptophan and seven phenylalanine residues which resembles what is termed the "greasy slide" of LamB, a region which is though to guide the diffusion of sugars. The aromatic ring of phenylalanine and tryptophan residues contribute to protein–sugar interactions in carbohydrate-binding proteins. The residues stack on one or both sides of the pyranoside rings of sugars, contributing to the Van der Waals contacts (Quiocho, 1989). OprB also has two cysteine residues (C148 and C156) which are fairly rare in porin proteins but are also found in LamB (Gerbl-Rieger *et al.*, 1991). No function has been attributed to these residues in LamB (Ferenci and Stretton, 1989), and the analysis of their function in OprB is presently underway.

Using interposon mutagenesis, *P. aeruginosa* strains were created which either lacked or overexpressed OprB (Wylie and Worobec, 1995). These strains were used to study the in vivo role of OprB in the diffusion of glucose and other carbohydrates across the outer membrane. These experiments demonstrated that OprB acts as a substrate-selective porin for a variety of different sugars, including mannitol, glycerol, and fructose. Indeed, Williams *et al.* (1994), found the production of OprB in conjunction with prolonged growth on glycerol. Thus, rather than acting exclusively as a glucose-selective proin, OprB facilitates the diffusion of a wide range of carbohydrates and thus has been labeled a carbohydrate-selective porin.

Other members of the *Pseudomonadaceae* rRNA group I carry a gene which hybridizes with the *P. aeruginosa oprB* structural gene and produce proteins with strong homology to OprB (based on SDS-PAGE mobility, immunochemical cross-reaction, and N-terminal sequence), when grown on glucose or other carbohydrates, such as xylose or maltose (Wylie and Worobec, 1994). This again reinforces the role of OprB as a general component of carbohydrate uptake in pseudomonads.

13. OprD

OprD was first identified as a modestly expressed outer membrane protein called D2 (Hancock and Carey, 1979). It was subsequently demonstrated that mutants of *P. aeruginosa*, which had become specifically resistant to a broad-spectrum β-lactam called imipenem, had lost OprD (Quinn *et al.*, 1986). Interestingly imipenem strongly resembles a dipeptide of a basic amino acid plus one other amino acid, so much so that a renal dipeptidase inhibitor, cilastatin, must be coadministered with imipenem to improve its stability in vivo (N.B: to some extent all β-lactams mimic the D-ala-D-Ala peptide in peptidoglycan precursors). Thus it seemed quite reasonable when Trias and Nikaido (1990a,b) demonstrated that OprD is a specific porin that recognizes basic amino acids, peptides containing basic amino acids, and imipenem and related carbapenem β-lactams (e.g., meropenem) as substrates. It was confirmed that these two classes of substrates bind to the same site on OprD (Trias and Nikaido, 1990b) and that high concentrations of basic amino acids substantatively increase the minimal inhibitory concentration (MIC) of imipenem and micropenem (Fukuoka *et al.*, 1991). Competition for binding to a common site within the OprD channel was definitively demonstrated by model membrane studies (Huang and Hancock, 1996; Nakae *et al.*, 1996; Trias and Nikaido, 1990b). Overexpression of OprD from the cloned gene was used to prove that the channel is quite specific for carbapenem β-lactams and does not permit passage of other β-lactams or quinolones (Huang and Hancock, 1993). This corrects some misleading conclusions in the literature (Michea-Hamzehpour *et al.*, 1991). Indeed the observation of cross-resistance between imipenem and fluoroquiinolones is more likely caused by regulatory or efflux mutants (Masuda and Ohya, 1995). Nevertheless, OprD overexpression studies indicate that gluconate can pass through this channel, possibly in a nonspecific manner (Huang and Hancock, 1993).

OprD is the first specific porin that has sequence alignment with the (generally) non-specific porin superfamily (Huang *et al.*, 1995). This has permitted constructing a membrane topology model (Fig. 1b). The model was tested with reasonable success by PCR-mediated site-specific deletion. By studying the resultant 4–8 amino acid deletion mutants, the following general conclusions were drawn: (1) loop 2 has a role in imipenem binding and (2) loops 5, 7, and 8 fold in to constrict the size of the channel and prevent high nonspecific movement of molecules through the OprD channel (Huang *et al.*, 1995; Huang and Hancock, 1996).

Recently Nakae and callaborators (1996) observed very weak protease activity (turnover number 10^{-5}/sec) for OprD that is inhibitable by imipenem. We propose that this "protease" site actually functions as a surface-binding site for basic amino acids and imipenem and is the loop 2 binding site. We further hypothesize that the binding site within the OprD channel is actually located in loop 3 (which for the known crystallized porins, both specific and nonspecific, usually forms the narrowest constriction zone of the porin channel).

14. OprE

OprE was first identified as a porin (Yoshihara and Nakae, 1989) with an exclusion limit of approximately a di- to trisaccharide (see Hancock *et al.*, 1990 for an overview), but it is now clear that it differs from outer membrane protein E described by Hancock and Carey (1979). Yamano *et al* (1993) demonstrated that it is a non-heat-modifiable protein with an overall 39% identity and 51% similarity with OprD. The protein was induced under anaerobic conditions and the upstream region of the *oprE* gene contains a weak FNR box sequence and a sigma 54-RNA polymerase-binding region. It should be noted that *Pseudomonas aeruginosa*, although termed an "obligate" aerobe, grows anaerobically in the presence of nitrate or arginine as terminal electron acceptors, and on occasion may grow in this fashion in vivo.

There is no prescribed function for this porin, although its homology to OprD suggests that it may be a specific porin with an unknown substrate. Yamano *et al.* (1993) suggested that it may be involved in the uptake of ferrous iron because anaerobic conditions favor the reduced form of iron and depress the synthesis of IROMPs. Interposon mutants lacking OprE did not change in antibiotic susceptibility, although these studies were somewhat ambiguous because they were not done under anaerobic conditions (Yoneyama *et al.*, 1995).

15. E2

Protein E2, formerly E, has not been well studied. It is present in reasonable concentrations in outer membrane samples grown under normal condition (Hancock and Carey, 1979), but it is the most prominent band in the whole cell protein gels (R. Hancock, unpublished), and thus may not be a true outer membrane protein. In experiments involving progressive solubilization of outer membrane proteins, protein E2 remained as the major detergent-insoluble protein in the residue, even after peptidoglycan digestion (Hancock *et al.*, 1981). In some antibiotic-resistant mutants, protein E levels were reduced (Yamano *et al*, 1993).

16. OprF

OprF, the most predominant outer membrane protein in *P. aeruginosa* under most media conditions, is present at around 200,000 copies/cell. It has a major role in maintaining correct cell structure and stability because mutants lacking OprF do not grow in a low osmolarity medium and are significantly shorter, giving them an almost rounded morphology. These properties and sequence similarity place OprF in the OmpA superfamily (Woodruff and Hancock, 1988). The sequence similarity to OmpA occurs entirely within the carboxy-terminal half and extends to more distantly related proteins, including OprL (the peptidolycan associated lipoprotein), pIII of *Neisseria gonorrhoeae* (the serum blocking protein), and *motB* of *Bacillus subtilis* (a non-outer membrane protein involved in motility). Interestingly, like OmpA, OprF can be progressively deleted down to an N-terminal core structure of 168 amino acids which it is predicted, forms an eight-stranded β-barrel where the eighth β-strand has high homology to the 16th (C-terminal β-strand) of members of the porin superfamily (Rawling, 1995). Deletions into this homologous β-strand, e.g., in PhoE (Bosch *et al.*, 1989) are nonpermissive and prevent a protein product from being formed. Therefore, this represents both a similarity and a difference between members of the OmpA and porin superfamilies.

Study of the above series of C-terminal deletion mutant of OprF has demonstrated that this region of OprF has a substantial role in determining both cell shape and the ability to grow in a low osmolarity medium. One explanation for these properties would be if OprF somehow determines the pattern and rate of peptidoglycan growth. With regard to this, OprF is noncovalently associated with the peptidoglycan, as demonstrated by cofractionation after solubilization of outer membrane/ peptidoglycan complexes in SDS at moderate temperature. It shares this property with OprL and OprI. Any deletion of the C-terminus, even one with as few as 36 amino acids, resulted in loss of peptidoglycan association defined in this way.

The folding of OprF in the membrane has evoked some controversy. First, the model for OmpA places the C-terminus in the periplasm, albeit using indirect evidence (Freudl *et al.*, 1986). Second, at least one of the members of the OmpA superfamily is not at all an outer membrane protein. Third, one might expect that the strong noncovalent association of OprF with the peptidoglycan would require that a significant portion of OprF is present in the periplasm. Fourth, only two regions in the C-terminus are strongly predicted to be transmembranous amphipathic β-strands as found in other outer membrane proteins. Despite this there is considerable evidence that several parts of the C-terminus of OprF are exposed on the surface of the outer membrane and that a significant portion of the OprF C-terminus (AAs 164–326) must be folded into β-sheets. This data involves (1) the mapping of linear, surface-localized, monoclonal antibody binding sites to regions involving amino acids 55 to 62, 237 to 244, and 297 to 314 (Rawling *et al.*, 1995; seven other antibodies with conformational epitopes were also mapped to larger amino acid stretches of the C-terminus); (2) the mapping of a surface-located protease cleavage site in a conserved region of *P. fluorescens* OprF to amino acids 188 to 197 (DeMot *et al.*, 1994); 93) the demonstration that 10 amino acid malarial epitopes inserted at residues 188, 196, 213, 215, 231, 290, and 310 can be localized to the surface by immunofluorescence (Wong *et al.*, 1995); and (4) the creation of surface-reactive antibodies using, amino acids 261 to 274, 305 to 318 (Hughes *et al.*, 1995), 188 to 216, and 308 to 326 as immunogens (von Specht *et al.*, 1995) (N.B: the first of these is disputed; see Hughes *et al.*, 1992 and von Specht *et al.*, 1995). In addition, circular dichroism spectra clearly indicate a β-strand content of approximatley 60% (Siehnel *et al.*, 1990), that can be accommodated only if one ensures that approximately 50% of the C-terminal residues are located in β-strands. Taking all of the above data together, a membrane topology model is presented in Figure 1c. One interesting part of this model is the cysteine disulfide region in its center. This insertion, missing in most *P. fluorescens* OprF sequences and in OmpA, has homology to a calcium-binding region in several eukaryotic proteins (De Mot and Vandeleyden, 1994).

In addition to its structural role, OprF functions as a porin. Although this was disputed for some time, there is now a consensus that OprF is the major porin of *P. aeruginosa* permitting nonspecific passage of substances larger than disaccharides (Bellido *et al.*, 1992; Nikaido *et al.*, 1991) and that it contributes substantially to, but is not the sole determinant of, the large exclusion limit and low outer membrane permeability of the *P. aeruginosa* outer membrane (Bellido *et al.*, 1992). These two properties of OprF and the outer membrane have led to considerable confusion because they indicate that even minor outer membrane proteins, or those with small channels, may contribute to the net permeation of substrates in *P. aeruginosa*. This is further complicated by efflux systems which are expected to influence net antibiotic uptake. Thus it is not surprising that defined OprF-deficient mutants have at most a two-to threefold increase in MIC to β-lactam antibiotics (Woodruff and Hancock, 1988; Yoneyama *et al.*, 1995). Nevertheless there is some evidence that OprF is the major uptake route for antibiotics, in that trained mutants selected for quinoline resistance acquire a multiple antibiotic resistance (MAR) phenotype (i.e., cross-resistance to other antibiotic classes) and OprF deficiency simultaneously (Piddock *et al.*, 1991; Zhanel *et al.*, 1994). Such MAR resistance can be reversed by complementing with the cloned *oprF* gene (G. Zhanel, personal communication). MAR mutants with OprF deficiencies can be isolated in the clinic (Piddock *et al.*, 1991).

In addition to these porin functions, evidence has been presented by comparing isogenic OprF-deficient mutants and wild-type that OprF is the route of uptake for ironsiderophore complexes, for which no specific IROMP exists (Meyer, 1992), and is possibly the route of toluene uptake (L. Li, personal communication).

17. OprG

OprG is a 25,000-Da protein which is expressed in the outer membrane of *P. aeruginosa* under a variety of growth conditions. OprG levels are affected by temperature, carbon sources, LPS content, the presence of iron, and growth in the stationary phase. OprG production has been linked to a low-affinity, iron-uptake system and to quinilone uptake. It was also reported that OprG levels are lowered in an OprN-overproducing, norfloxacin resistant *nfxC* mutant strain of *P. aeruginosa*, along with the decrease of OprD and possibly OprJ (see below). No definite function has yet been attributed to OprG.

18. OprH

OprH is a minor outer membrane protein that becomes dramatically overexpressed when cells are grown on media with low ($50\mu M$) Mg^{2+} concentrations (providing Ca^{2+}, Sr^{2+} and Mn^{2+} are also low). Under such conditions (and in mutant H181 which overexpresses OprH in a normal medium), cells become cross-resistant to polymyxins, aminoglycosides, and EDTA. Studies with an isogenic OprH-deficient interposon mutant have shown that OprH is necessary for these resistances (Young *et al.*, 1992). Overexpression of OprH from its cloned gene, however, does not fully replicate the resistance phenotype suggesting that an alteration in another outer membrane component (possibly LPS) is required. Presumably the constitutive OprH-overexpressing mutant H181 is actually a regulatory mutant, a suggestion consistent with cloning results (Bell and Hancock, 1989). We have hypothesized that OprH actually replaces divalent cations in the outer membrane, under conditions of medium insufficiency, to permit outer membrane stabilization (Nicas and Hancock, 1983).

OprH has no apparent porin activity (Young and Hancock, 1993). Our recent studies (Rehm and Hancock, manuscript in preparation), however, have caused us to reevaluate this conclusion. Circular dichroism studies indicate that OprH contains 50% β-sheet. A model of OprH as an eight-stranded β-barrel was created and tested using a combination of the methods utilized above for OprD and OprF. Deletion of external loops 1 and 3 had no effect on OprH's ability to form channels.

However, an OprH mutant with a 10 amino acid deletion of loop 4 formed large channels with a single-channel conductance of 1.3 nS. An analogous observation for the *E. coli* iron-regulated outer membrane protein FepA has led to the conclusion that FepA is a gated porin (Rutz *et al.*, 1992), and thus we suggest the same for OprH.

19. OprL

OprL is a constitutively produced, peptidoglycan-associated lipoprotein which contains covalently bound fatty acyl chains (Mizuno, 1979, 1981). The protein runs on SDS-PAGE with an apparent MW of 20,500 Da (Hancock and Carey, 1979). Like OprI, this protein is highly conserved among the pseudomonads found in the rRNA group I, as detected by immunochemical cross-reactivity studies (Mutharia and Hancock, 1985). The gene has been cloned and sequenced (Lim *et al.*, 1995) and demonstrates good homology to members of the PAL family of peptidoglycan-associated lipoproteins (and consequent C-terminal homology to OprF). Although no direct function has been assigned to OprL, it appears tobe involved in the structural integrity of the *P. aeruginosa* outer membrane (L. Ramos, personal communications).

20. OprI

Lipoprotein I (OprI) is a highly abundant, 8-KDa outer membrane protein which has become an important vaccine candidate for *P. aeruginosa* (Finke *et al.*, 1990, 1991; Rahner *et al.*, 1990; von Specht *et al.*, 1995). The *oprI* structural gene has been cloned and sequenced by three groups (Cornelis *et al.*, 1989; Duchene *et al.*, 1989; Saint-Onge *et al.*, 1992) and has many similarities to its *E. coli* counterpart, Braun's lipoprotein. The mature OprI protein has a molecular weight of 5,950 Da and a single, covalently linked fatty acid. The two proteins differ in amino acid sequence with only 30% identical residues. The majority of identical residues are located at the signal sequence cleavage site and at the C-terminus, which is the peptidoglycan linkage point of the *E. coli* lipoprotein. Using PCR and probes derived from the *P. aeruginosa oprI* gene, DeVos *et al.* (1993) extensively surveyed members of the *Pseudomondaceae* and close relatives to find that the gene could be amplified only in members of the rRNA group I, those species considered to be authentic pseudomonads. Much interest has been generated in using recombinant OprI, OprI–OprF fusion proteins, or OprI specific monoclonal antibodies as vaccine candidates for *P. aeruginosa*. OprI is antigenically cross-reactive among all 17 known serotypes of *P. aeruginosa*, according to the International Antigenic Typing Scheme (Finke *et al.*, 1990), and large quantities of recombinant OprI can be rapidly purified to homogeneity from *E. coli* (Toth *et al.*, 1994). Most recently, von Specht *et al.*, (1995) found that a glutathione S-transferase (GST)-linked fusion of OprF and OprI protects immunized animals 975-fold against *P. aeruginosa*.

ACKNOWLEDGMENTS. The financial assistance of the Medical Research Council of Canada and the Canadian Cystic Fibrosis Foundation in our research is gratefully acknowledged.

REFERENCES

Ankenbauer, R. G., 1992, Cloning of the outer membrane high-affinity Fe (III)-pyochelin receptor of *Pseudomonas aeruginosa*., *J. Bacteriol.* **174:**4401–4409.

Ankenbauer, R. G., and Quan, H. N., 1994, FptA, the Fe(III)-pyochelin receptor of *Pseudomonas aeruginosa*: A phenolate siderophore receptor homologous to hydroxamate siderophore receptors, *J. Bacteriol.* **176:** 307–319.

Bayer, A. S., Peter, J., Parr, T. R., Chan, L., and Hancock, R. E. W., 1987, In vivo development of ceftazidime resistance in an experimental *Pseudomonas aeruginosa* endocarditis model, *Antimicrob. Agents Chemother.* **31:** 253–258.

Bell, A., and Hancock, R. E. W., 1989, Outer membrane protein H1 of *Pseudomonas aeruginosa*: Expression, gene cloning, and nucleotide sequence, *J. Bacteriol.* **171:** 3211–3217.

Bellido, F., Martin, N. C., Siehnel, R. J. and Hancock, R. E. W., 1992, Reevaluation, using intact cells, of the exclusion limit and role of porin Opr in *Pseudomonas aeruginosa* outer membrane permeability, *J. Bacteriol.* **174:** 5196–5203.

Bitter, W., Marugg, J. D., de Weger, L. A., Tommassen, J. and Weisbeek, P. J., 1991, The ferric-pseudobactin receptor PupA of *Pseudomonas putida*: Homology to TonB-dependent *Escherichia coli* receptors and specificity of the protein, *Mol. Microbiol.* **5:** 647–655.

Bosch, D., Scholten, M., Verhagen, C. and Tommassen, J., 1989, The role of carboxy-terminal membrane-spanning fragment in the biogenesis of *Escherichia coli* K-12 outer membrane protein PhoE, *Mol. Gen. Genet.* **216:** 144–148.

Chamberland, S., Bayer, A. S., Schollaardt, T., Wong, S. A. and Bryan, L. E., 1989, Characterization of mechanisms of quinolone resistance in *Pseudomonas aeruginosa* strains isolated in vitro and in vivo during experimental endocarditis, *Antimicrob. Agents Chemother.* **33:** 624–634.

Chu, L., May, T. B., Chakrabarty, A. M. and T. K., 1991, Nucleotide sequence and expression of the algE gene involved in alginate biosynthesis by *Pseudomonas aeruginosa*, *Gene* **107:** 1–10.

Cornelis, P., Bouia, A., Belarbi, A., Guyonvarch, A., Kammerer, B., Hannaert, V. and Hubert, J. C., 1989, Cloning and analysis of the gene for the major outer membrane lipoprotein from *Pseudomonas aeruginosa*, *Mol. Microbiol.* **3:** 421–428.

Cornelis, P., Moguilevsky, N., Jacques, J. F., and Masson, P. L., 1987, Study of the siderophores and receptors in different clinical isolates of *Pseudomonas aeruginosa*, in: *Basic Research and Clinical Aspects of Pseudomonas aeruginosa* (G. Döring, I. A. Holder, and K. Botzenhart, eds.) S. Karger, Basel, pp. 290–306.

Cowan, S. W., Schirmer, T., Rummel, G., Steiert, M., Ghosh, R., Paupitt, P. A., Janosonius, J. N., and Rosenbusch, J. P., 1992, Crystal structures explain functional properties of two *Escherichia coli* porins, *Nature* (London) **358:** 727–733.

Cox, C. D., 1980, Iron uptake with ferripyochelin and ferric citrate by *Pseudomonas aeruginosa*, *J. Bacteriol.* **142:** 581–587.

Cox, C. D., and Adams, P., 1985, Siderophore activity of pyoverdine for *Pseudomonas aeruginosa*, Infect. Immun. **48:** 130–138.

Curtis, N. A. C., Eisenstadt, R. L., East, S. J., Cornford, R. J., Walker, L. A. and White, A. J., 1988, Iron-regulated outer membrane proteins of K-12 and mechanism of action of catechol-substituted cephalosporins, *Antimicrob. Agents Chemother.* **32:** 1879–1886.

Dean, C. R., and Poole, K., 1993, Cloning and characterization of the ferric enterobactin receptor gene (*pfeA*) of *Pseudomonas aeruginosa*, *J. Bacteriol.* **175:** 317–324.

De Mot, R., Schoofs, G., Roelandt, A., Declerk, P., Proost, P., van Damme, J., and Vanderleyden, J., 1994, Molecular characterization of the major outer membrane protein OprF from plant-root colonizing *Pseudomonas fluorescens*, *Microbiology* **140:** 1377–1387.

De Mot, R. and Vanderleyden, J., 1994, A conserved surface-exposed domain in outer membrane proteins of pathogenic *Pseudomonas* and *Branhamella* species shares sequence homology with the calcium-binding repeats of the eukaryotic extracellular matrix protein thrombospondin, *Mol. Microbiol.* **13:** 379–380.

De Vox, D., Lim, A., Jr., De Vox, P., Sarniguet, A., Kersters, K., and Cornelis, P., 1993, Detection of the outer membrane lipoprotein I and its gene in fluorescent and nonfluorescent pseudomonads: Implications for taxonomy and diagnosis, *J. Gen. Microbiol.* **139:** 2215–2223.

Duchene, M., Barron, C., Schweizer, A., von Specht, B.–U., and Domdey, H., 1989, *Pseudomonas aeruginosa* outer membrane lipoprotein I gene: Molecular cloning, sequence, and expression in *Escherichia coli*, *J. Bacteriol.* **171:** 4130–4137.

Earhart, C. F., 1987, Ferrienterobactin transport in *Escherichia coli*, in: *Iron Transport in Microbes, Plants, and animals* (G. Winkelmann, D. van der Helm, and J. B. Neilands, eds.), VCH Verlagsgellschaft mbH, Weinheim, Germany, pp. 67–84.

Ferenci, T. and Stretton, S., 1989, Cysteine-22 and cysteine-38 are not essential for the function of maltoporin (LamB protein), *FEMS Microbiol. Lett.* **61:** 335–340.

Finke, M., Duchene, M., Eckhardt, A., Domdey, H. and vonSpecht, B.–U., 1990, Protection against experimental *Pseudomonas aeruginosa* infection by recombinant *P. aeruginosa* lipoprotein I exxpressed in *Escherichia coli*, *Infect. Immun.* **58:** 2241–2244.

Finke, M., Muth G., Reichhelm, T., Thomas, M., Duchene, M., Hungerer, K.–D., Domdy, H., and von Specht, B. U., 1991, Protection of immunosuppressed mice against infection with *Pseudomonas aeruginosa* by recombinant *P. aeruginosa* by recombinant *P. aeruginosa* lipoprotein I and lipoprotein I-specific monoclonal antibodies, *Infect. Immun.* **59:** 1251–1254.

Freudl, R., MacIntyre, S., Degen, M., and Henning, U., 1986, Cell surface exposure of the outer membrane protein OmpA of *Eschericha coli* K12, *J. Mol. Biol.* **188:** 491–494.

Fukuda, H., Hosaka, M., Hirai, K., and Iyobe, S., 1990, New norfloxacin resistance gene in *Pseudomonas aeruginosa* PAO, *Antimicrob. Agents Chemother.* **34:** 1757–1761.

Fukuda, H., Hosaka, M., Iyobe, S., Gotoh, N., Nishino, T., and Hirai, K., 1994, *nfxC*-type quinolone resistance in a clinical isolate of *Pseudomonas aeruginosa*, *Antimicrob. Agents Chemother.* **39:** 790–792.

Fukuoka, T., Masuda, N., Takenouchi, T., Sekine, N., Iijima, M., and Ohya, S., 1991,

Increase in susceptibility of *Pseudomonas aeruginosa* to carbapenem antibiotics in low amino acid media, *Antimicrob. Agents Chemother.* **35:** 529–532.

Gensberg, K., Hughes, K., and Smith, A. W., 1992, Siderophore-specific induction of iron uptake in *Pseudomonas aeruginosa*, *J. Gen. Microbiol.* **138:** 2381–2387.

Gerbl-Rieger, S., Peters, J., Kellerman, J., Lottspeich, F., and Baumeister, W., 1991, Nucleotide and derived amino acid sequences of the major porin of *Comamonas acidovorans* and comparison of porin primary structures, *J. Bacteriol.* **173:** 2196–2205.

Glaser, P., Sakamoto, H., Bellalou, J., Ullmann, A., and Danchin, A., 1988, Secretion of cyclolysin, the calmodulin-sensitive adenylate cyclase-haemolysin bifunctional protein of *Bordatella pertussis, EMBO J.* **7:** 3997–4004.

Gotoh, N., Itoh, N., Tsujimoto, H., Yamagishi, J.-I., Oyamada, Y., and Nishino, R., 1994a, Isolation of OprM-deficient mutants of *Pseudomonas aeruginosa* by transposon insertion mutagenesis: Evidence of involvement in multiple antibiotic resistance, *FEMS Microbiol. Lett.* **122:** 267–274.

Gotoh, N., Itoh, N., Yamada, H., and Nishino, T., 1994b, Evidence for the location of OprM in the *Pseudomonas aeruginosa* outer membrane, *FEMS Microbiol. Lett.* **122:** 309–312.

Grabert, E., Wingender, J., and Winkler, U. K., 1990, An ouiter membrane protein chractterisitic of mucoid strains of *Pseudomonas aeruginosa, FEMS Microbiol. Lett.* **68:** 83–88.

Hancock, R. E. W., and Carey, A. M., 1979, Outer membrane of *Pseudomonas aeruginosa*: Heat- and 2-mercaptoethanol-modifiable proteins, *J. Bacteriol.* **140:** 901–910.

Hancock, R. E. W., and Carey, A. M., 1980, Protein D1 - a glucose inducible, pore-forming protein from the outer membrane of *Pseudomonas aeruginosa, FEMS Microbiol. Lett.* **8:** 699–707.

Hancock, R. E. W., Elgi, C., Benzy, R., and Siehnel, R. J., 1992, Overexpression in *Escherichia coli* and functional analysis of a novel PPi-selective porin, OprO, from *Pseudomonas aeruginosa, J. Bacteriol.* **174:** 471–476.

Hancock, R. E. W., Irvin, R. T., Costerton, J. W., and Carey, A. M., 1981, The outer membrane of *Pseudomonas aeruginosa*: Peptidoglycan associated proteins, *J. Bacteriol.* **145:** 628–631.

Hancock, R. E. W., Siehnel, R., and Martin, N., 1990, Outer membrane proteins of *Pseudomonas aeruginosa, Mol. Microbiol.* **4:** 1069–1075.

Heinrichs, D. E., and Poole, K., 1993, Cloning and sequence analysis of a gene (*pchR*) encoding an AraC family activator of pyochelin and ferripyochelin receptor synthesis in *Pseudomonas aeruginosa, J. Bacteriol.* **175:** 5882–5889.

Heinrichs, D. E., Young, L., and Poole, K., 1991, Pyochelin-mediated iron transport in *Pseudomonas aeruginosa*: Involvement of a high-molecular-mass outer membrane protein, *Infect. Immun.* **59:** 3680–3684.

Hirai, K., Suzue, S., Irikura, T., Iyobe, S., and Mitsuhashi, S., 1987, Mutations producing resistance to norflaxacin in *Pseudomonas aeruginosa, Antimicrob.* Agents Chemother, **31:** 582–586.

Hohnadel, D., and Meyer, J.–M., 1988, Specificity of pyoverdine-mediated iron uptake among fluorescent *Pseudomonas* strains, *J. Bacteriol.* **170—** 4865–4873.

Hosaka, M., Gotoh, N. and Nishino, R., 1995, Purification of a 54-kilodalton protein (OprJ) produced in NfxB mutants of *Pseudomonas aeruginosa* and production of a monoclonal antibody specific to OprJ, *Antimicrob. Agents Chemother.* **39:** 1731–1735.

Huang, H., and Hancock, R. E. W., 1996, Role of specific loop regions in determining the function of the imipenem specific pore-protein OprD of *Pseudomonas aeruginosa, J. Bacteriol.*, in press.

Huang, H., and Hancock, R. E. W., 1993, Genetic definition of the substrate selectivity of outer membrane protein OprD of *Pseudomonas aeruginosa, J. Bacteriol.* **175:** 7793–7800.

Huang, H., Jeanteur, D., Pattus, F., and Hancock, R. E. W., 1995, Membrane topology and site-specific mutagenesis of *Pseudomonas aeruginosa* porin OprD, *Mol. Microbiol.* **16:** 931–941.

Hughes, E. E., Gilleland, L. B., and Gilleland, H. E., 1992, Synthetic peptides representing epitopes of outer membrane protein F of *Pseudomonas aeruginosa* that elicit antibodies reactive with whole cells of heterologous immunotype strains of *P. aeruginosa, Infect. Immun.* **60:** 3497–3503.

Kropinski, A. M. B., Lewis, V., and Berry, D., 1987 Effect of growth temperature on the lipids, outer membrane proteins, and lipopolysaccharides of *Pseudomonas aertuginosa, J. Bacteriol.* **169:** 1960–1966.

Lee, H. S., Hancock, R. E. W., and Ingraham, J. L., 1989, Properties of a *Pseudomonas stutzeri* outer membrane channel-forming protein (NosA) required for production of copper containing N_2O reductase, . *Bacteriol.* **171:** 2096–2100.

Legakis, N. J., Tzouvelekis, L. S., Makris, A., and Kotsifaki, H., 1989, Outer membrane alterations in mutiresistant mutants of *Pseudomonas aeruginosa* selected by ciprofloxacin, *Antimicrob. Agents Chemother,* **33:** 124–127.

Létoffé, S., Delepelaire, P., and Wandersman, C., 1990, Protease secretion by *Erwinia chrysanthemi*: The specific secretion functions are analogous to those of *Escherichia coli* α-hemolysin, *EMBO J.* **9:** 1375–1382.

Li, X.-Z., Livermore, D. M., and Nikaido, H., 1994a, Role of efflux pump(s) in intrinsic resistance of *Pseudomonas aeruginosa*: Resistance to tetracycline, chloramphenicol, and norfloxacin, *Antimicrob. Agents Chemother.* **38:** 1732–1741.

Li, X.-Z., Nikaido, H., and Poole, K., 1995, Role of MexA-MexB-OprM in antibiotic efflux in *Pseudomonas aeruginosa, Antimicrob. Agents Chemother.* **39:** 1948–1953.

Lim, A., de Vox, D., Brauns, M., Gabella, A., Hamers, R., and Comelis, P., 1995, unpublished cited in Genbank release, Accession No. 250191.

Liu, P. V., and Shokrani, F., 1978, Biological activities of pyochelins: Iron-chelating agents of *Pseudomonas aeruginosa, Infect. Immun.* **22:** 878–890.

Lundrigan, M. D., and Kadner, R. J., 1986, Nucleotide sequence of the gene for the ferrienterochelin receptor FepA in *Escherichia coli*: Homology among outer membrane receptors that interact with TonB, *J. Biol., Chem.* **137:** 653–657.

Masuda, N., and Ohya, S., 1995, Outer membrane proteins responsible for multiple drug resistance in *Pseudomonas aeruginosa, Antimicrob. Agents Chemother.* **39:** 645–649.

Meyer, J. M., 1992, Exogenous siderophore-mediated iron uptake in *Pseudomonas aeruginosa*: Possible involvement of porin OprF in iron translocation, *J. Gen. Microbiol.* **138:** 951–958.

Meyer, J.-M., Hohnadel, D., Khan, A., and Cornelis, P., 1990, Pyoverdin-facilitated iron uptake in *Pseudomonas aeruginosa*: Immunological characterization of th ferripyoverdin receptor, *Mol. Microbiol.* **4:** 1401–1405.

Michea-Hamzehpour, M., Furet, Y. X., and Pechere, J. C., 1991, Role of protein D2 and lipopolysaccharide in diffusion of quinolones through the outer membrane of *Pseudomonas aeruginosa, Antimicrob. Agents Chemother.* **35:** 2091–2097.

Mizuno, T. 1979, A novel peptidoglycan-associated lipoprotein found in the cell envelope of *Pseudomonas aeruginosa* and *Escherichia coli, J. Biochem.* **86:** 991–1000.

Mizuno, T., 1981, A novel peptidoglycan-associated lipoprotein (PAL) found in the outer membrane of *Proteus mirabilis* and other gram-negative bacteria, *J. Biochem.* **89:** 1039–1049.

Mizuno, T., and Kageyama, M., 1979, Isolation and characterization of a major outer membrane protein of *Pseudomonas aeruginosa*. vidence for the occurence of a lipoprotein, *J. Biochem.* **85:** 115–122.

Morshed, S. R. M., Lei, Y., Yoneyama, H., and Nakae, T., 1995, Expression of genes associated with antibiotic extrusion in *Pseudomonas aeruginosa*, *Biochem. Biophys. Res. Commun.* **210:** 356–362.

Mutharia, L. M., and Hancock, R. E. W., 1985, Monoclonal antibody specific for an outer membrane lipoprotein of the *Pseudomonas fluorescens* group of the family *Pseudomonodaceae*, *Int. J. Syst. Bacteriol.* **35:** 530–532.

Nakae, T., Yoshihara, E., Yoneyama, H., and Ishii, J., 1996, Protein D2 of the outer membrane of *Pseudomonas aeruginosa*. In: *Abstracts 5th Int. Pseudomonas Meeting*, Tsukuba City, Japan, 1995.

Nicas, T. L., and Hancock, R. E. W., 1983, Alteration of susceptibility to EDTA polymyxin B and gentamicin in *Pseudomonas aeruginosa* by divalent cation regulation of outer membrane protein H1, *J. Gen. Microbiol.* **129:** 509–517.

Nicas, T. L., and Hancock, R. E. W., 1989, Outer membrane protein H1 of *Pseudomonas aeruginosa*: Involvement in adaptive and mutational resistance to ethylenediaminetetraacetate, polymisin B, and gentamycin, *J. Bacteriol.* **143:** 872–878.

Nikaido, H., Nikaido, T., and Harayama, S., 1991, Identification and characterization of porins in *Pseudomonas aeruginosa*, *J. Biol. Chem.* **266:** 770–779.

Nikaido, H., and Rosenberg, Y., 1990, cir and Fiu proteins in the outer membrane of *Escherichia coli* catalyse transport of monomeric catechols: Study with β-lactam antibiotics containing catechol and analogous groups, *J. Bacteriol.* **172:** 1361–1367.

Ohkawa, I., Shiga, S., and Kageyama, M., 1980, Effect of iron concentration in the growth medium on the sensitivity of *Pseudomonas aeruginosa* to pyocin S2, *J. Biochem.* **87:** 323–331.

Piddock, L. V. J., Hall, M. C. Bellido, F., Bains, M., and Hancock, R. E. W., 1992, A pleiotropic, posttherapy, enoxacin-resistant mutant of *Pseudomonas aeruginosa*, *Antimicrob. Agents Chemother.* **36:** 1057–1061.

Poole, K., Krebes, K., McNally, C., and Neshat, S., 1993, Multiple antibiotic resistance in *Pseudomonas aeruginosa*: Evidence for involvement of an efflux operon, *J. Bacteriol.* **175:** 7363–7372.

Poole, K., Neshat, S., and Heinrichs, D., 1991, Pyoverdine-mediated iron transport in *Pseudomonas aeruginosa*: Involvement of high-molecular-mass outer membrane protein, *FEMS Microbiol. Lett.* **78:** 1–6.

Poole, K., Neshat, S., Krebes, K., and Heinrichs, D., 1993, Cloning and nucleotide sequence analysis of the ferripyoverdine receptor gene *fpvA* of *Pseudomonas aeruginosa*, *J. Bacteriol.* **175:** 4597–4604.

Poole, K., Young, L., and Neshat, S., 1990, Enterobactin-mediated ironn transport in *Pseudomonas aeruginosa*, *J. Bacteriol.* **172:** 6991–6996.

Quinn, J. P., Dudeck, E. J., Divincenzo, C. A., Lucks, D. A., and Lerner, S. A., 1986, Emergence of resistance to imipenem during therapy for *Pseudomonas aeruginosa* infections, *J. Infect. Dis.* **154:** 289–294.

Quiocho, F. A., 1989, Protein-carbohydrate interactions: Basic molecular features, *Pure Appl. Chem.* **61:** 1293–1306.

Rahner, R., Eckhardt, A., Duchêne, MN., Domdey, H., and von Specht, B. U., 1990, Protection of immunosuppressed mice against infection with *Pseudomonas aeruginosa* by monoclonal antibodies to outer membrane protein OprI, *Infection* **18:** 242–254.

Rawling, E. G., 1995, Outer membrane protein OprF of *Pseudomonas aeruginosa*, Ph.D. Thesis, University of British Columbia, Vancouver.

Rawling, E. G., Martin, N. L., and Hancock, R. E. W., 1995, Epitope mapping of the

Pseudomonas aeruginosa major outer membrane protein OprF. *Infect. Immun.* **63**: 38–42.

Rehm, B., Boheim, G., Tommassen, J., and Winkler, U. K., 1994, Overexpression of *algE* in *Escherichia coli*: Subcellular localization, purification, and ion channel properties, *J. Bacteriol.* **176**: 5639–5647.

Rutz, J. M., Liu, J., Lyons, J. A., Goranson, J., Armstrong, S. K., McIntosh, M. A., Feix, J. B., and Klebba, P. E., 1992, Formation of a gated channel by a ligand-specific transport channel in the bacterial outer membrane, *Science* **258**: 471–=475.

Saint-Onge, A., Romeyer, F., Lebel, P. Masson, L., and Brousseau, R., 1992, Specificity of the *Pseudomonas aeruginosa* PAO1 lipoprotein I gene as a DNA probe and PCR target region with the *Pseudomonodaceae*, *J. Gen. Microbiol.* **138**: 733–741.

Saravolac, E. G., Taylor, N. F., Benz, R., and Hancock, R. E. W., 1991, Purification of glucose-inducible outer membrane protein OprB of *Pseudomonas putida* and reconstitution of glucose-specific pores, *J. Bacteriol.* **173**; 4970–4976.

Satake, S., Yoshihara, E., and Nakae, T., 1990, Diffusion of β-lactam antibiotics through liposome membranes reconstituted from purified porins of the outer membrane of *Pseudomonas aeruginosa, Antimicrob. Agents Chemother.* **34**: 685–690.

Schirmer, T., Keller, T. A., Wang, Y.–F., and Rosenbusch, J. P., 1995, Structural basis for sugar translocation through maltoporin channels at 3.1 Å resolution, *Science* **267**: 512–514.

Siehnel, R. J., Egli, C., and Hancock, R. E. W., 1992, Polyphospate-selective porin OprO of *Pseudomonas aeruginosa*: Expression purification and sequence, *Mol. Microbiol.* **6**: 2319–2326.

Siehnel, R. J., Martin, N. L., and Hancock, R. E. W., 1990, Function and structure of the porin proteins OprF and OprP of *Pseudomonas aeruginosa*, In: *Pseudomonas: Biotransformation, Pathogenesis and Evolving Biotechnology*, S. Silver, A. M. Chakrabarty, B. Iglewski, and S. Kaplan eds.) American Society for Microbiology, Washington, D.C., pp. 328–342.

Sokol, P. A., and Woods, D. E., 1983, Demonstration of an iron-siderophore-binding protein in the outer membrane of *Pseudomonas aeruginosa, Infect. Immun.* **40**: 665–669.

Sukhan, A., and Hancock, R. E. W., 1995, Insertion mutagenesis of the *Pseudomonas aeruginosa* phosphate-specific porin OprP, *J. Bacteriol.* **177**: 4914–4920.

Sukhan, A., and Hancock, R. E. W., 1996, The role of specific lysine residues in the passage of phosphate and chloride ions through the *Pseudomonas aeruginosa* phosphate starvation inducible porin OprP. *J. Biol. Chem.* **271**:21239–21242.

Thanassi, D. G., Suh, G. S., and Nikaido, H., 1995, Role of outer membrane barrier in efflux-mediated tetracycline resistance of *Escherichia coli, J. Bacteriol.* **177**: 998–1007.

Toth, A., Schödel, F., Duchêne, M., Massarrat, K., Blum, B., Schmitt, A., Domdey, H., and von Specht, B.–U., 1994, Protection of immunosuppressed mice against tanslocation of *Pseudomonas aerugginosa* from the gut by oral immunization with recombinant *Pseudomonas aeruginosa* outer membrane protein I expressing *Salmonella dublin, Vaccine* **12**: 1215–1221.

Trias, J., and Nikaido, H., 1990a, Outer membrane protein D2 catalyzes facilitated diffusion of carbapenems and penems through the outer membrane of *Pseudomonas aeruginosa, Antimicrob. Agents Chemother.* **34**: 52–57.

Trias, J., and Nikaido, 1990b, Protein D2 channel of the *Pseudomonas aeruginosa* outer membrane has a binding site for basic amino acids and peptides, *J. Biol. Chem.* **265**: 15680–15684.

Trias, J., Rosenberg, E. Y., and Nikaido, H., 1988, Specificity of the glucose channel formed by protein D1 of *Pseudomonas aeruginosa, Biochem. Biophys. Acta* **938**: 493–496.

VonSpecht, B.-U., Knapp, B., Muth, G., Bröker, M., Hungerer, K.–D., Diehl, K.–D.,

Massarrat, K., Seemann, A., and Domdey, H., 1995, Protection of immunocompromised mice against lethal infection with *Pseudomonas aeruginosa* by active or passive immunization with recombinant *P. aeruginosa* outer membrane protein F and outer membrane protein I fusion proteins, *Infect. Immun.* **63:** 1855–1862.

Williams, S. G., Greenwood, J. A., and Jones, C. W., 1994, The effect of nutrient limitation on glycerol uptake and metabolism in continuous cultures of *Pseudomonas aeruginosa*, *Microbiology* **140:** 2961–2969.

Wong, R. S. Y., Wirtz, R. A., and Hancock, R. E. W., 1995, *Pseudomona aeruginosa* outer membrane protein OprF as an expression vector for foreign epitopes: The effects of positioning and length on the antigenicity of the epitope, *Gene* **158:** 55–60.

Woodruff, W. A., and Hancock, R. E. W., 1982, Construction and characterization of *Pseudomonas aeruginosa* protein F-deficient mutant after in vivo and in vitro insertion mutagenesis of the cloned gene, *J. Bacteriol.* **170:** 2592–2598.

Wylie, J. L., 1994, Characterization of the OprB porin of *Pseudomonas aeruginosa*, Ph.D. Thesis, University of Manitoba, Winnipeg, Manitoba, Canada.

Wylie, J. L., Bernegger-Egli, C., O'Neil, J. D.J., and Worobec, E. A., 1993, Biophysical characterization of OprB, a glucose-inducible porin of *Pseudomonas aeruginosa*, *J. Bioenerg. Biomembr.* **25:** 547–556.

Wylie, J. L., and Worobec, E. A., 1994, Cloning and nucleotide sequence of the *Pseudomonas aeruginosa* glucose-selective OprB porin gene and distribution of OprB within the family *Pseudomonadaceae*, *Eur. J. Biochem.* **220:** 505–512.

Wylie, J. L., and Worobec, E. A., 1995, The OprB porin plays a central role in carbohydrate uptake in *Pseudomonas aeruginosa*, *J. Bacteriol.* **177:** 3021–3026.

Yamano, Y., Nishikawa, T. and Komastusu, Y., 1993, Cloning and nucleotide sequence of anaerobically induced porin protein E1 (OprE) of *Pseudomonas aeruginosa* PAO1, *Mol. Microbiol.* **8:**993–1004.

Yamano, Y., Nishkawa, T., and Komatsu, Y., 1994, Ferric iron transport system of *Pseudomonas aeruginosa* PAO1 that functions as the uptake pathway of a novel catechol-substituted chephalosporin, S-9096, *Appl. Microbiol. Biotechnol.* **40:** 892–897.

Yates, J. M., Morris, G., and Brown, M. R. W., 1989, Effect of iron concentration and growth rate on the expression of protein G in *Pseudomonas aeruginosa*, *FEMS Microbiol. Lett.* **58:** 259–262.

Yoneyama, H., Yamano, Y., and Nakae, T., 1995, Role of porins in the antibiotic susceptibility of *Pseudomonas aeruginosa*: Construction of mutants with deletions in the multiple porin genes, *Biochem. Biophys. Res. Commun.* **213:** 88–95.

Yoshida, T., Muratani, T., Iyobe, S., and Mitsuhashi, S., 1994, Mechanisms of high-level resistance to quinolones in urinary tract isolates of *Pseudomonas aeruginosa*, *Antimicrob. Agents Chemother.* **38:** 1466–1469.

Yoshihara, E., and Nakae, T., 1989, Identification of porins in the outer membrane of *Pseudomonas aeruginosa* that form small diffusion pores, *J. Biol. Chem.* **264:** 6297–6301.

Young, M. L., Bains, M., Bell, A., and Hancock, R. E. W., 1992, Role of *Pseudomonas aeruginosa* outer membrane protein OprH in polymyxin and gentamicin resistance: Isolation of an OprH-deficient mutant by gene replacement techniques, *Antimicrob. Agents Chemother.* **36:** 2566–2568.

Young, M. L., and Hancock, R. E. W., 1992, Fluoroquinolone supersusceptibility mediated by Opr overexpression in *Pseudomonas aeruginosa*: Evidence for involvement of a non-porin pathway, *Antimicrob. Agents Chemother.* **36:** 2365–2369.

Zhanel, G. G., Karlowsky, J. A., Saunders, M. H., Davidson, R. J., Hoban, D. J., Hancock, R. E. W., McLean, I., and Nicolle, L. E., 1994, Development of multiple antibiotic-resistant (MAR) mutants of *Pseudomonas aeruginosa* after serial exposure to fluoroquinolones, *Antimicrob. Agents Chemother.* **39:** 489–495.

Transport Systems in *Pseudomonas*

6

TOSHIMITSU HOSHINO

1. INTRODUCTION

Nutrient acquisition and waste excretion are important functions of all living cells including bacteria. Cellular membranes composed of lipid bilayers are inherently impermeable to most nutrients and wastes because of their hydrophilicity. Therefore, living organisms have developed a variety of transport systems that accumulate or excrete a particular solute across the cytoplasmic membrane. These transport systems either equilibrate solutes across the membrane (facilitated diffusion) or use energy to concentrate solutes (active transport). In bacteria, most solutes are translocated across the cytoplasmic membrane by active transport systems, which allow accumulation of solute in chemically unmodified form against a concentration gradient. Such transport systems can be classified into two classes according to the mode of energy coupling (Harold, 1972; Konings *et al.*, 1989): primary transport systems that are directly coupled to biochemical reaction to translocate solutes across the cytoplasmic membrane; and secondary transport systems that utilize electrochemical energy of some compound (mostly electrochemical potential of protons or Na^+ ions) generated by metabolic reaction. Another important mechanisms for accumulation of solute employed by bacteria is group translocation, which couples the translocation of a solute across the membrane to the chemical modification of the solute. Phosphoenolpyruvate (PEP): carbohydrate phosphotransferase systems (PTSs) are representative of the group translocation mechanisms and are widely distributed among bacteria (Postma *et al.*, 1993).

TOSHIMITSU HOSHINO • Mitsubishi Kasei Institute of Life Sciences, Machida-Shi, Tokyo 194, Japan.

Pseudomonas, edited by Montie. Plenum Press, New York, 1998.

Transport systems in gram-negative bacteria are also categorized according to their response to osmotic shock (Neu and Heppel, 1965) into shock-sensitive and shock-resistant transport systems. The transport systems in the latter category are usually retained by membrane vesicle preparations and thus are called membrane-bound transport systems. The shock-sensitive transport systems, on the other hand, are lost in membrane vesicle preparations and require periplasmic components that are called periplasmic binding proteins. Another important characteristic used to distinguish between these two categories of transport systems is in the nature of the energy-coupling mechanism. Shock-resistant transport systems are driven by an electrochemical potential of protons or Na^+ ions (secondary coupling) (Berger and Heppel, 1974; Kaback, 1983; Wright *et al.*, 1986); whereas it has long been postulated that shock-sensitive systems are fueled by ATP or other compounds with high-energy phosphate bond (primary coupling) (Berger and Heppel, 1974; Hong *et al.*, 1979). Recently it has been shown that some of the shock-sensitive transport systems are directly energized by ATP (Bishop *et al.*, 1989; Davidson and Nikaido, 1990; Hoshino *et al.*, 1992). Extensive studies with recombinant DNA technology in the past 15 years have elucidated the molecular basis of many transport systems in bacteria and have revealed that transport systems in the same category have a common molecular feature. Shock-resistant transport systems, represented by the *E. coli* lactose (*lacY*) transport system (Buchel *et al.*, 1980; Kaback, 1983), are mediated by transporters/carriers encoded by single genes. Shock-sensitive transport systems, on the other hand, are multicomponent systems, requiring several membranous components including ATP-binding proteins in addition to periplasmic binding proteins (Ames, 1986; Doige and Ames, 1993).

In this chapter, I overview the current status of knowledge of various transport systems identified in *Pseudomonas,* particularly in the context of the energy-coupling mechanism and the molecular architecture of the transport systems. In the past 30 years, transport systems for various compounds, such as sugars, amino acids, and ions have been identified in *Pseudomonas* mainly by kinetic analysis of intact cells. Most of the transport systems, however, have not yet been clarified at the genetic and molecular level except for branched-chain amino acid transport systems in *Pseudomonas aeruginosa,* which are among best characterized transport systems in bacteria, including *Escherichia coli,* the most well-studied organism in microbiology. Thus, I present this review with strong emphasis on the structure and function of branched-chain amino acid transport systems in *P. aeruginosa.*

2. SUGARS

Pseudomonas cannot metabolize most hexoses to triose phosphate via the Embden–Meyerhof pathway (glycolysis), because they lack 6-phosphofructokinase. This defect makes *Pseudomonas* considerably different in mechanisms for sugar transport from obligate and facultative anaerobic bacteria. In *E. coli,* for example, PTSs are the main routes used for the uptake of various sugars. In *Pseudomonas,* on the other hand, the only PTS reported is that for fructose (Baumann and Baumann, 1975; Sawyer *et al.,* 1977), which is catabolized via the Entner–Doudoroff pathway and also via the Embden–Meyerhof pathway. The unavailability of PTS for other hexoses not metabolized by the Embden–Meyerhof pathway may reflect the low yield of PEP, the energy source for PTS, by the Entner–Doudoroff pathway. This pathway generates only one mole of PEP from one mole of hexose, whereas the Embden–Meyerhof pathway generates two moles of PEP from one mole of hexose.

2.1. Glucose and Gluconate

Glucose in *Pseudomonas* is catabolized by the Entner–Doudoroff pathway via 6-phosphogluconate, a key intermediate. There are two pathways for converting glucose to 6-phosphogluconate (Lessie and Phibbs, 1984). One is the oxidative pathway which includes direct oxidation of glucose to gluconate and then to 2-ketogluconate before phosphorylation. Another is the phosphorylative pathway which includes direct phosphorylation of glucose to glucose-6-phosphate before oxidation. A number of investigators have studied glucose metabolism including uptake systems in *Pseudomonas* with intact cells or membrane vesicle preparations. Cumulative evidence obtained from the studies with *P. aeruginosa* shows that glucose is oxidized to gluconate extracellularly and further to ketogluconate by dehydrogenases localized in the cytoplasmic membrane and that a transport system dependent on a periplasmic glucose-binding protein (GBP) is involved in the phosphorylative pathway, whereas transport systems for gluconate and 2-ketogluconate are requisite for the oxidative pathway.

Phibbs and Eagon (1970) first reported that *P. aeruginosa* possesses a glucose transport system, which is specifically induced by glucose. This transport system shows saturation kinetics with an apparent K_m of 10 µM and is inhibited by energy poisons, such as uncouplers (Eagon and Phibbs, 1971). The PEP:glucose PTS activity was not detected in *P. aeruginosa* cells grown on glucose (Phibbs and Eagon, 1970), suggesting that glucose is transported by the transport system as the free sugar. The glucose transport system is also specific for methyl α-glucoside and 2-de-oxy-D-glucoside, nonmetabolizable glucose analogs, although the affinities for these analogs are much lower than that for glucose (Guymon and Eagon, 1974; Midgley and Dawes, 1973; Mukkada *et al.*, 1973). Using these analogs as a transport substrate, it was demonstrated that the glucose transport system is an active transport system.

Midgley and Dawes (1973) identified an alternate, more complex mechanism for the transport of glucose in *P. aeruginosa* cells grown on glucose. This system involves extracellular conversion of glucose to gluconate by membrane-bound glucose dehydrogenase, followed by transport of gluconate across the cell membrane. Thus, the system is rather called a gluconate transport system. The gluconate transport system has an apparent K_m of 40 μM for gluconate and is induced in *P. aeruginosa* cells grown on glucose or gluconate (Guymon and Eagon, 1974). The inducer for this transport system is not glucose but gluconate, however, because the transport system is induced in mutants defective in glucose dehydrogenase by growth on gluconate but not on glucose (Whiting *et al.*, 1976). This contrasts with the glucose transport system which is induced by growth on glucose but not by growth on gluconate (Midgley and Dawes, 1973).

Eagon and his colleagues developed a method for preparing transport-active membrane vesicles from *P. aeruginosa* cells and found that the vesicles retain the ability to transport gluconate against its concentration gradient (Guymon and Eagon, 1974; Stinnett *et al.*, 1973). Gluconate transport by the vesicles requires exogeneous energy donors, such as ascorbate-phenazinemethosulfate, and shows saturation kinetics with an apparent K_m of 20 μM, which closely agrees with the K_m for gluconate transport by intact cells. These findings indicate that the gluconate transport system is a carrier-mediated transport system coupled to the proton motive force. The membrane vesicles, however, do not retain the ability to transport free glucose (Stinnett *et al.*, 1973), suggesting that some component of the glucose transport system is lost during preparation of the vesicles. Stinson *et al.* (1976) identified GBP in the osmotic shock fluid of *P. aeruginosa* cells and found that GBP was coinduced with the glucose transport system. Mutants unable to grow on glucose were isolated from *P. aeruginosa* strains with defects in the glucose oxidation pathway, including glucose dehydrogenase (Cuskey and Phibbs, 1985; Stinson *et al.*, 1977). These mutants were defective in both GBP production and glucose transport, but revertants that grow on glucose simultaneously regain both activities. These findings strongly suggest that the glucose transport system is an obligatory step in the phosphorylative pathway for glucose utilization and that GBP is an essential component of the glucose transport system, which explains why the glucose transport system is not retained by the membrane vesicles. The mutations (*glcT*, later designated *gltB*) responsible for this phenotype were mapped near a cluster of carbohydrate catabolic genes on the *P. aeruginosa* PAO chromosome (Cuskey and Phibbs, 1985; Sly *et al.*, 1993; Temple *et al.*, 1990). The *gltB* gene might be the structural gene for GBP or a putative regulatory gene required for induction by glucose (Cuskey and Phibbs, 1985). Recent characterization of *gltB* by cloning and determination of nucleotide sequence strongly suggests that the latter possibility is the case (Sage *et al.*, 1993). Thus, isolating a *P. aeruginosa* mutant with a defect in the structural gene for GBP may be necessary to clarify further the role of GBP in the glucose transport system.

2.2. Fructose and Mannitol

Fructose is the only sugar in *Pseudomonas* that is transported by vectorial phosphorylation (Baumann and Baumann, 1975; Sawyer *et al.*, 1997). As in other bacteria, the sugar is transported by a PEP:fructose PTS and accumulates as fructose-1-phosphate. Thus, fructose can be metabolized by *Pseudomonas* via the Entner–Doudoroff pathway and also via the Embden–Meyerhof pathway. The PEP:fructose PTS is induced by fructose (Baumann and Baumann, 1975). *P. aeruginosa* mutants were isolated, which did not grow on fructose but grow normally on mannitol utilized after conversion to fructose (Roehl and Phibbs, 1982). These mutants were defective in the PEP:fructose PTS activity and exhibited no fructose uptake. A *P. doudoroffi* mutant with similar properties has also been reported (Baumann and Baumann, 1975). These facts suggest that the PTS is the only route of entry for fructose in most *Pseudomonas*. On the other hand, *P. cepacia* appears to use another type of transport system because utilization of fructose in this species depends exclusively on an inducible fructokinase and the following Entner–Doudoroff pathway (Allenza *et al.*, 1982). Durham and Phibbs (1982) suggested that the *P. aeruginosa* PTS consists of only two components, a soluble 72-kDa enzyme I and a membrane-associated enzyme II complex. The *Pseudomonas* PTS seems atypical in lacking an HPr-like low molecular weight phospho-carrier protein.

Mannitol is metabolized by *Pseudomonas* via the Entner-Doudoroff pathway after conversion to fructose (Allenza *et al.*, 1982; Phibbs *et al.*, 1978; Stanier *et al.*, 1966). Induction of a transport system for mannitol by growth on mannitol has been shown in *P. cepacia* and *P. aeruginosa* (Allenza *et al.*, 1982; Phibbs *et al.*, 1978). Eisenberg and Phibbs (1982) identified a mannitol-binding activity in osmotic shock fluid from mannitol-grown *P. aeruginosa* cells and purified the mannitol-binding protein from the shock fluid. The protein is highly specific for mannitol and found only in mannitol-grown cells. The osmotic shock treatment causes a considerable decrease in mannitol uptake by mannitol-grown cells (Eisenberg and Phibbs, 1982). Although these findings suggest that the mannitol-binding protein is involved in the mannitol transport system, further genetic and biochemical characterization is required to assess the physiological function of this binding protein.

3. AMINO ACIDS

3.1. Branched-Chain Amino Acids

Transport of branched-chain amino acids in *P. aeruginosa* has been extensively studied by Hoshino and co-workers in recent years. Three transport systems, designated LIV-I, LIV-II, and LIV-III, are involved in the transport of branched-chain amino acids in *P. aeruginosa*. All of the genes required for these transport systems (*bra* genes) have been located on the *P. aeruginosa* PAO chromosome, cloned, and sequenced. All of the gene products (Bra proteins) have also been identified. The biochemical properties of these transport systems have been elucidated by studies with intact cells, membrane vesicles, reconstituted proteoliposomes, and purified proteins. The overall characteristics of the branched-chain amino acid transport systems in *P. aeruginosa* are illustrated in Figure 1.

P. aeruginosa cells show monophasic kinetics for leucine transport in the absence of Na+ with an apparent K_m of 0.4 μM, whereas the cells show biphasic kinetics for leucine transport in the presence of Na+, giving apparent K_m values of 0.4 and 10 μM (Hoshino, 1979; Hoshino and Nishio, 1982). The transport system with low K_m is designated LIV-I and the transport system with higher K_m, which is detected only in the presence of Na+, LIV-II. Competition experiments with or without Na+ (Hoshino, 1979) show that LIV-II is specific for leucine and isoleucine (K_m = 10–15 μM) with lower affinity for valine (K_m = 130 μM), whereas LIV-I is specific for branched-chain amino acids (K_m = 0.2–0.4 μM) and also for alanine and threonine (K_m = 3–5 μM). The LIV-III transport system that was identified later in *P. aeruginosa* (Hoshino *et al.*, 1991), on the other hand, is specific for isoleucine and valine (K_m = 12 μM) with lower affinity for leucine (K_m = 150 μM).

In bacteria, it is not rare that multiple systems function in the trans-

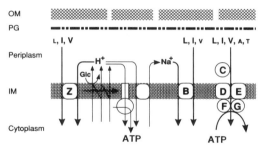

Figure 1. Schematic representation of the branched-chain amino acid transport systems, LIV-I, LIV-II, and LIV-III, in *P. aeruginosa*. OM, PG, and IM stand for outer membrane, peptidoglycan layer, and inner (cytoplasmic) membrane, respectively. The LIV-I system encoded by five genes, *braCDEFG*, is a periplasmic binding protein (BraC)-dependent transport system for branched-chain amino acids and also for alanine and threonine with lower affinities. This transport system has an ATPase activity coupled to substrate translocation. The LIV-II system encoded by the *braB* gene is a transporter (carrier) specific for leucine and isoleucine with lower affinity for valine and driven by an electrochemical gradient of Na^+ generated by a putative Na^+/H^+ antiporter. The LIV-III transport system encoded by the *braZ* gene is a transporter for isoleucine and valine with lower affinity for leucine and probably driven by the proton motive force generated by oxidation of respiratory substrates as glucose via the electron transfer chain. The proton motive force may also be used by F_0F_1-type H^+-translocating ATPase to synthesize ATP, although no study has been reported on the ATPase in *P. aeruginosa*.

port of a specific substrate, which often causes difficulties in precisely characterizing a particular transport system which utilizes this substrate. In the case of branched-chain amino acid transport by *P. aeruginosa*, however, this difficulty is reduced considerably by taking the above properties of the LIV transport systems into consideration as follows. Transport activities of LIV-I, LIV-II, and LIV-III in *P. aeruginosa* can be assayed separately even with the same whole cell preparation: LIV-I by leucine uptake in the absence of Na^+; LIV-II with excess alanine in the presence of Na^+; and LIV-III by valine uptake with excess alanine in the absence of Na^+. These distinct assay conditions may be advantageous to researchers in characterizing the LIV transport systems in *P. aeruginosa* with overlapping substrate specificities. DL-5,5,5-Trifluoroleucine (TFL), a leucine analog that is a potent inhibitor of *P. aeruginosa* growth, is taken up by *P. aeruginosa* cells via LIV-I and LIV-II (Hoshino and Kageyama, 1982), whereas DL-4-azaleucine, another leucine analog used as a growth inhibitor, is taken up specifically via LIV-I (Hoshino and Nishio, 1982). Thus, the use of these inhibitors is also instrumental in the study, particularly in genetic analysis of the LIV transport systems in *P. aeruginosa*.

3.1.1. LIV-I

The high-affinity LIV-I transport system is sensitive to osmotic shock (Hoshino, 1979) and is lost in membrane vesicle preparations (Hoshino and Kagayama, 1979), suggesting that LIV-I is a periplasmic binding protein-dependent transport system. In fact, a binding protein for branched-chain amino acids was identified and purified from the osmotic shock fluid of *P. aeruginosa* cells (Hoshino and Kagayama, 1980). This protein (LIVAT-BP) exhibits the substrate specificity and affinities for its substrates characteristic of the LIV-I transport system. Mutants defective in the production of LIVAT-BP were identified among those deficient in LIV-I, and revertants from the mutants to LIV-I-positive phenotype simultaneously recover normal levels of the binding protein (Hoshino and Kagayama, 1980), strongly suggesting the involvement of LIVAT-BP in the LIV-I transport system. A *P. aeruginosa* PAO mutant defective in LIV-I with an altered LIVAT-BP was further isolated by selection for resistance to azaleucine (Hoshino and Nishio, 1982). The mutation (*braC310*) was mapped in a region near *fla* genes on the *P. aeruginosa* PAO chromosome (Hoshino *et al.*, 1983; Hoshino and Nishio, 1982). The gene for LIVAT-BP was later isolated by screening the *P. aeruginosa* PAO genomic library for the production of LIVAT-BP with an antibody raised against the purified LIVAT-BP (Hoshino and Kose, 1989). The gene encodes a protein with a signal peptide for secretion, and the amino acid composition of the mature product predicted from the nucleotide sequence agree well with that of the purified LIVAT-BP. The plasmid carrying the LIVAT-BP gene restores the LIV-I transport system in the *braC* mutant (Hoshino and Kose, 1989). This fact clearly demonstrates that *braC* is the structural gene for LIVAT-BP, giving conclusive evidence for the requirement of LIVAT-BP in the LIV-I transport system.

P. aeruginosa PAO mutants defective in the LIV-I system alone are preferentially isolated by the selection for resistance to azaleucine (Hoshino and Kose, 1990b; Hoshino *et al.*, 1983; Hoshino and Nishio, 1982). All of the mutations thus selected are transductionally linked to the region II *fla* genes, suggesting that the genes required for LIV-I constitute a cluster on the *P. aeruginosa* PAO chromosome. A DNA fragment of *P. aeruginosa* PAO that confers the LIV-I transport activity on *P. aeruginosa* MT1562 with an extensive chromosomal deletion around the *braC* gene has been isolated (Hoshino and Kose, 1990a). The fragment contains the *braC* gene and the 4-kb DNA segment adjacent to the 3' region of *braC*. The nucleotide sequence of the 4-kb segment shows that the segment contains four open reading frames, designated *braD, braE, braF,* and *braG*. All of the mutants defective in LIV-I are complemented by a plasmid (pKTH24) harboring the *braC, braD, braE, braF*, and *braG* genes. Five cistrons corresponding to the *bra* genes were identified by complementation analysis of the mutants with various derivatives of pKTH24, confirming that the *braD, braE, braF*, and *braG* genes, in addition to *braC*, are required by the LIV-I transport system (Hoshino and Kose, 1990b). The complementation analysis further shows that the *bra* genes are organized as an operon and are cotranscribed in the order *braC-braD-braE-braF-braG* from a promoter located in the 5′-flanking region of the *braC* gene. This type of gene organization is typical of those for periplasmic binding protein-dependent transport systems in *E. coli* and *S. typhimurium* (Ames, 1986). Recently, genes for the high-affinity branched-chain amino acid transport systems in *E. coli* and *S. typhimurium*, both of which are periplasmic binding protein-dependent transport systems, have been characterized (Adams *et al.*, 1990; Matsubara *et al.*, 1992). As illustrated in Figure 2, the gene organization of these transport systems in *E. coli* and *S. typhimurium* are similar to that in *P. aeruginosa*, suggesting that the high-affinity branched-chain amino acid transport systems in these organisms have the same ancestral origin.

The *braD* and *braE* genes specify highly hydrophobic proteins with about 10 membrane-spanning segments, and *braF* and *braG* specify proteins containing amino acid sequences typical of proteins with ATP-binding sites (Hoshino and Kose, 1990a). When BraD, BraE, BraF, and BraG are overproduced together in *E. coli*, all of the Bra proteins are localized in the cytoplasmic membrane (Hoshino and Kose, 1990a). Immunoprecipitation analysis with antibodies raised against the Bra proteins (Hoshino *et al.*, 1992) shows that the Bra proteins form a complex in the *E. coli* membrane and are solubilized as the complex from the membrane with octyl glucoside, a detergent used to solubilize integral membranous proteins in a functional form. Using the Bra proteins overexpressed in *E. coli*, the LIV-I transport system was reconstituted into proteoliposomes (Hoshino *et al.*, 1992). In this reconstituted system, branched-chain amino acid transport depends completely on the presence of all five Bra components, including BraC added externally and ATP loaded internally to the proteoliposomes. The reconstituted proteoliposomes take up alanine and threonine as effectively as branched-chain amino acids, reflecting the substrate specificity of BraC. A concentration gradient of at least 50-fold is established by the reconstituted proteoliposomes, showing that LIV-I is an active transport system. Consumption of loaded ATP and concomitant production of orthophosphate were further observed only when BraC and a substrate for LIV-I were added together to the proteoliposomes, indicating clearly that the

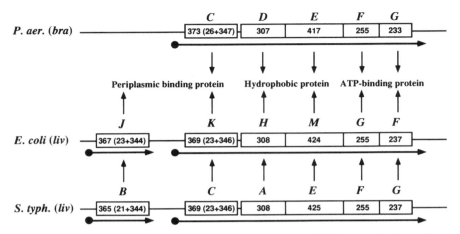

Figure 2. Genetic structure of operons for the periplasmic binding protein-dependent branched-chain amino acid transport systems in *P. aeruginosa* (*bra* operon), *E. coli* (*liv* operon), and *S. typhimurium* (*liv* operon). The arrows with solid circles indicate the transcriptional units and the directions of transcription. The numbers of the amino acid residues of the gene products are also shown in this figure. The numbers in each parentheses represent those of amino acid residues of the signal peptide and the mature product for the periplasmic binding protein. Each polycistronic operon consists of five genes: one gene for a periplasmic binding protein; two genes for integral membrane proteins; and two genes for putative nucleotide-binding proteins. The branched-chain amino acid transport system in *E. coli* or *S. typhimurium* contains two periplasmic binding proteins with different substrate specificities, LivJ and LivK or LivB and LivC, respectively. The *P. aeruginosa* LIV-I transport system, on the other hand, contains BraC alone as the periplasmic binding protein for branched-chain amino acids. The genes for *E. coli* LivJ and *S. typhimurium* LivB proteins which have substrate specificities very similar to *P. aeruginosa* BraC protein form monocistronic transcriptional units. Instead, *livK* and *livC*, genes encoding the leucine-specific binding proteins of *E. coli* and *S. typhimurium*, respectively, are involved in polycistronic operons.

LIV-I transport system is an ATPase coupled to translocation of branched-chain amino acids across the membrane. Maltose and histidine transport systems, both of which are the periplasmic binding protein-dependent transport systems in *E. coli* and *S. typhimurium*, respectively, have also been reconstituted into proteoliposomes and have ATPase activities similar to those of reconstituted LIV-I (Bishop *et al.*, 1989; Davidson and Nikaido, 1990). Early studies of binding protein-dependent transport systems suggested that transport was energized by the hydrolysis of ATP (Berger and Heppel, 1974) or some other compound with a high-energy phosphate bond (Hong *et al.*, 1979), whereas some data suggest that the transport systems are energized by proton motive force

(Plate, 1979) or directly by electron flow (Richarme, 1985). Thus, the recent reconstitution studies, including that of the *P. aeruginosa* LIV-I system (Bishop *et al.*, 1989; Davidson and Nikaido, 1990; Hoshino *et al.*, 1992), have solved a long-standing controversy over the energy coupling mechanism of periplasmic binding protein-dependent transport systems.

3.1.2. LIV-II

The low-affinity LIV-II transport system of *P. aeruginosa* is resistant to osmotic shock and requires Na^+ to function (Hoshino, 1979). The LIV-I and LIV-II systems are mutationally separable (Hoshino *et al.*, 1983; Hoshino and Kagayama, 1980), showing that LIV-II is genetically distinct from the LIV-I transport system. The properties of the LIV-II system, including Na^+ dependence, are similar to those of the branched-chain amino acid transport system identified in membrane vesicles (Hoshino, 1979; Hoshino and Kagayama, 1979). This suggests that LIV-II is a membrane-bound, carrier-mediated transport system. Membrane vesicles prepared from a LIV-II-defective mutant show no leucine uptake, whereas the vesicles prepared from a mutant with an enhanced LIV-II activity exhibit much higher leucine uptake than the vesicles prepared from a wild-type strain (Hoshino and Kagayama, 1980), confirming that LIV-II is a membrane-bound transport system. Imposition of an inwardly directed concentration gradient of Na^+ ions drives leucine uptake in nonenergized membrane vesicles (Hoshino and Kagayama, 1979), suggesting that LIV-II is a Na^+/substrate cotransport system. Data also suggested that Li^+ would also serve as a coupling cation for the LIV-II transport system (Hoshino, 1979). Uratani *et al.* (1989) showed that Na^+ and Li^+ are taken up by the *P. aeruginosa* cells when an inwardly directed concentration gradient of leucine, isoleucine, or valine is imposed on the cells. No uptake of ions is induced by such a gradient in the LIV-II-defective mutant, whereas an enhanced uptake of ions is detected in the mutant with enhanced LIV-II activity, confirming that LIV-II is a $Na^+(Li^+)$/substrate cotransport system. Comparison of the LIV-II transport systems between *P. aeruginosa* PAO and PML strains (Uratani and Hoshino, 1989) demonstrated a difference in the Na^+ dependence of the systems. The apparent K_m values for Na^+ with respect to leucine transport were $3\mu M$ for the PAO strain and $95\ \mu M$ for the PML strain. The LIV-II transport systems in PAO and PML strains gave the same V_{max} value for leucine transport, however, suggesting that the difference is caused by an alteration in the gene encoding the LIV-II carrier protein.

The mutation, *braB*, in a strain defective in LIV-II function is located in the *nar-9011-puuC10* region on the *P. aeruginosa* PAO chromosome (Hoshino *et al.*, 1983). The *braB* genes of *P. aeruginosa* PAO and PML strains were isolated by selection for Hrb⁺ (recovery of growth at low leucine concentration) in a leucine-requiring strain of *P. aeruginosa* PAO carrying *braB* and *braC* mutations (Hoshino *et al.*, 1990; Uratani and Hoshino, 1989). Cloned *braB* genes confer the LIV-II function typical of their original strains on the LIV-II-defective mutant of either strain, indicating clearly that *braB* is the structural gene for the LIV-II carrier protein. Determination of nucleotide sequences of the *braB* genes showed that *braB* encodes a very hydrophobic protein of 437 amino acid residues with 12 membrane-spanning segments and revealed that the substitution of the amino acid at position 292 of the LIV-II carrier (threonine for PAO and alanine for PML) causes a difference in the Na^+ dependence of the carriers of the PAO and PML strains.

The method for reconstituting the *P. aeruginosa* LIV-II transport system into proteoliposomes was established by Uratani (1985). This reconstituted system exhibited Na^+-dependent counterflow and Na^+ gradient-driven transport of branched-chain amino acids, whose kinetic properties are consistent with those of LIV-II in intact cells. Reconstitution studies with the solubilized LIV-II carrier and synthetic phospholipids (Uratani and Aiyama, 1986; Uratani *et al.*, 1987) suggest that the LIV-II carrier requires phospholipids with appropriate polar head groups and particular acyl chain length for its maximal activity. The BraB protein of *P. aeruginosa* PML has recently been overexpressed in *E. coli*, solubilized, and purified to homogeneity by immunoaffinity chromatography with an antibody raised against the synthetic C-terminal peptide of BraB (Uratani, 1992). Proteoliposomes reconstituted with the purified BraB protein exhibit Na^+ or Li^+ concentration gradient-driven transport of leucine, isoleucine, and valine, clearly demonstrating that the LIV-II carrier is a $Na^+(Li^+)$/substrate cotransporter (symporter) for branched-chain amino acids, composed of BraB alone.

3.1.3. LIV-III

The LIV-III transport system was accidentally identified in *P. aeruginosa* PAO in the course of cloning the *braB* gene for the LIV-II carrier. Hoshino *et al.* (1991) screened the *P. aeruginosa* PAO genomic library for Hrb⁺ in *P. aeruginosa* PAO3536 (*leu braB braC*) on a medium containing a high concentration of Na^+ ions. They failed, however, to isolate the *braB* gene using the medium. Instead, a gene which conferred a branched-chain amino acid transport activity independent of Na^+ on the strain was isolated and mapped to the *cys-54* region of the *P. aeruginosa* PAO chromosome. This gene is designated as *braZ* and the transport system coded for by *braZ* as LIV-III. The LIV-III transport system is very efficient in isoleucine and valine uptake. The transport system is inefficient, however, in leucine uptake due to its very low affinity for leucine, which contrasts with LIV-I and LIV-II. This inefficiency of LIV-III in leucine uptake explains well why the *leu* strain defective in both LIV-I and LIV-II shows Hrb^- at low leucine concentrations, although the strain still retains the LIV-III transport system. Isoleucine and valine uptake by the *P. aeruginosa* PAO strain defective in both LIV-I and LIV-II show monophasic kinetics in a wide range of substrate concentrations, strongly suggesting that no branched-chain amino acid transport system other than LIV-I, LIV-II, and LIV-III exists in *P. aeruginosa* PAO.

The *braZ* gene encodes a highly hydrophobic protein of 437 amino acid residues with a 12 membrane-spanning segments (Hoshino *et al.*, 1991) typical of transport carrier proteins (Botfield *et al.*, 1992; Büchel *et al.*, 1980; Hoshino *et al.*, 1990). Comparison of the amino acid sequences of BraZ, BraB, and BrnQ, the carrier for the LIV-II transport system in *S. typhimurium* (Ohnishi *et al.*, 1988), reveals extensive homology among the proteins. BraZ and BrnQ are particularly homologous, showing 67% identity. This higher homology may reflect specificities for coupling cations rather than substrate specificities of the transport systems because the LIV-III (BraZ) system is quite different in substrate specificity from the LIV-II systems (BraB and BrnQ) of *P. aeruginosa* (Hoshino, 1979) and *S. typhimurium* (Matsubara *et al.*, 1988), both of which are specific for leucine and isoleucine and less specific for valine. A study suggests that the *E. coli* LIV-II transport system (BrnQ/HrbA), homologous to *S. typhimurium* LIV-II, is a H^+-coupled transport system (Yamato and Anraku, 1977). The LIV-III transport system may also be coupled to H^+ ions (Hoshino *et al.*, 1991). As described above, the difference in the Na^+ requirement of the LIV-II carrier between PAO and PML strains results from the substitution of an amino acid at position 292, suggesting that the amino acid residue at this position is important in determining specificity for coupling cations. In this sense, it seems worth noting that both BraZ and BrnQ contain valine at this position. This amino acid is different from those of the PAO and PML BraB proteins (threonine and alanine, respectively), whereas the homology of the amino acid sequence around position 292 is retained among the BraB, BraZ, and BrnQ proteins.

3.2. Basic Amino Acids (Arginine, Lysine, Histidine)

Pseudomonas utilizes basic amino acids as sole sources of carbon and nitrogen. Transport systems for basic amino acids in *Pseudomonas* have been studied in the context of the metabolism of these amino acids as carbon or nitrogen sources. Rodwell and his colleagues (Fan *et al.*, 1972; Miller and Rodwell, 1971) found that two lysine transport systems are induced in *P. putida* grown on lysine as a sole carbon and nitrogen source. Competition experiments suggested that the high K_m (7.3 μM) transport system is specific for D and L isomers of lysine, arginine, and ornithine and the low K_m (0.4 μM) system is specific for lysine and ornithine. They found that the lysine transport by lysine-grown cells is completely abolished by KCN or NaN_3, suggesting that the transport is an active process. A transport system specific for L-arginine alone (K_m = 0.05 μM) has also been found in *P. putida* cells grown on L-arginine (Fan *et al.*, 1972). Later, Weill-Thevenet *et al.* (1979) reexamined lysine transport by *P. putida* using a wild-type strain and a mutant strain lacking lysine oxygenase activity. They found that two transport systems for lysine (K_m = 1 and 50 μM) function even in succinate-grown cells of either strain. The high-affinity system was specific for arginine and ornithine in addition to lysine. The relationship between the lysine transport systems identified by Rodwell's group and those by Weill-Thevenet's group is not clear at present, because no genetic or molecular basis for the transport systems has yet been clarified.

P. acidovorans, unable to utilize lysine as a sole source of carbon or nitrogen, on the other hand, has a single transport system for lysine with a K_m of 1 μM (Weill-Thevenet *et al.*, 1979). This transport system was constitutively expressed and is highly specific for L-lysine. No other transport system is induced in this organism by lysine added to the culture medium.

P. aeruginosa metabolizes arginine under anaerobic, nitrate-free conditions by the arginine deiminase pathway and generates ATP for motility and growth (Armitage *et al.*, 1983; Vander Wauben *et al.*, 1984). Genetic and cloning studies (Luthi *et al.*, 1990; Vander Wauben *et al.*, 1984) demonstrate that another gene, named *arcD*, is required for anaerobic arginine metabolism in addition to *arcABC* encoding the enzymes for the arginine deiminase pathway. The *arcD* mutations rendered the cells unable to utilize extracellular arginine under anaerobic conditions although the enzymes of the arginine deiminase pathway were expressed in the mutants (Vander Wauben *et al.*, 1984). The nucleotide sequence of *arcD* suggests that *arcD* encodes a hydrophobic 52-kDa protein with 13 membrane-spanning segments (Luthi *et al.*, 1990).

When *arcD* is expressed in *E. coli*, the ArcD protein is detected in the membrane fraction (Luthi *et al.,* 1990). These facts suggest that ArcD is involved in a specific transport system that delivers arginine to the arginine deiminase pathway under oxygen-limited conditions. Analyses of arginine transport by membrane vesicles and proteoliposomes prepared from *E. coli* cells expressed for ArcD (Verhoogt *et al.,* 1992) clearly show that this is the case. The membrane vesicles exhibit a marked uptake of arginine and ornithine in an ArcD-dependent manner when the proton motive force is imposed on the vesicles. The study with ionophores suggests that the membrane potential and not ΔpH acts as the main driving force for the transport by ArcD. Ornithine accumulated in the vesicles was released by an excess of lysine, arginine, or ornithine but not citrulline, suggesting that the ArcD protein transports arginine, ornithine, and lysine and catalyzes their rapid exchange. The study of the ArcD-dependent arginine transport reconstituted into proteoliposomes reveals that ArcD mediates the stoichiometric exchange between arginine and ornithine when the proteoliposomes are preloaded with ornithine, suggesting that the ArcD protein is the arginine–ornithine antiporter. The arginine–ornithine antiport reaction showed monophasic saturation kinetics and gave an apparent K_m value of 11 μM for arginine. The antiport reaction was several orders of magnitude faster than the membrane potential-driven transport. This property may be advantageous to *P. putida* grown on arginine under anaerobic, nitrate-free conditions because the cells can economize the energetic costs for arginine uptake by using the concentration gradient of ornithine, a product of the arginine deiminase pathway.

3.3. Proline

It has been suggested that *P. aeruginosa* has at least two transport systems for proline (Kay and Gronlund, 1969). The double-reciprocal plot of proline uptake by *P. aeruginosa* shows a biphasic kinetics, giving apparent K_m values of 1 and 20 μM. The transport systems are highly specific for proline because 18 commonly occurring amino acids did not inhibit proline uptake at 100-fold excess concentration over proline. *P. aeruginosa* cells grown on glucose with added proline showed much higher proline uptake than those without proline, suggesting that a proline transport system is induced by proline. The inducer seems to be proline and not a degradation product thereof because a mutant strain unable to degrade proline can still induce the transport system. Kinetic analysis at low proline concentrations gave the same K_m value (1 μM) for the transport by cells grown with or without proline. Kay and

Gronlund (1969) isolated a strain with a point mutation, which could not utilize proline as a carbon and nitrogen source. This mutant had a severe defect in the inducible proline transport system, whereas the mutant had normal levels of the enzymes for proline degradation. *P. aeruginosa* mutants with similar properties were also reported by Meile *et al.* (1982). The mutations carried by the mutants were mapped in the very late region of the *P. aeruginosa* chromosome and found cotransducible with those in the enzymes for proline degradation. These findings may imply that the low K_m system is required for growth on proline as a sole carbon and nitrogen source and is coinduced by proline together with the enzymes required for proline degradation.

3.4. Aromatic Amino Acids

Kay and Gronlund (1971) studied the transport of aromatic amino acids (phenylalanine, tyrosine, and tryptophan) by *P. aeruginosa*. They suggested the existence of two high-affinity transport systems for the amino acids in *P. aeruginosa* based on competition studies, although only a K_m value (approximately 1 μM) was obtained for the transport of each amino acid. The first system was specific for phenylalanine and tyrosine and rather less specific for tryptophan. The second system, on the other hand, was most specific for tryptophan. These two systems were mutationally separable, showing that their suggestion is highly likely.

P. acidovarans also has a transport system for tryptophan (Rosenfeld and Feigelson, 1969) which is inducible by tryptophan or kinurenine. This system differs considerably from those of *P. aeruginosa* because the transport system is inducible by tryptophan or kynurenine and competed by kynurenine and formyl kynurenine, which are not substrates for *P. aeruginosa* aromatic amino acid transport systems.

3.5. Methionine

P. fluorescens UK1 has an inducible methionine transport system with a K_m of 25 μM (Mantsala *et al.*, 1974) which has broad specificity for amino acids (Laakso, 1976). Membrane vesicles prepared from methionine-grown *P. fluorescens* UK1 exhibit methionine uptake in an energy-dependent manner, suggesting that the transport system is a carrier-mediated transport system (Mantsala *et al.*, 1974). *P. aeruginosa* also has a high-affinity, energy-requiring methionine transport system with a K_m of 0.2–0.3 μM (Montie and Montie, 1979). This system has a high degree of specificity, however, because 14 common amino acids tested exhibited no inhibition of methionine transport at 100-fold excess concentration over methionine.

4. INORGANIC IONS

4.1. Cations

Very few studies have been done on the transport of K^+, Na^+, Mg^{2+}, and Ca^{2+} in *Pseudomonas* spp., which was reviewed several years ago by Cervantes and Silver (1990). Since that time no additional information has been obtained about these cation transport systems in *Pseudomonas*. Iron transport, on the other hand, has been studied considerably in fluorescent *Pseudomonas, P. putida* and *P. aeruginosa*. Recent progress in the study of iron transport in *P. putida* and *P. aeruginosa* is well documented in the review of iron transport in microorganisms by Guerinot (1994). Iron metabolism in *Pseudomonas* is also extensively reviewed by Meyer and Stintzi in Chapter 7 of this book. Readers who are interested in cation transport in *Pseudomonas* are referred to these reviews.

4.2. Anions

4.2.1. Phosphate

The transport of inorganic phosphate has been extensively studied in *E. coli,* where two transport systems have been identified (Rosenberg, 1987). One is a high-affinity, periplasmic binding protein-dependent transport system (Pst system) which is derepressed by phosphate limitation. Another is a low-affinity, carrier-mediated transport system (Pit system) which is constitutively expressed.

Phosphate transport by *P. aeruginosa* shows biphasic kinetics with the apparent K_m values of 1.1 and 10 μM (Lacoste *et al.*, 1981), suggesting that two distinct transport systems for phosphate function in *P. aeruginosa* as in *E. coli*. Osmotic shock treatment dramatically reduces phosphate transport at low phosphate concentrations in *P. aeruginosa* (Lacoste *et al.*, 1981), suggesting the involvement of a binding protein in the high-affinity phosphate transport system in *P. aeruginosa*. When *P. aeruginosa* cells are starved for phosphate, a periplasmic, 37-kDa protein is induced (Hancock *et al.*, 1982). The 37-kDa protein, later purified, bound phosphate with a K_d of 0.34 μM (Poole and Hancock, 1984). The phosphate-binding protein is a component of the high-affinity phosphate transport system because the high-affinity transport system is lost in a *P. aeruginosa* mutant lacking the binding protein (Poole and Hancock, 1984). No studies have been done on the cytoplasmic membrane components of the high-affinity transport system in *P. aeruginosa*. In the *E. coli* Pst system, on the other hand, three inner membrane components have been identified: PstA and PstC, integral membrane proteins, and PstB, a peripheral inner membrane component with an ATP-binding motif (Surin *et al.*, 1985). The *P. aeruginosa* high-affinity transport system may be composed of proteins analogous to these Pst proteins, because considerable similarity in overall properties of phosphate metabolism and regulation of genes involved in the metabolism has been observed between *P. aeruginosa* and *E. coli* (Hancock *et al.*, 1982; Lacoste *et al.*, 1981; Poole and Hancock, 1983).

The low-affinity phosphate transport system in *P. aeruginosa* is resistant to osmotic shock (Lacoste *et al.*, 1981) and is retained in the mutant lacking the phosphate-binding protein (Poole and Hancock, 1984). The transport system is also sensitive to inhibitors of the electron transfer chain (Lacoste *et al.*, 1981). These facts suggest that the low-affinity transport system may be mediated by a carrier-type transporter similar to the Pit system of *E. coli* (Rosenberg, 1987).

4.2.2. Nitrate

Nitrate is used by various bacteria as a terminal electron acceptor under anaerobic conditions. Kristjansson *et al.* (1978) suggested that the reduction site of nitrate to nitrite is on the inner surface of the cytoplasmic membrane in *P. denitrificans*. They further suggested that nitrate uptake by anaerobically grown *P. denitrificans* is mediated by facilitated diffusion. The continuous diffusion of nitrate into the cells is maintained by rapid intracellular reduction of nitrate to nitrite, which then diffuses out of the cells. Betlach *et al.* (1981) suggested, on the other hand, that *P. fluorescens* could take up nitrate actively when the cells were grown with tungstate which was added as an inhibitor of the nitrate reductase. Using a *P. stutzeri* mutant defective in nitrite (not nitrate) reductase, Dias *et al.* (1990) showed saturation kinetics for nitrate uptake with an apparent K_m of 306 μM and a stoichiometric relationship between nitrate import and nitrite export, suggesting a nitrate/nitrite antiport mechanism for nitrate transport in *P. stutzeri*. Hernandez *et al.* (1991) also found that *P. aeruginosa* cells grown with tungstate, which inhibits the generation of nitrite from nitrate, incorporate only nitrate in the presence of exogenously added nitrite, supporting the nitrate/ nitrite antiport mechanism. Because the nitrate incorporated is converted very rapidly to nitrite (Betlach *et al.*, 1981), determination of intracellular nitrate concentration is very difficult, which limits extensive investigation of nitrate transport by whole cells. Therefore, study of nitrate transport in membrane vesicles or in mutants lacking nitrate reductase is necessary to confirm the transport mechanisms for nitrate transport in the *Pseudomonas* species described above. Isolation of mutants defective in nitrate transport would be more favorable for better understanding of nitrate transport although Sias and Ingraham (1979) did not find such mutants among those defective in nitrate assimilation in *P. aeruginosa*. As an anaerobic process, denitrification is regulated by oxygen (Stouthamer, 1976). Inhibition of nitrate uptake of oxygen has been suggested as a primary mechanism for the regulation of denitrification in most bacterial species including *P. aeruginosa* (Hernandez and Rowe, 1988).

4.2.3. Sulfate

Inorganic sulfate is utilized as the sole source of sulfur by a wide variety of sulfate-assimilating microorganisms (Peck, 1961). *Pseudomonas aeruginosa* also utilizes many inorganic sulfur compounds as sole sources of sulfur for its growth (Schook and Berk, 1978). Ohtake *et al.* (1987) showed that *P. fluorescens* has a saturable sulfate transport system with a K_m of 6–7 μM. The sulfate transport system is repressed by growth on cysteine and derepressed by growth on djenkolic acid (Ohtake *et al.*, 1987), similar to the *S. typhimurium* sulfate transport system which requires at least three components encoded by genes in the cysA region in addition to the sulfate-binding protein (Dreyfuss, 1964; Ohta *et al.*, 1971; Sirko *et al.*, 1990). The *P. fluorescens* sulfate transport system is specific for chromate because sulfate uptake is competitively inhibited by chromate with a K_i of 12–13 μM. This transport system may be responsible for the sensitivity of *P. fluorescens* to chromate, a toxic contaminant in the natural environment (Ohtake *et al.*, 1987).

5. OTHER COMPOUNDS

5.1. Compounds Catabolized by the β-Ketoadipate Pathway

5.1.1. Benzoate

Benzoate is dissimilated in *P. putida* via the catechol branch of the ketoadipate pathway, and mandelate is also catabolized via this pathway when converted to benzoate. These compounds are weak and relatively hydrophobic acids and were initially presumed to enter the cells by diffusion (Rottenberg, 1975). Later, a saturable transport system was identified in benzoate-grown *P. putida* cells (Thayer and Wheelis, 1976). The transport system is specific for benzoate alone because benzoate uptake is not inhibited by mandelate, *cis,cis*-muconate or adipate which is added at 10-fold the concentration of benzoate. The K_m and V_{max} values of benzoate uptake by wild-type cells are 20 μM and 30 nmol/min/mg of dry weight, respectively, which are high enough to support growth on benzoate (Thayer and Wheelis, 1982). It seems that benzoate is actively transported because the substrate accumulate in the cells against its concentration gradient and transport is dissipated by various inhibitors of energy metabolism such as NaN_3 and carbonyl cyanide m-chloro-phenylhydrazone (CCCP) (Thayer and Wheelis, 1982). No benzoate-binding activity was detected in osmotic shock fluid, suggesting that the benzoate transport system is mediated by a carrier-type transporter (Thayer and Wheelis, 1982). A mutant unable to grow on benzoate was isolated and characterized (Thayer and Wheelis, 1976). The mutant spontaneously reverted to wild-type, indicating that the mutation is a point mutation. The gene, designated *benP*, is cotransducible with *catB* and *catC* genes. The expression of *benP*, is independent of that of *catB* and *catC* because benzoate transport activity is detectable in a *catR* mutant which is required to induce *catB* and *catC*. The *benP* mutant grew normally on mandelate, suggesting that *benP* most likely defines a gene for benzoate transport, not genes for enzymes for benzoate dissimilation. No data is available, however, to determine if *benP* is the structural gene for the benzoate transporter or a regulatory gene required for the expression of the transporter.

Data also suggest that two other compounds metabolized via the catechol branch of the β-ketoadipate pathway in *P. Putida*, mandelate (Higgins and Mandelstam, 1972) and β-carboxy-*cis, cis*-muconate (Meagher *et al.*, 1972), induce their own transport systems. However, very little has been shown about the characteristics of these transport systems.

5.1.2. 4-Hydroxybenzoate

4-Hydroxybenzoate is catabolized via the protocatechuate branch of the β-ketoadipate pathway. Harwood *et al.* (1994) recently isolated a transposon mutant defective in 4-hydroxybenzoate chemotaxis and cloned a chromosomal DNA segment which complemented the mutation. Their sequencing, mutational analysis, and complementation studies show that three genes are involved in this segment. One of the genes, designated *pcaK*, encodes a 47-kDa protein with homology to a large number of carrier-type transport proteins (Marger and Saier, 1993). A gene with strong homology to *pcaK* at the amino acid level was isolated from *Acinetobacter calcoaceticus* (Kowalchuk *et al.*, 1994) although its function in this organism has not yet been elucidated. The *pcaK* mutant has no obvious defect in growth with 4-hydroxybenzoate as a carbon source except at high pH (pH 8.1). The mutant cells grown on 4-hydroxybenzoate, however, show a severe defect in 4-hydroxybenzoate transport. The PcaK protein was expressed in *E. coli* by a T7 RNA polymerase/promoter system. The *E. coli* cells overexpressing PcaK protein accumulate 4-hydroxybenzoate at a substantial rate, whereas the cells do not accumulate benzoate. These findings by Harwood *et al.* (1994) indicate that *pcaK* encodes a carrier-type transporter specific for 4-hydroxybenzoate. A role is suggested for the *pcaK* gene in chemotaxis to 4-hydroxybenzoate in addition to its function in transport (Harwood *et al.*, 1994) although it remains to be solved whether or not PcaK itself is a chemoreceptor for 4-hydroxybenzoate.

5.1.3. β-Ketoadipate

Pseudomonas putida has a transport system for β-ketoadipate which is a common intermediate of benzoate and 4-hydroxybenzoate catabolism. The properties of the β-ketoadipate transport system were studied with adipate, a nonmetabolizable analog in *P. putida,* as a transport substrate. The β-ketoadipate transport system was induced by 4-hydroxybenzoate or β-ketoadipate (Ornston and Parke, 1976). The transport system was induced by β-ketoadipate even in a mutant defective in β-ketoadipate succinyl CoA transferase, suggesting that β-ketoadipate itself induces the expression of the transport system. Constitutive mutants of the β-ketoadipate transport system constitutively expressed three catabolic enzymes that give rise to β-ketoadipate from the metabolic precursor β-carboxy-*cis, cis*-muconate. Thus a single regulatory gene appears to govern the expression of these enzymes and the transport system. Intracellular adipate concentration exceeded the external concentration by 25-fold in the wild-type strain and by 200-fold in the constitutive mutants. The addition of electron transport inhibitors or proton conductors such as NaN_3 or CCCP, inhibits the adipate uptake and also causes a rapid loss of accumulated adipate, confirming the energy dependence of the transport system (Ondrako and Ornston, 1980). Maximal transport was observed at the relatively low pH of 5.5, suggesting the contribution of a high inward proton gradient to the driving of transport. The K_m for adipate transport is 230 μM, and the K_i for β-ketoadipate with respect to adipate transport is 40 μM. Thus, it appears that the transport system has a much higher affinity for β-ketoadipate, the natural substrate of this system, than for adipate, a substrate analog. Succinate is a competitive inhibitor of adipate transport with a K_i of 1.3 mM. The β-ketoadipate transport system is resistant to an osmotic shock treatment, and no adipate binding was detected in the shock fluid. Membrane vesicles prepared from the constitutive mutant cells accumulate adipate effectively when a respiratory substrate, ascorbate plus phenazine methosulphate, is added to the vesicles. The K_m for adipate transport by the vesicles is 140 μM, which is close to the K_m for the whole cells. These findings by Ondrako and Ornston (1980) strongly suggest that the β-ketoadipate transport system is mediated by a carrier-type transporter driven by an electrochemical potential of protons.

Expression of the β-ketoadipate transport system is not induced in the wild-type *P. putida* to a level sufficient to utilize exogeneous β-ketoadipate added as a growth substrate (Ornston and Parke, 1976). The transport system is widely distributed, however, among divergent fluorescent *Pseudomonas* species (Ondrako and Ornston, 1980), suggesting that the system has a selective function for the organisms. The involvement of the β-ketoadipate transport system in chemotaxis toward β-ketoadipate is suggested for such a function (Karimian and Ornston, 1981). β-Ketoadipate may serve as a simple chemical signal for change attracting fluorescent *Pseudomonas* to environments in which the microbial breakdown of complex aromatic polymers is underway.

5.2. *Myo*-inositol

Myo- and *scyllo*-inositol are polyalcohols that are structurally related to sugars and are secreted in soil by plant roots in a free or phosphorylated form. Although the cyclitols are ubiquitous in the natural environment, only a restricted number of bacterial species use the cyclitols as a carbon source for growth.

A *Pseudomonas* sp. (formerly classified as *P. putida*) which utilizes cyclitols as a carbon source was isolated from soil and characterized (Reber *et al.*, 1977). A transport system for *myo*-inositol and *scyllo*-inositol is induced in the strain when grown on cyclitols. This transport system is not retained by membrane vesicles prepared from the strain and is sensitive to osmotic shock treatment. *Myo*-inositol binding activity was detected in the shock fluid of *myo*-inositol-grown cells. These facts suggest that the *Pseudomonas* sp. has a periplasmic binding protein-dependent transport system for *myo* and *scyllo* isomers of cyclitol, which is inducible by these cyclitols. The *myo*-inositol transport system of *Pseudomonas* sp. contrasts with that of *Klebsiella aerogenes* (*Enterobacter aerogenes*), which is mediated by a membrane-bound H^+ symporter (Deshusses, 1985). A binding protein for these cyclitols was later purified from the shock fluid of *myo*-inositol-grown *Pseudomonas* cells (Deshusses and Belet, 1984). The protein was not detected in the shock fluid of glucose-grown cells. The protein shows substrate specificity and affinities for substrates similar to those of the *myo*-inositol transport system. Pulse-chase experiments (Feiss *et al.*, 1984) showed that the binding protein is synthesized as a 32-kDa precursor protein and then processed to the 30-kDa matured form. These findings suggest that the *myo*-inositol-binding protein is a periplasmic-binding protein which is involved in the *myo*-inositol transport system in the *Pseudomonas* sp. Mutants defective in the *myo*-inositol transport system were identified among those which failed to grow on *myo*-inositol (G.-Feiss *et al.*, 1985). Genomic clones which restore both the growth on *myo*-inositol and *myo*-inositol transport to the mutants were isolated from a gene bank of wild-type *Pseudomonas* sp. constructed in pMMB34, a broad host-range vector (G.-Feiss *et al.*, 1985). pJD2, one of these clones, contains an 11.5-kb DNA segment. Subcloning of fragments generated by digestion with restriction enzymes suggests that most of the 11.5-kb segment is necessary to restore *myo*-inositol transport in the mutant.

5.3. Steroids

Pseudomonas testosteroni grows on C_{19} and C_{21} steroids in the absence of other suitable carbon sources. Induction of steroid-binding activity (Watanabe *et al.*, 1973) and steroid dehydrogenase activities (Talalay, 1965) was found in *P. testosteroni* grown on steroid as a carbon source. Detection of steroid uptake by the whole cells is difficult, however, because the steroid taken up by the cells is metabolized very rapidly to CO_2 and H_2O (Watanabe *et al.*, 1973). Watanabe and Po (1974) identified and characterized a testosterone transport activity in *P. testosteroni* using membrane vesicle preparations. Membrane vesicles prepared from cells grown on testosterone show considerable testosterone uptake, whereas vesicles from cells grown without testosterone do not show any testosterone uptake, suggesting that the transport system is inducible. The uptake of testosterone by the vesicles is saturable, giving a K_m of 2 μM. Under optimal conditions, the intravesicular steroid concentration reached 800 times the steroid concentration in the medium. Transport is inhibited by various inhibitors of the electron transfer chain. These facts suggest that testosterone transport in membrane vesicles is an energy-dependent process. An unusual feature of this transport system, however, is the requirement for NAD^+, an electron acceptor, and its independence from externally added energy donors such as succinate. It is known that NAD$^+$ is required for oxidation of testosterone to androstenedione by 17β-hydroxysteroid dehydrogenase (Talalay, 1965). If such a dehydrogenase is required for testosterone transport, two mechanisms are plausible: (1) the dehydrogenase couples the oxidation and transport of testosterone in a group translocational manner; (2) the dehydrogenase oxidizes testosterone to androstenedione, and the resultant androstenedione is transported by a transporter specific for androstenedione. The group translocational mechanism seems favorable for explaining the above observations. Testosterone used as a transport substrate was converted intravesicularly to androstenedione, whereas no transport of androstenedione was detected in the membrane vesicles prepared from testosterone-grown cells (Watanabe and Po, 1976). The involvement of a membrane-bound 3β and 17β-hydroxysteriod dehydrogenase for testosterone transport was further suggested by experiments with inhibitors and antibodies for this dehydrogenase (Lefebvre *et al.*, 1979). Watanabe and his colleagues (Francis and Watanabe, 1981, 1982, 1983; Watanabe *et al.*, 1979) identified several binding proteins for steroids in the shock fluid and also in the membrane fraction of testosterone-grown *P. testosteroni* and partially purified the binding proteins. No evidence is obtained, however, that shows involvement of these binding proteins in the transport process of steroids in *P. testosteroni.* Francis *et al.* (1985) extracted a steroid binding protein from *P. testosteroni* membrane with an organic solvent system and reconstituted steroid transport activity into liposomes with the extracted protein. The addition of valinomycin, an electrogenic K$^+$ ionophore, caused testosterone transport by proteoliposomes loaded with potassium phosphate but not by those loaded with sodium phosphate, suggesting that the transport system reconstituted into proteoliposomes is mediated by a carrier-type transporter. This transport system may be different from that identified in membrane vesicles because the reconstituted system does not require NAD$^+$ nor 3β and 17β-hydroxysteroid dehydrogenase.

ACKNOWLEDGMENTS. I thank Dr. Y. Uratani for valuable suggestions and critical reading of the manuscript.

REFERENCES

Adams, M. D., Wagner, L. M., Graddis, T. J., Landick, R., Antonucci, T. K., Gibson, A. L., and Oxender, D. L., 1990, Nucleotide sequence and genetic characterization reveal six essential genes for the LIV-I and LS transport systems of *Escherichia coli, J. Biol. Chem.* **265:** 11436–11443.

Allenza, P., Lee, Y. N., and Lessie, T. G., 1982, Enzymes related to fructose utilization in *Pseudomonas cepacia, J. Bacteriol.* **150:** 1348–1356.

Ames, G. F.-L., 1986, Bacterial periplasmic transport systems: Structure, mechanism, and function, *Ann. Rev. Biochem.* **55:** 397–425.

Armitage, J. P., and Evans, M. C. W., 1983, The motile and tactic behaviour of *Pseudomonas aeruginosa* in anaerobic environments, *FEBS Lett.* **156:** 113–118.

Baumann, P., and Baumann, L., 1975, Catabolism of D-fructose and D-ribose by *Pseudomonas doudoroffi, Arch. Microbiol.* **105:** 225–240.

Berger, E. A., and Heppel, L. A., 1974, Different mechanisms of energy coupling for the shock-sensitive and shock-resistant amino acid permeases of *Escherichia coli, J. Biol. Chem.* **249:** 7747–7755.

Betlach, M. R., Tieje, J. M., and Firestone, R. B., 1981, Assimilatory nitrate uptake in *Pseudomonas fluorescens* studied using nitrogen-13, *Arch. Microbiol.* **129:** 135–140.

Bishop, L., Agbayani, R., Jr., Ambudkar, S. V., Maloney, P. C., and Ames, G. F.-L., 1989, Reconstitution of a bacterial periplasmic permease in proteoliposomes and demonstration of ATP hydrolysis concomitant with transport, *Proc. Natl. Acad. Sci. USA* **86:** 6953–6957.

Botfield, M. C., Naguchi, K., Tsuchiya, T., and Wilson, T. H., 1992, Membrane topology of the melibiose carrier of *Eschericia coli, J. Biol. Chem.* **267:** 1818–1822.

Büchel, D. E., Gronenborn, B., and Muller–Hill, B., 1980, Sequence of the lactose permease gene, *Nature* **283:** 541–545.

Cervantes, C., and Silver, S., 1990, Inorganic cation and anion transport systems of *Pseudomonas,* in: *Pseudomonas: Biotransformations, Pathogenesis, and Evolving Biotechnology* (S. Silver, A. M. Chakrabarty, B. Iglewski, and S. Kaplan, eds.), American Society of Microbiology, Washington, D.C., pp. 359–372.

Cuskey, S. M., and Phibbs, P. V., Jr., 1985, Chromosomal mapping of mutations affecting glycerol and glucose catabolism in *Pseudomonas aeruginosa* PAO, *J. Bacteriol.* **162:** 872–880.

Davidson, A. L., and Nikaido, H., 1990, Overproduction, solubilization, and reconstitution of the maltose transport system from *Escherichia coli, J. Biol. Chem.* **265:** 4245–4260.

Deshusses, J. P., 1985, *Myo*-inositol transport in bacteria: H^+ symport and periplasmic binding protein dependence, *Ann. N. Y. Acad. Sci.* **456:** 351–360.

Deshusses, J., and Belet, M., 1984, Purification and properties of the *myo*-inositol-binding protein from a *Pseudomonas sp., J. Bacteriol.* **159:** 179–183.

Dias, F. M., Ventullo, R. M., and Rowe, J. J., 1990, Regulation and energization of nitrate transport in a halophilic *Pseudomonas stutzeri, Biochem. Biophys. Res. Commun.* **166:** 424–430.

Doige, C. A., and Ames, G. F.-L., 1993, ATP-dependent transport systems in bacteria and humans: Relevance to cystic fibrosis and multidrug resistance, *Ann. Rev. Microbiol.* **47:** 291–319.

Dreyfuss, J., 1964, Characterization of a sulfate- and thiosulfate-transporting system in *Salmonella typhimurium, J. Biol. Chem.* **239:** 2292–2297.

Durham, D. A., and Phibbs, P. V., Jr., 1982, Fractionation and characterization of the phosphoenolpyruvate:fructose 1-phosphotransferase system from *Pseudomonas aeruginosa, J. Bacteriol.* **149:** 534–541.

Eagon, R. G., and Phibbs, P. V., Jr., 1971, Kinetics of transport of glucose, fructose, and mannitol by *Pseudomonas aeruginosa, Can. J. Biochem.* **49:** 1031–1041.

Eisenberg, R. C., and Phibbs, P. V., Jr., 1982, Characterization of an inducible mannitol-binding protein from *Pseudomonas aeruginosa*, *Curr. Microbiol.* **7:**229–234.

Fan, C. L., Miller, D. L., and Rodwell, V. W., 1972, Metabolism of basic amino acids in *Pseudomonas putida:* Transport of lysine, ornithine, and arginine, *J. Biol. Chem.* **247:** 2283–2288.

Feiss, D., Belet, M., and Deshusses, J., 1984, Precursor of the *myo*-inositol-binding protein of a *Pseudomonas species, FEBS Lett.* **170:** 165–168.

G.-Feiss, D., Frey, J., Belet, M., and Deshusses, J., 1985, Cloning of genes involved in *myo*-inositol transport in a *Pseudomonas sp.*, *J. Bacteriol.* **162:** 324–327.

Francis, M. M., and Watanabe, M., 1981, Membrane-associated steroid-binding proteins of *Pseudomonas testosteroni, Can. J. Microbiol.* **27:** 1290–1297.

Francis, M. M., and Watanabe, M., 1982, Partial purification and characterization of a membrane-associated steroid-binding protein from *Pseudomonas testosteroni, Can. J. Biochem.* **60:** 798–803.

Francis, M. M., and Watanabe, M., 1983, Purification and characterization of a membrane-associated testosterone-binding protein from *Pseudomonas testosteroni, Can. J. Biochem. Cell Biol.* **61:** 307–312.

Francis, M. M., Kowalsky, N., and Watanabe, M., 1985, Extraction of a steroid transport system from *Pseudomonas testosteroni* membranes and incorporation into synthetic liposomes, *J. Steroid Biochem.* **23:** 523–528.

Guerinot, M. L., 1994, Microbial iron transport, *Ann. Rev. Microbiol.* **48:** 743–772.

Guymon, L. F., and Eagon, R. G., 1974, Transport of glucose, gluconate, and methyl α-D-glucoside by *Pseudomonas aeruginosa*, *J. Bacteriol.* **117:** 1261–1269.

Hancock, R. E. W., Poole, K., and Benz, R., 1982, Outer membrane protein P of *Pseudomonas aeruginosa:* Regulation by phosphate deficiency and formation of small anion-specific channels in lipid bilayer membranes, *J. Bacteriol.* **150:** 730–738.

Harold, F. M., 1972, Conservation and transformation of energy by bacterial membranes, *Bacteriol. Rev.* **36:** 172–230.

Harwood, C. S., Nichols, N. N., Kim, M.-K., Ditty, J. L., and Parales, R. E., 1994, Identification of the *pcaRKF* gene cluster from *Pseudomonas putida:* Involvement in chemotaxis, biodegradation, and transport of 4-hydroxybenzoate, *J. Bacteriol.* **176:** 6479–6488.

Hernandez, D., and Rowe, J. J., 1988, Oxygen inhibition of nitrate uptake is a general regulatory mechanism in nitrate respiration, *J. Biol. Chem.* **263:** 7937–7939.

Hernandez, D., Dias, F. M., and Rowe, J. J., 1991, Nitrate transport and its regulation by O_2 in *Pseudomonas aeruginosa*, *Arch. Biochem. Biophys.* **286:** 159–163.

Higgins, S. J., and Mandelstam, J., 1972, Evidence for induced synthesis of an active transport factor for mandelate in *Pseudomonas putida, Biochem. J.* **126:** 917–922.

Hong, J.-H., Hunt, A. G., Masters, P. S., and Lieberman, M. A., 1979, Requirement of acetyl phosphate for the binding protein-dependent transport systems in *Escherichia coli, Proc. Natl. Acad. Sci. USA* **76:** 1213–1217.

Hoshino, T., 1979, Transport systems for branched-chain amino acids in *Pseudomonas aeruginosa, J. Bacteriol.* **139:** 705–712.

Hoshino, T., and Kagayama, M., 1979, Sodium-dependent transport of L-leucine in membrane vesicles prepared from *Pseudomonas aeruginosa*, *J. Bacteriol.* **137:** 73–81.

Hoshino, T., and Kagayama, M., 1980, Purification and properties of a binding protein for branched-chain amino acids in *Pseudomonas aeruginosa*, *J. Bacteriol.* **141:** 1055–1063.

Hoshino, T., and Kagayama, M., 1982, Mutational separation of transport systems for branched-chain amino acids in *Pseudomonas aeruginosa*, *J. Bacteriol.* **151:** 620–628.

Hoshino, T., and Kose, K., 1989, Cloning and nucleotide sequence of *braC*, the structural gene for the leucine-, isoleucine-, and valine-binding protein of *Pseudomonas aeruginosa* PAO, *J. Bacteriol.* **171:** 6300–6306.

Hoshino, T., and Kose, K., 1990a, Cloning, nucleotide sequences, and identification of the products of the *Pseudomonas aeruginosa* PAO *bra* genes, which encode the high-affinity branched-chain amino acid transport system, *J. Bacteriol.* **172:** 5531–5539.

Hoshino, T., and Kose, K., 1990b, Genetic analysis of the *Pseudomonas aeruginosa* PAO high-affinity branched-chain amino acid transport system by use of plasmids carrying the *bra* genes, *J. Bacteriol.* **172:** 5540–5543.

Hoshino, T., and Nishio, K., 1982, Isolation and characterization of a *Pseudomonas aeruginosa* PAO mutant defective in the structural gene for the LIVAT-binding protein, *J. Bacteriol.* **151:** 729–736.

Hoshino, T., Tsuda, M., Iino, T., Nishio, K., and Kagayama, M., 1983, Genetic mapping of *bra* genes affecting branched-chain amino acid transport in *Pseudomonas aeruginosa*, *J. Bacteriol.* **153:** 1272–1281.

Hoshino, T., Kose, K., and Uratani, Y., 1990, Cloning and nucleotide sequence of the gene *braB* coding for the sodium-coupled branched-chain amino acid carrier in *Pseudomonas aeruginosa* PAO, *Mol. Gen. Genet.* **220:** 461–467.

Hoshino, T., Kose-Terai, K., and Uratani, Y., 1991, Isolation of the *braZ* gene encoding the carrier for a novel branched-chain amino acid transport system in *Pseudomonas aeruginosa* PAO, *J. Bacteriol.* **173:** 1855–1861.

Hoshino, T., Kose-Terai, K., and Sato, K., 1992, Solubilization and reconstitution of the *Pseudomonas aeruginosa* high-affinity branched-chain amino acid transport system, *J. Biol. Chem.* **267:** 21313–21318.

Kaback, H. R., 1983, The *lac* carrier protein in *Escherichia coli, J. Membr. Biol.* **76:** 95–112.

Karimian, M., and Ornston, L. N., 1981, Participation of the β-ketoadipate transport system in chemotaxis, *J. Gen. Microbiol.* **124:** 25–28.

Kay, W. W., and Gronlund, A. F., 1969, Proline transport by *Pseudomonas aeruginosa, Biochim. Biophys. Acta* **193:** 444–455.

Kay, W. W., and Gronlund, A. F., 1971, Transport of aromatic amino acids by *Pseudomonas aeruginosa, J. Bacteriol.* **105:** 1039–1046.

Konings, W. N., Pooleman, B., and Driessen, A. J. M., 1989, Bioenergetics and solute transport in *Lactococci, CRC Crit. Rev. Microbiol.* **16:** 419–476.

Kowalchuk, G. A., Hartnett, G. B., Benson, A., Houghton, J. E., Ngai, K.-L., and Ornston, L. N., 1994, Contrasting patterns of evolutionary divergence within the *Acinetobactor calcoaceticus pca* operon, *Gene* **146:** 23–30.

Kristjansson, J. K., Walter, B., and Hollocher, T. C., 1978, Respiration-dependent proton translocation and the transport of nitrate and nitrite in *Paracoccus denitrificans* and other denitrifying bacteria, *Biochemistry* **17:** 5014–5019.

Laakso, S., 1976, The relationship between methionine uptake and demethiolation in a methionine-utilizing mutant of *Pseudomonads fluorescens* UK1, *J. Gen. Microbiol.* **95:** 391–394.

Lacoste, A.-M., Cassaigne, A., and Neuzil, E., 1981, Transport of inorganic phosphate in *Pseudomonas aeruginosa, Curr. Microbiol.* **6:** 115–120.

Lefebvre, Y. A., Lefebvre, D. D., Schulz, R., Groman, E. V., and Watanabe, M., 1979, The effects of specific inhibitors and an antiserum of 3β- and 17β-hydroxysteroid dehydrogenase on steroid uptake in *Pseudomonas testosteroni, J. Steroid Biochem.* **10:** 519–522.

Lessie, T. G., and Phibbs, P. V., Jr., 1984, Alternative pathways of carbohydrate utilization in *Pseudomonas, Ann. Rev. Microbiol.* **38:** 359–387.

Luthi, E., Bauer, H., Gamper, M., Brunner, F., Villeval, D., Mercenier, A., and Haas, D., 1990, The *arc* operon for anaerobic arginine catabolism in *Pseudomonas aeruginosa* contains an additional gene, *arcD*, encoding a membrane protein, *Gene* **87:** 37–43.

Mantsala, P., Laakso, S., and Nurmikko, V., 1974, Observations on methionine transport in *Pseudomonas fluorescens* UK1, *J. Gen. Microbiol.* **84:** 19–27.

Marger, M. D., and Saier, M. H., Jr., 1993, A major superfamily of transmembrane facilitators that catabolize uniport, symport, and antiport, *Trends Biochem. Sci.* **18:** 13–19.

Matsubara, K., Ohnishi, K., and Kiritani, K., 1992, Nucleotide sequences and characterization of *liv* genes encoding components of the high-affinity branched-chain amino acid transport system in *Salmonella typhimurium*, *J. Biochem.* **112:** 93–101.

Meagher, R. B., McCorkle, G. M., Ornston, M. K., and Ornston, L. N., 1972, Inducible uptake system for β-carboxy-*cis, cis*-muconate in a permeability mutant of *Pseudomonas putida*, *J. Bacteriol.* **111:** 465–473.

Meile, L., Soldati, L., and Leisinger, T., 1982, Regulation of proline catabolism in *Pseudomonas aeruginosa* PAO, *Arch. Microbiol.* **132:** 189–193.

Midgley, M., and Dawes, E. A., 1973, The regulation of transport of glucose and methyl α-glucoside in *Pseudomonas aeruginosa*, *Biochem. J.* **132:** 141–154.

Miller, D. L., and Rodwell, V. W., 1971, Metabolism of basic amino acids in *Pseudomonas putida*, *J. Biol. Chem.* **246:** 1765–1771.

Montie, T. C., and Montie, D. B., 1979, Methionine transport in *Pseudomonas aeruginosa*, *Can. J. Microbiol.* **25:** 1103–1107.

Mukkada, A. J., Long, G. L., and Romano, A. H., 1973, The uptake of 2-deoxy-D-glucose by *Pseudomonas aeruginosa* and its regulation, *Biochem. J.* **132:** 155–162.

Neu, H. C., Heppel, L. A., 1965, The release of enzymes from *Escherichia coli* by osmotic shock and during the formation of spheroplasts, *J. Biol. Chem.* **240:** 3685–3692.

Ohnishi, K., Hasegawa, A., Matsubara, K., Date, T., Okada, T., and Kiritani, K., 1988, Cloning and nucleotide sequence of the *brnQ* gene, the structural gene for a membrane-associated component of the LIV-II transport system for branched-chain amino acids in *Salmonella typhimurium*, *Jpn. J. Genet.* **63:** 343–357.

Ohta, N., Galsworthy, P. R., and Pardee, A. B., 1971, Genetics of sulfate transport by *Salmonella typhimurium*, *J. Bacteriol.* **105:** 1053–1062.

Ohtake, H., Cervantes, C., and Silver, S., 1987, Decreased chromate uptake in *Pseudomonas fluorescens* carrying a chromate resistance plasmid, *J. Bacteriol.* **169:** 3853–3856.

Ondrako, J. M., and Ornston, L. N., 1980, Biological distribution and physiological role of the β-ketoadipate transport system, *J. Gen. Microbiol.* **120:** 199–209.

Ornston, L. N., and Parke, D., 1976, Properties of an inducible uptake system for β-ketoadipate in *Pseudomonas putida*, *J. Bacteriol.* **125:** 475–488.

Parke, D., and Ornston, L. N., 1976, Constitutive synthesis of enzymes of the protocatechuate pathway and of the β-ketoadipate uptake system in mutant strains of *Pseudomonas putida*, *J. Bacteriol.* **126:** 272–281.

Peck, H. D., Jr., 1961, Enzymatic basis for assimilatory and dissimilatory sulfate reduction, *J. Bacteriol.* **82:** 933–939.

Phibbs, P. V., Jr., and Eagon, R. G., 1970, Transport and phosphorylation of glucose, fructose, and mannitol by *Pseudomonas aeruginosa*, *Arch. Biochem. Biophys.* **138:** 470–482.

Phibbs, P. V., Jr., McCowen, S. M., Feary, T. W., and Blevins, W. T., 1978, Mannitol and fructose catabolic pathways of *Pseudomonas aeruginosa* carbohydrate-negative mutants and pleiotropic effects of certain enzyme deficiencies, *J. Bacteriol.* **133:** 717–728.

Plate, C. A., 1979, Requirement for membrane potential in active transport of glutamine by *Escherichia coli*, *J. Bacteriol.* **137:** 221–225.

Poole, K., and Hancock, R. E. W., 1983, Secretion of alkaline phosphatase and phospholipase C is specific and does not involve an increase in outer membrane permeability, *FEMS Microbiol. Lett.* **16:** 25–29.

Poole, K., and Hancock, R. E. W., 1984, Phosphate transport in *Pseudomonas aeruginosa:* Involvement of a periplasmic phosphate-binding protein, *Eur. J. Biochem.* **144:** 607–612.

Postma, P. W., Lengeler, J. W., and Jacobsen, G. R., 1993, Phosphoenolpyruvate:carbohydrate phosphotransferase systems in bacteria, *Microbiol Rev.* **57:** 543–594.

Reber, G., Belet, M., and Deshusses, J., 1977, *Myo*-inositol transport system in *Pseudomonas putida, J. Bacteriol.* **131:**872–875.

Richarme, G., 1985, Possible involvement of lipoic acid in binding protein-dependent transport systems in *Escherichia coli, J. Bacteriol.* **162:** 286–293.

Roehl, R. A., and Phibbs, P. V., Jr., 1982, Characterization and genetic mapping of fructose phosphotransferase mutations in *Pseudomonas aeruginosa, J. Bacteriol.* **149:** 897–905.

Rosenberg, H., 1987, Phosphate transport in prokaryotes, in: *Ion Transport in Prokaryotes* (B. P. Rosen and S. Silver, eds.), Academic Press, San Diego, pp. 205–248.

Rosenfeld, H., and Feigelson, P., 1969, Product inhibition in *Pseudomonas acidovorans* of a permease system which transports L-tryptophan, *J. Bacteriol.* **97:** 705–714.

Rottenberg, H., 1975, The measurement of transmembrane electrochemical proton gradients, *J. Bioenerg.* **7:** 61–74.

Sage, A., Temple, L., Christie, G. B., and Phibbs, P. V., Jr., 1993, Nucleotide sequence and expression of the glucose catabolism and transport genes in *Pseudomonas aeruginosa,* Program and Book of Abstracts for *4th International Symposium on Pseudomonas,* August 1993, p. 105.

Sawyer, M. H., Baumann, P., Baumann, L., Berman, S. M., Canovas, J. L., and Berman, R. H., 1977, Pathways of D-fructose catabolism in species of *Pseudomonas, Arch. Microbiol.* **112:** 49–55.

Schook, L. B., and Berk, R. S., 1978, Nutritional studies with *Pseudomonas aeruginosa* grown on inorganic sulfur sources, *J. Bacteriol.* **133:** 1377–1382.

Sias, S. R., and Ingraham, J. L., 1979, Isolation and analysis of mutants of *Pseudomonas aeruginosa* unable to assimilate nitrate, *Arch. Microbiol.* **122:** 263–270.

Sirko, A., Hryniewicz, M., Hulanicka, D., Bock, A., 1990, Sulfate and thiosulfate transport in *Escherichia coli* K12: Nucleotide sequence and expression of the *cysTWAM* gene cluster, *J. Bacteriol.* **172:** 3351–3357.

Sly, L. M., Worobec, E. A., Perkins, R. E., and Phibbs, P. V., Jr., 1993, Reconstitution of glucose uptake and chemotaxis in *Pseudomonas aeruginosa* glucose transport defective mutants, *Can. J. Microbiol.* **39:** 1079–1083.

Stanier, R. Y., Palleroni, N. J., and Doudoroff, M., 1966, The aeruobic pseudomonads: A taxonomic study, *J. Gen. Microbiol.* **43:** 149–171.

Stinett, J. D., Guymon, L. F., and Eagon, R. G., 1973, A novel technique for the preparation of transport-active membrane vesicles from *Pseudomonas aeruginosa:* Observations on gluconate transport, *Biochem. Biophys. Res. Commun.* **52:** 284–290.

Stinson, M. W., Cohen, M. A., and Merrick, J. M., 1976, Isolation of dicarboxylic acid- and glucose-binding proteins from *Pseudomonas aeruginosa, J. Bacteriol.* **128:** 573–579.

Stinson, M. W., Cohen, M. A., and Merrick, J. M., 1977, Purification and properties of the periplasmic glucose-binding protein of *Pseudomonas aeruginosa, J. Bacteriol.* **131:** 672–681.

Stouthamer, A. H., 1976, Biochemistry and genetics of nitrate reductase in bacteria, *Adv. Microbiol. Physiol.* **14:** 315–375.

Surin, B. P., Rosenberg, H., and Cox, G. B., 1985, Phosphate-specific transport system of *Escherichia coli:* Nucleotide sequence and gene-polypeptide relationships, *J. Bacteriol.* **161:** 189–198.

Talalay, P., 1965, Enzymatic mechanisms in steroid biochemistry, *Ann. Rev. Biochem.* **34:** 347–380.

Temple, L., Cuskey, S. M., Perkins, R. E., Bass, R. C., Morales, N. M., Christie, G. E., Olsen, R. H., and Phibbs, P. V., Jr., 1990, Analysis of cloned structural and regulatory genes for carbohydrate utilization in *Pseudomonas aeruginosa* PAO, *J. Bacteriol.* **172:** 6396–6402.

Thayer, J. R., and Wheelis, M. L., 1976, Characterization of a benzoate permease mutant of *Pseudomonas putida, Arch. Microbiol.* **110:** 37–42.

Thayer, J. R., and Wheelis, M. L., 1982, Active transport of benzoate in *Pseudomonas putida, J. Gen. Microbiol.* **128:** 1749–1753.

Uratani, Y., 1985, Solubilization and reconstitution of sodium-dependent transport system for branched-chain amino acids from *Pseudomonas aeruginosa, J. Biol. Chem.* **260:** 10023–10026.

Uratani, Y., 1992, Immunoaffinity purification and reconstitution of sodium-coupled branched-chain amino acid carrier of *Pseudomonas aeruginosa, J. Biol. Chem.* **267:** 5177–5183.

Uratani, Y., and Aiyama, A., 1986, Effect of phospholipid composition on activity of sodium-dependent leucine transport system in *Pseudomonas aeruginosa, J. Biol. Chem.* **261:** 5450–5454.

Uratani, Y., and Hoshino, T., 1989, Difference in sodium requirement of branched-chain amino acid carrier between *Pseudomonas aeruginosa* PAO and PML strains is due to substitution of an amino acid at position 292, *J. Biol. Chem.* **264:** 18944–18950.

Uratani, Y., Wakayama, N., and Hoshino, T., 1987, Effect of lipid acyl chain length on activity of sodium-dependent leucine transport system in *Pseudomonas aeruginosa, J. Biol. Chem.* **262:** 16914–16919.

Uratani, Y., Tsuchiya, T., Akamatsu, Y., and Hoshino, T., 1989, Na+(Li+)/branched-chain amino acid cotransport in *Pseudomonas aeruginosa, J. Membr. Biol.* **107:** 57–62.

Vander Wauben, C., Pierard, A., Kley–Raymann, M., and Haas, D., 1984, *Pseudomonas aeruginosa* mutants affected in anaerobic growth on arginine: Evidence for a four-gene cluster encoding the arginine deiminase pathway, *J. Bacteriol.* **160:** 928–934.

Verhoogt, H. J. C., Smit, H., Abee, T., Gamper, M., Driessen, A. J. M., Haas, D., and Konings, W. L., 1992, *arcD*, the first gene of the *arc* operon for anaerobic arginine catabolism in *Pseudomonas aeruginosa*, encodes an arginine-ornithine exchanger, *J. Bacteriol.* **174:** 1568–1573.

Watanabe, M., and Po, L., 1974, Testosterone uptake by membrane vesicles of *Pseudomonas testosteroni, Biochim. Biophys. Acta* **345:** 419–429.

Watanabe, M., and Po, L., 1976, Membrane bound 3β- and 17β-hydroxysteroid dehydrogenase and its role in steroid transport in membrane vesicles of *Pseudomonas testosteroni, J. Steroid Biochem.* **7:** 171–175.

Watanabe, M., Phillips, K., and Watanabe, H., 1973, Induction of steroid-binding activity in *Pseudomonas testosteroni, J. Steroid. Biochem.* **4:** 622–631.

Watanabe, M., Sy, L. P., Hunt, D., and Lefebvre, Y., 1979, Binding of steroids by a partially purified periplasmic protein from *Pseudomonas testosteroni, J. Steroid Biochem.* **10:** 207–213.

Weill-Thevenet, N. J., Hermann, M., and Vandecasteele, J.-P., 1979, Lysine transport systems in *Pseudomonas* in relation to their physiological function, *J. Gen. Microbiol.* **111:** 263–269.

Whiting, P. H., Midgley, M., and Dawes, E. A., 1976, The regulation of transport of glucose, gluconate, and 2-oxogluconate and of glucose catabolism in *Pseudomonas aeruginosa, Biochem. J.* **154:** 659–668.

Wright, J. K., Seckler, R., and Overath, P., 1986, Molecular aspects of sugar:ion cotransport, *Ann. Rev. Biochem.* **55:** 225–248.

Yamato, I., and Anraku, Y., 1977, Transport of sugars and amino acids in bacteria. XVIII. Properties of isoleucine carrier in the cytoplasmic membrane vesicles of *Escherichia coli, J. Biochem.* **81:** 1517–1523.

Iron Metabolism and Siderophores in *Pseudomonas* and Related Species

7

JEAN-MARIE MEYER and ALAIN STINTZI

1. INTRODUCTION

With the exception of mainly lactic acid bacteria (Archibald, 1983; Pandey *et al.*, 1994), iron is required as an oligoelement by all living organisms and particularly by aerobic bacteria, such as *Pseudomonas*. Because of its redox potentialities, this element is involved in many, if not all, primary biological functions, e.g., electron transport, carbon metabolism, nitrogen fixation, and nucleic acid biosynthesis. Although abundant in nature, iron is, however, not readily available for bacteria because of its profound tendency to hydrolyze and polymerize at physiological pH under aerobiosis. Thus, solubilization of environmental iron by excretion of powerful iron-chelating secondary metabolites, i.e., the siderophores, is the most common way used by microorganisms to sustain their iron requirement.

Pyoverdine, the typical yellow-green fluorescent pigment of the so-called fluorescent pseudomonads (Elliott, 1958) and the first siderophore recognized among *Pseudomonas* (Meyer and Abdallah, 1978; Meyer and Hornsperger, 1978) is one of the most sophisticated structures among siderophores (Teintze *et al.*, 1981). The multiple variations around a unique theme played by a number of strains make the pyoverdine system a fascinating world. Furthermore, it gained in interest when evidence for a strong correlation between iron, pyoverdine, and the plant-beneficial effects of the *Pseudomonas fluorescens-putida* group was

JEAN-MARIE MEYER and ALAIN STINTZI • Laboratoire de Microbiologie et de Génétique, Unité de Recherche Associée au Centre National de la Recherche Scientifique No. 1481, Université Louis-Pasteur, 67000 Strasbourg, France.

Pseudomonas, edited by Montie. Plenum Press, New York, 1998.

raised in the beginning of the 1980s (Kloepper *et al.*, 1980a,b). Another good reason to be interested in that molecule is the important effect that iron metabolism has on bacterial pathogenicity (Weinberg, 1978, 1993) which justifies special focus on *Pseudomonas aeruginosa,* a species involved in human pathogenicity, especially in cystic fibrosis. Indeed, the discovery of pyoverdine as a siderophore raised the question of siderophores in nonfluorescent pseudomonads, among them species presenting interesting biotechnological features. The wide siderophore diversity already found could be associated with the heterogeneity encountered among nonfluorescent pseudomonads, which justified a recent reclassification of some of these strains (Yabuuchi *et al.*, 1992). Thus, siderophore systems evolved by both fluorescent and nonfluorescent *Pseudomonas* and by formerly *Pseudomonas* strains are reviewed in this chapter.

2. SIDEROPHORES AT THE BENCH

2.1. Optimization of Siderophore Production

Siderophores are synthesized in response to cellular iron deficiency. Thus, two iron-related growth parameters should be considered for microbial siderophore production; (1) the available iron content of the growth medium; and (2) the iron requirement of the microorganism. Precise quantification of growth and siderophore production as a function of the concentration of added iron in a given iron-poor medium defines the iron level for optimal siderophore production. This value varies depending on the microorganism and on the growth conditions. Indeed, the level of contaminating iron brought into a culturing medium together with the various ingredients of the growth medium should be kept to a minimum by choosing the purest quality chemicals. Moreover, special care should be taken with regard to the cleanliness of the glassware used for growth experiments, although the HCl-rinsing program with which modern dishwashing machines are equipped is usually efficient enough to remove traces of iron from the glassware.

Among the growth parameters which may affect the synthesis of siderophores, the choice of the carbon source is of particular importance because it has a strong influence on the iron requirement of the bacteria. For instance, to reach full growth, citrate-grown *P. fluorescens* cells need half the amount of iron required by succinate-grown cells (Meyer *et al.*, 1987). Thus, the contaminating iron level in a citrate medium is high enough to sustain maximal growth and, consequently, no siderophore is produced in a citrate medium. A succinate medium, which has a contaminating iron content similar to a citrate medium, allows reduced, iron-limited cellular growth with concomitant siderophore production because of increased cellular demand for iron in such a medium, i.e., for synthesizing a high level of the iron-containing enzymes, succinate and fumarate dehydrogenases. Decreasing the contaminating iron level in a citrate medium by filtration on Chelex-100 resin or by treatment with 8-hydroxyquinoline (Waring and Werkman, 1942), effectively results in iron-starved cells and siderophore production, just as does growth in a succinate medium.

Some other growth factors may also affect the synthesis of siderophores. The stimulating effect of zinc or other metals (Höfte *et al.*, 1993) or the inhibiting effect of phosphates (Barbhaiya and Rao, 1985) on pyoverdine production by *P. aeruginosa* can be seen as an indirect effect of these ions on overall iron metabolism. Other compounds, especially amino acids, may increase the production of siderophores which involve amino acyl residues or amino acid derivatives in their structures, e.g., pyoverdines (Kisaalita *et al.*, 1993), ornibactins (Meyer *et al.*, 1995), and desferriferrioxamines (Azelvandre, 1993). Finally, it should be remembered that the temperature of incubation may drastically affect siderophore synthesis and that no siderophore is produced at temperatures close to the maximal temperature tolerance limit for growth, i.e., 30 °C for *P. fluorescens*, 37 °C for *P. chlororaphis*, and 40 °C for *P. aeruginosa* (Garibaldi, 1971; Meyer *et al.*, 1996). The reason for this temperature effect on siderophore biosynthesis, which appears to be a feature common to all bacteria and not specific to *Pseudomonas* siderophores (Garibaldi, 1972) and how bacteria fulfill their iron requirements, when growing at siderophore-inhibiting temperatures, are presently unknown.

2.2. Detection of Siderophores

Chemical methods for detecting catecholate (Arnow, 1937) or hydroxamate (Csaky, 1948) siderophores are of interest for detecting siderophores in culture supernatants and initially to gain knowledge of the chemical nature of the compounds produced. Description of these tests and comments on their usefulness and limits are well elaborated (Gillam *et al.*, 1981; Neilands and Nakamura, 1991). *Pseudomonas* siderophores are so diverse in structure that a general method of detection, independent of the nature of the ligand, is usually more appropriate than specific chemical reactions. Two rapid procedures are presently available for this. The simplest consists of directly visualizing the colored siderophore-iron(III) complex which is formed when a small volume (1–2 ml) of the culture supernatant is supplemented with iron chloride (0.5 μl of a commercial $FeCl_3$ solution). The color developed by the *Pseudomonas* siderophore-iron(III) complexes varies from yellow-orange (ornibactins, cepabactin) to purple (salicylic acid), and brown to red-brown for pyoverdines, pyochelin, and desferriferrioxamines. An interesting feature of this method is that it quantitatively estimates the siderophore produced by measuring the absorbance at the wave length λ_{max} reached in the visible range, usually between 400 to 450 nm depending on the compound.

The second method for siderophore detection is the chromeazurol-S (CAS) assay developed by Schwyn and Neilands (1987). It is based on the removal of the ferric iron contained in the bright-blue ternary complex of CAS-iron(III)-HDTMA (hexadecyltrimethyl ammonium bromide) by the siderophore, resulting in a change in color from blue to orange, the color of the deferrated dye. Although some factors frequently encountered in bacterial cultures may severely interfere (like the presence of phosphates in the growth medium or an acidic pH developed during the culture), the method is particularly useful for detecting production of siderophores at low levels because it is much more sensitive than the iron chloride test. The reader is advised to refer to the original publication (Schwyn and Neilands, 1987) for technical considerations and to Neilands and Nakamura (1991) for specific comments. Interestingly, the CAS-assay can be extended to agar-solidified media, resulting in an easy way to screen mutants affected in the biosynthesis and transport of siderophores. Unfortunately, the method is usually not applicable when the bacteria produce more than one siderophore under iron-starvation, a feature common to *Pseudomonas* and related strains. For example, *P. aeruginosa* ATCC 15692 produces pyoverdines together with pyochelin (Cox and Graham, 1979; Cox and Adams, 1985), *P. fluorescens* CHA0 produces pyoverdines and salicylic acid (Meyer *et al.*, 1992), and *B. cepacia* ATCC 25416 produces pyochelin, cepabactin, and ornibactins (Meyer *et al.*, 1995).

Recently, a promising procedure for detecting siderophores in culture supernatants has been described which is based on isoelectrophoresis of the siderophores in a culture supernatant coupled with an overlay of CAS-agar for detection (Koedam *et al.*, 1994). The method originally demonstrated with pyoverdines of fluorescent pseudomonads is applicable to siderophores of other microbial origins, as we have recently shown for *B. cepacia* strains. It is particularly useful when analyzing large collections of strains for comparison.

2.3. Purification of Siderophores

Because of the great diversity of siderophore structures, there is no standardized method for purification. The three methods described in

Figure 1 allow purification with good yields of all known siderophores produced by *Pseudomonas* (see Table I and Fig. 2). First the chloroform (or ethyl acetate) extraction method involves acidification of the culture supernatant with HCl to pH 2 to 3. Then the acidified supernatant is vigorously shaken with one-fifth volume of organic solvent. The extrac-

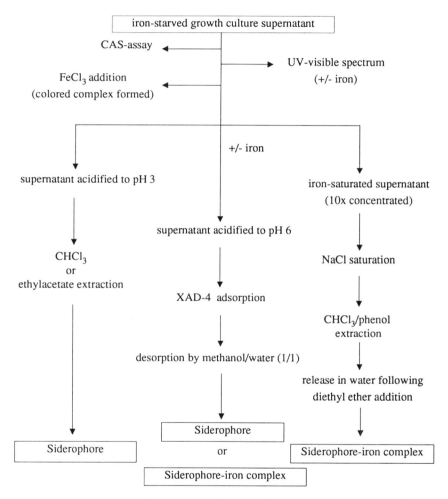

Figure 1. Detection and purification procedures of siderophores. Subsequent purification steps could be CM- or DEAE-Sephadex chromatography for water-soluble compounds, LH-20 Sephadex chromatography for compounds preferentially soluble in methanol, or HPLC.

Table I. Siderophores of *Pseudomonas* and Related Species

Strains	Siderophores	Structures	References
I. Fluorescent pseudomonads			
P. aeruginosa	Pyoverdines	Table II, Fig. 2(a)	see Table II
P. fluorescens	as main siderophores for all strains,		
P. chlororaphis	pyochelin and/or salicylic acid		Cox et al., 1981
P. putida	as secondary siderophores for some strains	Fig. 2(b) and 2(h)	Meyer et al., 1992
			Visca et al., 1993
II. Nonfluorescent pseudomonads and related species			
P. stutzeri ATCC 17588	Desferrioxamine E	Fig. 2(j)	Meyer and Abdallah, 1980
	Desferrioxamine D2	Fig. 2(k)	Azelvandre, 1993
P. stutzeri RC7	Catechol type	2-3 DHB, Arg	Chakraborty et al., 1990
Pseudomonas sp.	Aerobactin	Fig. 2(f)	Buyer et al., 1991
B. cepacia	Ornibactins	Fig. 2(d)	Meyer et al., 1995
	Pyochelin	Fig. 2(b)	Sokol, 1986
	Cepabactin	Fig. 2(c)	Meyer et al., 1989
	Salicylic acid	Fig. 2(h)	Sokol et al., 1992; Visca et al., 1993
	Tropolone	Fig. 2(e)	Korth et al., 1982
B. solanacearum	Schizokinen	Fig. 2(g)	Münziger, 1995
B. pseudomallei	Malleobactin	Uncharacterized hydroxamate	Yang et al., 1991
B. vietnamiensis	Ornibactins	Fig. 2(d)	Meyer et al., 1995
Pseudomonas sp.	Tropolone	Fig. 2(e)	Lindberg et al., 1980
B. plantarii			Azegami et al., 1988
P. fragi			Champomier-Vergès et al., 1996
P. mendocina	No siderophore detected		Meyer, J. M., unpublished
P. diminuta			Meyer, J. M., unpublished
P. stutzeri YPL-1			Lim et al., 1991

tion is usually repeated two or three times. The pooled organic phases are washed with distilled water, filtered on anhydrous sodium sulphate, and evaporated under vacuum to dryness. Subsequent purification set-ups, including thin-layer chromatography or liquid (methanol) chromatography on LH-20 Sephadex or HPLC, may be used for analytical purposes or for further purification, as required. This method, which has been widely used for purifying aromatic (catecholate) siderophores (Neilands, 1981), has been successful in purifying various *Pseudomonas* siderophores, e.g., pyochelin from *P. aeruginosa* (Cox and Graham, 1979; Liu and Shokrani, 1978); pyochelin and other related phenyl-thiazol derivatives from *B. cepacia* strains (Bukowits *et al.,* 1982; Meyer *et al.,* 1989; Sokol, 1986); cepabactin from *Pseudomonas alcaligenes* (Barker *et al.,* 1979) and from *B. cepacia* strains (Itoh *et al.,* 1979; Meyer *et al.,* 1989); and salicylic acid from *P. fluorescens* CHA0 (Meyer *et al.,* 1992) and other *Pseudomonas* spp. (Visca *et al.,* 1993).

Hydroxamate siderophores are not extractable by chloroform but rather by a mixture of chloroform and phenol. In this procedure, the spent medium is first concentrated 10 times in vacuo, saturated with NaCl, and then mixed with vigorous shaking with one-half volume of the chloroform/phenol solution (1:1, v:w). The organic phase (free of the aqueous phase) is recovered by sedimentation and, if necessary, by centrifugation, and the siderophore it contains is resolubilized in water by shaking with three volumes of diethyl ether and a small volume of water. This procedure allows the purification of the desferriferriox-amines produced by *P. stutzeri* (Azelvandre, 1993; Meyer and Abdallah, 1980). For other siderophores, e.g., the ornibactins of *B. cepacia* or *B. vietnamiensis* (Meyer *et al.,* 1995) or most of the pyoverdines from fluores-cent pseudomonads, the procedure is more efficient when applied to the siderophore-iron complexes. In these cases, the growth supernatant is first saturated with ferric chloride. Then the iron-free form of the side-rophores is obtained by decomplexing the purified iron complexes (Meyer and Abdallah, 1978; Wiebe and Winkelmann, 1975). The XAD procedure which is an easy and fast way of purifying these compounds from large scale cultures (Budzikiewicz, 1993) is especially of interest for pyoverdines. The culture supernatant brought to pH 6 is filtered through an Amberlite XAD-4 column (40 × 6 cm for 10 l of medium) or treated with the resin in a batch procedure. Washings with water fol-lowed by an elution with methanol-water (1:1, v:v) usually allows maxi-mal recovery of the siderophores (or their iron-complexes if the super-natant was first saturated with iron). For miniscale preparations, another simple way, which is especially useful when analyzing a large number of samples, is to filter the culture supernatant (from a few ml to not more

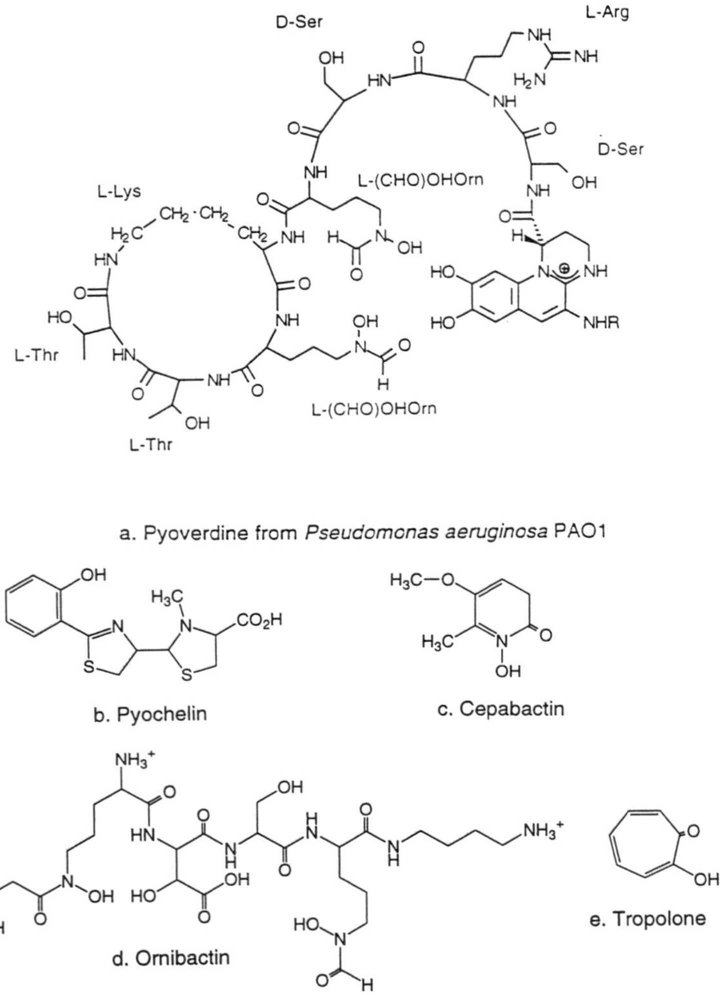
D-Ser
L-Arg
D-Ser
L-(CHO)OHOrn
L-Lys
L-Thr
L-Thr
L-(CHO)OHOrn
a. Pyoverdine from *Pseudomonas aeruginosa* PAO1
b. Pyochelin
c. Cepabactin
d. Ornibactin
e. Tropolone

Figure 2. Siderophores of *Pseudomonas* and related species. The iron-transport properties of schizokinen and tropolone remain to be determined (see text). Desferriferrioxamine B has been added for comparison with desferriferrioxamine E and desferriferrioxamine D2.

than 50 ml) through a commercially available SepPack cartridge (Waters). Pyoverdines bind very tightly to the support and, thus, can be obtained free of other medium components by eluting with acetonitrile after washing with water. More detailed purification procedures for pyoverdines have been recently described, involving XAD and CM-Sep-

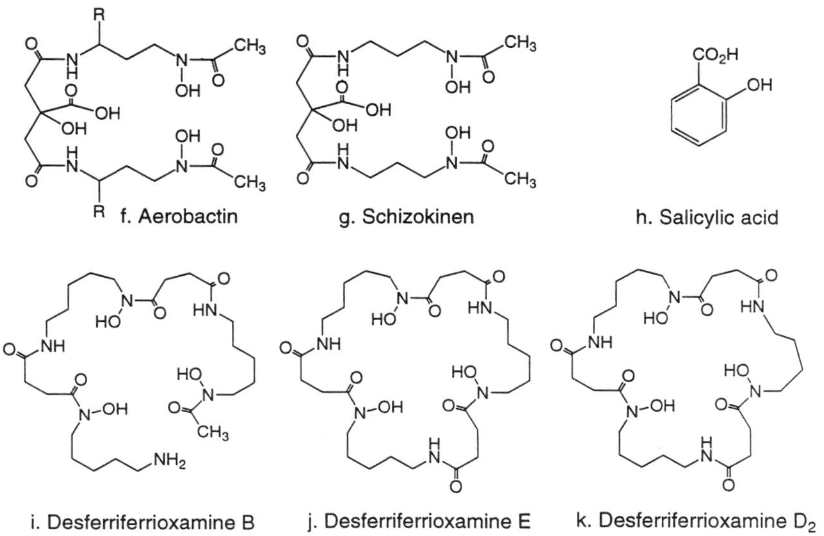

f. Aerobactin   g. Schizokinen   h. Salicylic acid

i. Desferriferrioxamine B   j. Desferriferrioxamine E   k. Desferriferrioxamine D$_2$

Figure 2. (*Continued*).

hadex chromatography followed by HPLC (Nowak-Thompson and Gould, 1994a). A very different approach involves copper-chelate chromatography (Xiao and Kisaalita, 1995).

3. SIDEROPHORE-MEDIATED IRON UPTAKE SYSTEMS IN FLUORESCENT *PSEUDOMONAS*

3.1. Fluorescent *Pseudomonas* Siderophores

3.1.1. The Pyoverdines

Pyoverdines are chromopeptides composed of a quinoleinic chromophore bound together with a peptide and an acyl chain (see Fig. 2(a) as an example and Table 1). The chromophore, identical for all the pyoverdines so far identified, is (1S)-5-amino-2,3-dihydro-8,9-dihydroxy-1H-pyrimido [1,2-a] quinoline-1-carboxylic (Michels *et al.*, 1991). One exception, however, has been recently described for a strain of *P. putida* which has a pyoverdine with a carboxylic group attached to the C$_3$ instead of the C$_1$ of the quinoline cyclic ring (Jacques *et al.*, 1993, 1995). Because of the chromophore which confers its characteristics UV-visible spectrum and its strong fluorescence on the molecule, pyoverdine has been known since Pasteur's time (Schroeter, 1870). It was first described as the yellow-green fluorescent, water-soluble pigment characteristic of the so-called fluorescent pseudomonads (Turfitt, 1936, 1937; Turfreijer *et al.*, 1938). First introduced by Elliot (1958), the term pyoverdine advantageously replaced the too confusing "fluorescein" or "bacterial fluorescein" used previously (King *et al.*, 1954).*

The acyl chain of pyoverdine, which binds at the NH_2 group of the chromophore, can be succinic acid, succinamide, α-ketoglutaric acid, glutamic acid, malic acid or its amide form, depending on the strain and growth conditions. Thus, it is usual in an iron-starved growth supernatant to find several forms of pyoverdines varying in amounts which are easily separated by ion exchange chromatography, HPLC, or electrophoresis. In young cultures, a predominance of the succinamide pyoverdine is usually found, which decreases with a concomitant increase of the succinic form when cultures age because of a spontaneous chemical hydrolysis of the amide group. The rationale for the multiplicity of pyoverdine molecular species, most of which are excreted by bacteria, is so far not understood, but they can be viewed as minor variations arising from a unique precursor during the biosynthesis of pyoverdine (apparently glutamic acid, see paragraph 3.4.1.). As shown for the succinic, succinamide and α-ketoglutaric acid forms of the *P. aeruginosa* ATCC 15692 pyoverdine, modification of the acyl chain does not affect the biological function of these compounds, i.e., their iron uptake (Meyer *et al.*, 1987).

Table II details some examples of complete or fairly complete structures of pyoverdine peptides so far identified (more than 20), originating from various *Pseudomonas* strains belonging to the main species, i.e., *P. aeruginosa*, *P. fluorescens*, and *P. putida*, and to some other species. Complete or partial structures have also been published for pyoverdines produced by *P. syringae* strains (Cody and Gross, 1987; Torres *et al.*, 1986), *P. mildenbergii* (Newkirk and Hulcher, 1969), *P. ovalis* (Elliot,

---

*Unfortunately, confusion arose again when the name pseudobactin was introduced to describe pyoverdines produced by pseudomonads of plant or soil origin (Kloepper *et al.*, 1980b). Furthermore, other authors (Cox and Adams, 1985) turned the name pyoverdine into "pyoverdin," with very debatable arguments in favor of removing the last "e." Thus, presently three different names are used in the literature to describe similar compounds. Our own stand is to respect the valid name first introduced, which is pyoverdine (Elliott, 1958). The appelation "pyoverdine" has also been used (Menhart *et al.*, 1991) to describe the closely related yellow-green fluorescent pigments produced by *Azotobacter* strains (Bulen and LeComte, 1962) and so far called azotobactins in regard to their siderophore functions. The choice of only one name, pyoverdine, to designate compounds belonging to the same family by their closely related structures and their identical biological function is strongly supported by the recent finding of one *Pseudomonas* strain which produces both pyoverdine-type and azotobactin-type related pigments (Hohlneicher *et al.*, 1995).

Table II. Peptidic Sequences of Various Pyoverdines

Strain	Sequence of the pyoverdine peptide[a]	Number of a.a. residues	References
P. aeruginosa ATCC 15692	(Chr)-Ser-Arg-Ser-FoOHOrn-Lys-FoOHOrn-Thr-Thr-	8	Briskot et al., 1989
P. aeruginosa ATCC 27853	(Chr)-Ser-FoOHOrn-Orn-Gly-aThr-Ser-cOHOrn	7	Tappe et al., 1993
P. aeruginosa PaR or Pa6[b]	(Chr)-Ser/DAB-FoOHOrn-Gln-Gln-Gly-FoOHOrn	6	Gipp et al., 1991
P. putida 589	(Chr)-Asp-Lys-OHAsp-Ser-Thr-Ala-Glu-Ser-cOHOrn[c]	9	Persmark et al., 1990
P. putida WCS358	(Chr)-Lys-OHAsp-Ser-Thr-Ala-aThr-Lys-Asp-cOHOrn[c]	9	van der Hofstad et al., 1986
P. putida C	(Chr)-Asp-OHbutOHOrn-Dab-Thr-Gly-Ser-Ser-OHAsp-Thr	9	Seinsche et al., 1993
P. fluorescens ATCC 13525	(Chr)-Ser-Lys-Gly-FoOHOrn-Lys-FoOHOrn-Ser-[c]	7	Linget et al., 1992
P. fluorescens strain 12	(Chr)-Ser-Lys-Gly-FoOHOrn-Ser-Ser-Gly-Lys-FoOHOrn-Glu-Ser-	11	Geisen et al., 1992
P. fluorescens strain ii	(Chr)-Ala-Lys-Gly-Gly-OHAsp-Gln-Ser-Ala-Ala-Ala-Ala-cOHOrn	12	Mohn et al., 1990
P. chlororaphis ATCC 9446	(Chr)-Ser-Lys-Gly-FoOHOrn-Lys-FoOHOrn-Ser-	7	Hohlneicher et al., 1995
P. tolaasii NCPPB 2192	(Chr)-Ser-Lys-Ser-Ser-Thr-Ser-AcOHOrn-Thr-Ser-cOHOrn	10	Demange et al., 1990
P. aptata strain 4a	(Chr)-Ala-Lys-Thr-Ser-AcOHOrn-cOHOrn	6	Budzikiewicz et al., 1992
Pseudomonas B10	(Chr)-Lys-threo-OHAsp-Ala-aThr-Ala-cOHOrn	6	Teintze et al., 1981
Pseudomonas 7SR1	(Chr)-Ser-Gly-Ala-AcOHOrn-Ser-Thr-OHAsp-Ser-[c]	8	Yang and Leong, 1984

[a]Abbreviations: Chr, chromophore; OHOrn, ∂N-hydroxy Orn; cOHOrn, cyclo-OHOrn; OHAsp, threo-β-hydroxy Asp; aThr, alloThr; FO(Ac,OHbut)OHOrn, ∂N-formyl (acetyl, β-hydroxy butyryl) OHOrn; Dab, diaminobutyric acid. D-amino acids are underlined. The pyoverdine of strain Pa6 has an identical structure (R. Tappe and H. Budzikiewicz, personal communication).

[b]Structure determined for the pyoverdine of strain P. aeruginosa PaR.

[c]Stereochemistry partially determined or unknown.

1958), and from natural isolates, usually *P. fluorescens* or *P. putida* strains (Andriollo *et al.*, 1992; Buyer *et al.*, 1986; Gwose and Taraz, 1992, Hancock and Reeder, 1993; Poppe *et al.*, 1987; Taraz *et al.*, 1991a; Teintze *et al.*, 1981).

The peptide chain is the most fascinating part of the pyoverdine molecule. As shown in Table II, great diversity in structure exists among pyoverdines from strains which belong to different species and from strains within a single species. This diversity, which is illustrated by more than ten different structures so far described for *P. fluorescens* strains, presents, however, some limits for other species. For example, *P. aeruginosa* strains could be divided into three groups, depending on the pyoverdine they produce (Cornelis *et al.*, 1989; Meyer *et al.*, 1993). In contrast, the same pyoverdine has been described for two strains belonging to two different species, i.e., *P. fluorescens* ATCC 13525 and *P. chlororaphis* ATCC 9446 (see Table II and Hohlneicher *et al.*, 1995).

The number of amino acyl residues forming the peptide chain varies from 6 to 12 for the pyoverdines so far analyzed. Remarkably unusual amino acids are involved, e.g., ∂N-acyl(acetyl, formyl, or β-hydroxybutyryl)-∂N-hydroxyornithine, cyclized ∂N-hydroxyornithine, β-hydroxy aspartic acid, β-hydroxyhistidine, diaminobutyric acid, and as common amino acids but some of them with a D-configuration. The hydroxy amino acids, together with the catechol-like group of the chromophore participate in iron complexation, whereas the other amino acids are involved in the recognizing and binding the iron complex to the outer membrane ferripyoverdine receptor. The usually strict strain specificity observed for pyoverdine-mediated iron transport (Buyer and Leong, 1986; Cornelis *et al.*, 1989; Hohnadel and Meyer, 1988) correlates well with the diversity observed in pyoverdine structures. A few exceptions, however, have been noted, e.g., *P. aeruginosa* ATCC 15692 which recognizes, indeed, its own (ferri)pyoverdine, but also the (ferri)pyoverdine of *P. fluorescens* ATCC 13525 (Hohnadel and Meyer, 1988). Both pyoverdines, although different in their peptide chain (Table 2), have in common another special feature encountered in some pyoverdines, an internal loop within the peptide chain. Whether this loop participates in the recognition and binding of the ferripyoverdine complex to its receptor is a question which remains to be answered.

Due to a high affinity for iron(III) on the order of 10^{32} M^{-1} (Meyer and Abdallah, 1978), pyoverdines efficiently chelate and solubilize iron usually present in the bacterial environment as insoluble salts or hydroxides. Pyoverdines also may successfully compete for iron with siderophores produced by other microorganisms in the same environment (see paragraph 5.1), or with iron proteins like transferrin, lactoferrin, and ferritin, which represent the sole potential sources of iron for pathogenic bacteria developing in vivo (see paragraph 5.2).

3.1.2. Pyochelin and Salicylate

Apart from pyoverdine, some fluorescent pseudomonads, e.g., *P. aeruginosa*, synthesize an additional iron-chelating compound named pyochelin (Cox and Graham, 1979; Liu and Shokrani, 1978), which differs totally in structure and in physicochemical properties from pyoverdine. Pyochelin, which is 2-(2-*o*-hydroxyphenyl-2-thiazolin-4-yl)-3-methylthiazolidine-4-carboxylic acid [Fig. 2(b)], can be considered a condensation product of salicylic acid and two cysteinyl residues (Cox *et al.*, 1981). Its biosynthesis requires salicylic acid (Ankenbauer and Cox, 1988), but more details on its complete biosynthetic pathway are still lacking. Structurally related compounds, previously isolated from the pyochelin-producers *P. aeruginosa* and *P. cepacia*, e.g., aeruginoic acid (2-(*o*-hydroxyphenyl)-4-thiazole carboxylic acid) and derivatives (Bukowits *et al.*, 1982; Yamada *et al.*, 1970) may be intermediates. Their relationship to iron metabolism has not been studied so far.

That pyochelin acts as a siderophore is apparently debatable considering its low production (less than 10 mg per liter of *P. aeruginosa* ATCC 15692 growth medium compared to 200–300 mg for pyoverdine), its low solubility in water, and its low affinity for iron (5.10^5 M^{-1}, compared to 10^{32} M^{-1} for pyoverdine). However, when involved in iron uptake experiments, pyochelin demonstrates good capacities, which are lower than those of pyoverdine for the wild-type *P. aeruginosa* ATCC 15692, but increased and even better compared with pyoverdine, for pyoverdine-deficient mutants (Ankenbauer *et al.*, 1988; Cox, 1980; Meyer, 1992). Moreover, pyochelin has been involved, although to a lesser extent compared with pyoverdine, in the in vivo development and virulence of *P. aeruginosa* (Ankenbauer *et al.*, 1985; Cox, 1982; Meyer *et al.*, 1996; Sriyosachati and Cox, 1986). Interestingly, it has been recently proposed that pyochelin has a more general function in mineral nutrition in cobalt and molybdenum delivery (Visca *et al.*, 1992b).

Salicylic acid [Fig. 2(h)] was first described as a siderophore for *P. fluorescens* CHA0 (Meyer *et al.*, 1992), a strain which produces pyoverdine but apparently not pyochelin. Salicylic acid was also recently described [also under the name of azurechelin (Sokol *et al.*, 1992)] in the pyochelin-producers *P. aeruginosa* and *B. cepacia* (Visca *et al.*, 1993). It is presently unclear whether salicylic acid should be considered a biologically active by-product of the pyochelin biosynthesis pathway using the same receptor as ferripyochelin or an independent siderophore having its own outer membrane receptor, as suggested for *P. fluorescens* CHA0 which cannot incorporate iron chelated by pyochelin (Meyer *et al.*, 1992).

3.2. Internalization of Iron

3.2.1. Translocation through Outer and Inner Membranes

Siderophore-iron complexes are internalized in the bacterial cell first through specific iron-regulated outer membrane proteins (IROMPs, Williams *et al.*, 1990) acting as ferrisiderophore receptors (Neilands, 1982). Iron-deficient fluorescent *Pseudomonas*, like other gram (−) bacteria, have such IROMPs with apparent molecular masses in the range of 75–92 kDa (de Weger *et al.*, 1986; Meyer *et al.*, 1979, 1990; Poole *et al.*, 1991) except for the *P. aeruginosa* ATCC 15692 ferripyochelin-receptor which has been described as a 14-kDa protein (Sokol and Woods, 1983). Recent reports, however, favor a 75-kDa IROMP as the ferripyochelin receptor of *P. aeruginosa* ATCC 15692 (Ankenbauer, 1992; Heinrichs *et al.*, 1991).

The parallelism between pyoverdines and their corresponding IROMPs is remarkable. Electrophoretic IROMP-profiles usually differ in strains producing different pyoverdines, whereas strains producing identical pyoverdines have identical IROMP-profiles (Cornelis *et al.*, 1989; Meyer *et al.*, 1979). Although identical biological functions involve closely structurally related ligands, ferripyoverdine receptors present astonishing immunological specificities when comparing strains producing different pyoverdines: A polyclonal antiserum raised against the ferripyoverdine-receptor of *P. aeruginosa* ATCC 15692 showed no cross-reaction with its counterpart in *P. aeruginosa* ATCC 27853 and vice versa (Cornelis *et al.*, 1989). Thus, marked immunospecificity characterizes each receptor protein, a conclusion which is also suggested when comparing the deduced amino acid sequences of the three ferripyoverdine-receptor genes *fpvA* (Poole *et al.*, 1993), *pupA* (Bitter *et al.*, 1991) and *pupB* (Koster *et al.*, 1993) already determined. Interestingly, the ferripyoverdine uptake system of *P. aeruginosa* has also been associated with the sensitivity of cells to pyocin Sa (Smith *et al.*, 1992).

The active transport mechanism by which iron is translocated through membranes and the periplasm for the most part remains to be elucidated. However, the ExbB, ExbD and, to a lesser extent, TonB proteins recently characterized in *P. putida* WCS358 (Bitter *et al.*, 1993) showed enough similarities with their *E. coli* counterparts to support the hypothesis that an energy-coupled system similar to, although not fully compatible with, that found in *E. coli* (Braun and Hantke, 1991) should act in fluorescent *Pseudomonas*.

3.2.2. Intracellular Release of Iron from Ferrisiderophores

The ultimate step in siderophore-mediated iron metabolism is relevant to the release of iron from the internalized siderophore-iron complex. The extreme difference in the affinity of siderophores for ferrous and ferric ions favors the hypothesis of a reductive mechanism, as already suggested by Neilands during his pioneer work (Neilands, 1957). It is agreed, however, that this last step is separate from the overall siderophore genetic system because siderophore-reductase activities are never iron-regulated. The purification of a *P. aeruginosa* cytoplasmic protein, which releases iron from ferripyoverdine in an in vitro system, allowed a better understanding of the biological iron-siderophore reduction process (Hallé and Meyer, 1992a). In fact, the purified protein demonstrates NADH/FMN oxido-reductase activity whereas $FMNH_2$, the product of the reaction, chemically reduces the iron of many, if not all, ferrisiderophores, providing that an iron(II)-trap is present, for example, ferrozine in the in vitro system (Hallé and Meyer, 1992b). A similar NAD(P)H:flavin oxidoreductase was isolated by Coves and Fontecave (1993) as a component of the ribonucleotide-reductase complex of *E. coli*. This protein demonstrates in vitro ferrisiderophore reductase activity which is also effective in assays where ferrozine is replaced by the iron-free form of the ribonucleotide reductase subunit R2. Thus, in situ, the iron-free forms (apo-proteins) of the potential iron targets may regulate their own iron release from the ferrisiderophores. Such a system would present the advantage of not allowing the accumulation inside the cell of excess free iron which could generate damaging free hydroxyl radicals. In this way, the Fur protein, which is involved in regulating siderophore biosynthesis (Bagg and Neilands, 1987a,b; Prince *et al.*, 1991) should be the last apo-protein to react as an iron(II) trap.

P. aeruginosa contains a ferritin-like protein (Moore *et al.*, 1986). Although one would predict that a biological relationship should exist, only a few links have presently been established between ferrisiderophores and such a protein which may act in an iron storage capacity (Mielczarek *et al.*, 1992).

3.3. Iron Uptake from Other Iron Sources

Beside their own siderophores, fluorescent *Pseudomonas* usually have many other ways to pick up iron from their environment. (Ferri)siderophores of foreign origin, produced by bacteria or fungi, may be used directly as iron sources. These compounds could be closely related to the indigenous siderophores and, thus, could use the same indigenous translocation pathways, as expected from competition studies for incorporating iron mediated by the *P. fluorescens* ATCC 13525 pyoverdine in *P. aeruginosa* ATCC 15692 (Meyer, unpublished data). However, the existence of additional outer membrane receptors, which specifically recognize (ferri)siderophores of foreign origin, as they exist in *E. coli* (Braun *et al.*, 1987), should be the most common way of translocating such compounds, as demonstrated in some studies (de Weger *et al.*, 1988; Morris *et al.*, 1992). Because one frequently finds several IROMPs in *Pseudomonas* iron-starved cells, it could be expected, although not yet demonstrated, that similar features should occur for efficient iron incorporation mediated by desferriferrioxamine B, ferrichrome, or cepabactin in various fluorescent *Pseudomonas* (Cornelis *et al.*, 1987; Jurkevitch *et al.*, 1992; Meyer, 1992). Siderophores with low efficiency in iron uptake, however, may have a nonspecific way of incorporation by utilizing porins, as demonstrated with OprF in *P. aeruginosa* ATCC 15692 (Meyer, 1992).

Another original feature found in several *Pseudomonas* strains is the inducibility of some of the additional ferrisiderophore uptake systems. *P. aeruginosa* ATCC 15692, grown in presence of the *E. coli* siderophore enterobactin, induces an enterobactin-mediated iron transport (Poole *et al.*, 1990). This apparently occurs by the synthesis of a two-component regulatory system utilizing a supplementary IROMP which acts as a (ferri)enterobactin receptor (PfeA; Dean and Poole, 1993a,b). PfeA could also be involved in the iron transport mediated by dihydroxybenzoyl-serine, a breakdown product of enterobactin (Screen *et al.*, 1995). Similar inducible systems have been described in *P. putida* WCS358 for pyoverdines for foreign origin (Koster *et al.*, 1993, 1994) and also in a nonfluorescent pseudomonad, *P. fragi*, for the pyoverdine of *P. aeruginosa* ATCC 15692 (Champomier-Vergès *et al.*, 1996). Moreover, indigenous siderophores themselves modulate the expression of their respective receptors (Gensberg *et al.*, 1992).

Nonsiderophore iron-chelating compounds may also be efficiently used as iron sources, as shown for citrate (Cox, 1980; Harding and Royt, 1990), myoinositol hexakisphosphate (Smith *et al.*, 1994) and, more surprisingly, for the xenobiotic compound nitrilotriacetate (Meyer and Hohnadel, 1992). Entrance pathways for these compounds remain unknown. Finally, in addition to these well-demonstrated iron transport systems, other hypothetical methods could involve, for example, membrane-associated iron chelators (Royt, 1988) or even less-defined, low-affinity systems (Neilands *et al.*, 1987) to explain the ability of bacteria to grow well in an iron-rich environment without apparently producing any siderophores.

3.4. Biosynthesis, Genetic Organization, and Regulation of Siderophores

3.4.1. Biosynthetic Pathway of Pyoverdine

Presently the biochemical pathway of pyoverdine biosynthesis is poorly understood. When considering the sophisticated structure of pyoverdine and its subdivision into three distinct moieties, namely, the chromophore, the peptide chain, and the acyl chain, it is not surprising that the first assumptions included independent biosynthetic pathways for the chromophore and acyl chain and for the peptide chain, with an expected final condensation reaction between the different elements (Budzikiewicz, 1994; Marugg *et al.*, 1985). Indeed, independent chemical syntheses of the pyoverdine chromophore, obtained from dihydroxyphenylalanine (DOPA) as a precursor, and of a pyoverdine peptide chain have been described (Kolasa and Miller, 1990; Okonya *et al.*, 1995). It should be mentioned, however, that mutants producing only the chromophore instead of pyoverdine have never been obtained, whereas mutants which produce iron-regulated nonfluorescent peptides harboring hydroxamate functions have already been described (Marugg *et al.*, 1985; Stintzi, 1993; Visca *et al.*, 1992a). From labeled experiments, it has been concluded that phenylalanine and/or tyrosine (Maksimova *et al.*, 1992; Nowack-Thomson and Gould, 1994b; Stintzi, 1993) are the potent precursors of the quinoleinic chromophore.* An hypothetical scheme of synthesis (Longerich *et al.*, 1993), which proceeds from phenylalanine to the chromophore backbone, includes several steps, such as hydroxylation reactions with tyrosine, DOPA, and TOPA (trihydroxyphenylalanine) as intermediates and a condensation of the latter with diaminobutyrate (see Fig. 3). Interestingly, the synthesis pathway of pseudoverdine, a coumarin-derivative produced by a *P. aeruginosa* PAO pyoverdine-deficient mutant harboring a 10.8-kb chromosomal DNA fragment of PAO (Longerich *et al.*, 1993; Stintzi *et al.*, 1996), could be easily derivatized from this scheme as the product of an internal condensation of TOPA. Labeled tyrosine has been recognized as a biosynthetic precursor for the pyoverdine chromophore and for pseudoverdine (Stintzi, 1993). Another biosynthetic pathway, however, has been proposed for the chromophore with dihydroorotate as a precursor (Maksimova *et al.*, 1993).

Synthesis of the peptide portion is not likely to be through the

---

*Incorporation of ^{14}C within the chromophore of pyoverdine (azotobactin) of *Azotobacter vinelandii* has been observed for ^{14}C-phenylalanine- and ^{14}C-DOPA-supplemented, iron-starved growth media (Fukasawa and Goto, 1973).

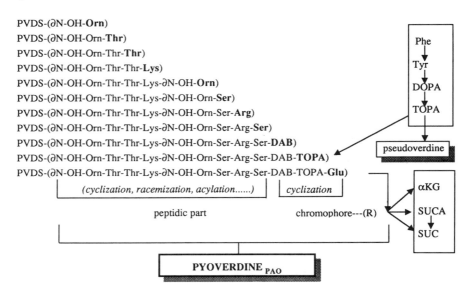

Figure 3. Hypothetical biosynthetic pathway of pyoverdine of *P. aeruginosa* ATCC 15692. Biosynthesis starts (top left) with the attachment of a ∂N-hydroxyornithine (∂N-OH-Orn) to the pyoverdine synthetase (PVDS) and continues step-by-step for the other amino acids or amino acid derivatives. Phenylalanine (top right) is hydroxylated in several steps to trihydroxyphenylalanine (TOPA) which is included after diaminobutyric acid (DAB) within the peptide chain in the wild-type strain PAO1 or is transformed in pseudoverdine in the mutant PAO6624 harboring the plasmid pPYP180 (see Longerich *et al.*, 1993, and Stintzi *et al.*, 1996). The acyl chain (R) attached to the chromophore [see Fig. 2(a)] arises from the transformation of the last amino acyl residue in the peptide chain (Glu) in β-ketoglutarate (βKG0, or succinamide (SUCA), or succinic acid (SUC).

classical ribosomal pathway of peptide synthesis. The size of the peptide chain (6 to 12 aminoacids depending on the pyoverdines; see paragraph 3.1.1 and Table II), together with the presence of uncommon amino acids or amino acid derivatives and D-amino acids in it, and the internal loops which characterize some of the pyoverdine peptides, are all peculiarities shared with microbial peptidic antibiotics, e.g., gramicidin, tyrocidin, and actinomycin (Kleinkauf and von Döhren, 1987) and syringotoxins and syringomycins produced by strains of *P. syringae* (Morgan and Chatterjee, 1988; Xu and Gross, 1988). Therefore, a multienzyme thiotemplate mechanism involving peptide synthetases, as demonstrated for the biosynthesis of the peptidic antibiotics (Kleinkauf and von Döhren, 1990) is most probable for the synthesis of the peptidic siderophores (pyoverdines and ornibactins, see paragraph 4.2). This concept is also supported by evidence showing that *pvdD*, a pyoverdine gene of *P. aeruginosa*, has high similarity to antibiotic peptide synthetase genes (Merriman *et al.*, 1995). Secondly, the detection of iron-repressed cytoplasmic proteins (IRCPs) with unusual, apparent high molecular mass values (M_r ranging from 180 to 600 kDa according to SDS-gel electrophoresis patterns), comparable to those of antibiotic peptide synthetases, has been reported (Georges and Meyer, 1995). Furthermore, IRCP-gel profiles differ for strains producing different pyoverdines but are identical for strains producing identical pyoverdines (Georges and Meyer, 1995). High molecular mass cytoplasmic proteins have also been detected in *P. syringae* pv. syringae. These are associated with pyoverdine biosynthesis because they are visible only in iron-deficient cells (Morgan and Chatterjee, 1988). The partial characterization of amino acyl-AMP transferase activity correlating with the synthesis of pyoverdine in *Azotobacter vinelandii* (Menhart and Viswanatha, 1990) and the recent isolation of the pyoverdine biosynthetic gene *pbsC* in *Pseudomonas* strain M114 (Adams *et al.*, 1994; see Table III) also strengthen the proposal of a nonribosomal pathway for pyoverdine biosynthesis.

A fully hypothetical scheme for the biosynthetic pathway of a pyoverdine, e.g., the pyoverdine of *P. aeruginosa* PAO, as illustrated in Figure 3, could be constructed by taking into account the data described above. In this scheme, the molecule is synthesized from amino acids step-by-step with the integration during the process of the expected precursors of the chromophore, TOPA and DAB, as occurs for other amino acids. Indeed, secondary reactions like racemization, acylation, cyclization, and conversion of the last amino acyl residue (glutamic acid, from which the acyl chain linked to the chromophore is donated), could occur during the process at steps which are presently unknown. ∂N-hydroxy-ornithine, however, should be integrated as such, because the corresponding hydroxylase and gene have been characterized (Visca *et al.*, 1994). Another secondary reaction not indicated on the scheme could be a transitory sulfonation which has recently been postulated in consequence of the isolation of dihydropyoverdine sulfonic acids from *P. aptata* grown supernatants (Schröder *et al.*, 1995). The scheme also does not take into account the multiplicity of expected peptide synthetases observed as IRCPs (five for *P. aeruginosa* PAO; Georges and Meyer, 1995). Another point of the proposed scheme is that it could explain the formation of pyoverdine-related minor compounds besides pyoverdine which could be isolated, i.e., ferribactin (Holneicher *et al.*, 1992, 1995; Maurer *et al.*, 1968) and dehydropyoverdine (pseudobactin A, Budzikiewicz *et al.*, 1992; Teintze and Leong, 1981). Errors at the level of amino acid incorporation during antibiotic synthesis through peptide synthetases have already been observed (H. Kleinkauf, personal communication). Similarly, if tyrosine is erroneously integrated during the process instead of TOPA, then ferribactin instead of pyoverdine becomes the final product. In this view, ferribactin should be seen as a side-product of pyoverdine rather than as a pyoverdine precursor, as postulated by some authors (Budzikiewicz, 1994; Taraz *et al.*, 1991b). Dehydropyoverdine, isopyoverdine (Jacques *et al.*, 1993, 1995), and the recently described pyoverdine, which contains an azotobactin-type chromophore (Hohlneicher *et al.*, 1995), may be considered similarly.

3.4.2. Genetics and Regulation

Approximately thirty siderophore-related genes have already been cloned and eventually sequenced in various fluorescent *Pseudomonas* strains, most of them related to the pyoverdine system of *P. aeruginosa* ATCC 15692 (PAO1 strain). Two natural isolates associated with interesting features in plant-related biocontrol, *P. putida* WCS358 and *Pseudomonas* M114, have also been intensively studied in regard to their respective pyoverdines. The genes presently identified are mainly outer membrane (ferri)pyoverdine-receptor genes, and also regulatory genes with only a few genes so far, whose products are involved in the biosynthesis of pyoverdines (Table III).

The genetic organization of pyoverdine-related genes reflects the complexity of the system. A minimum of 12 to 15 genes organized in four to five gene clusters have been recognized in various *Pseudomonas* strains as revealed by complementation studies (Loper *et al.*, 1984; Marugg *et al.*, 1985, 1988; Moores *et al.*, 1984; O'Sullivan *et al.*, 1990). Precise locations on the updated *P. aeruginosa* PAO chromosomal map (Holloway *et al.*, 1994; O'Hoy and Krishnapillai, 1987) are at 23 min (Hohnadel *et al.*, 1986) and 47 min (Ankenbauer *et al.*, 1986; Hohnadel *et al.*, 1986; Visca *et al.*, 1992a) as revealed by conjugation experiments, whereas a third location at 66–70 min was recently found by physical mapping for genes involved in pyoverdine chromophore synthesis (Stintzi *et al.*, 1996). The other biosynthetic genes so far described in PAO are *pvdA* coding for a L-ornithine N^5-oxygenase (Visca *et al.*, 1994) and *pvdD* which has strong similarity with the antibiotic peptide synthetase genes (Merriman *et al.*, 1995). These genes are located within the 47 min region of the chromosome which also contains *fpvA*, coding for the outer membrane (ferri)pyoverdine receptor (Poole *et al.*, 1993c), and a regulatory gene, *pvdS* (Cunliffe *et al.*, 1995; Miyazaki *et al.*, 1995). A 103-kb DNA fragment containing all of these genes has been cloned

Table III. Siderophore-Related Genes in *Pseudomonas* Species

	Gene	Function	Strains	References
Receptor genes	*fptA*	Ferripyochelin receptor	*P. aeruginosa* PAO	Ankenbauer and Quan, 1994
	fpvA	Ferripyoverdine receptor	*P. aeruginosa* PAO	Poole *et al.*, 1993c
	pfeA	Receptor of ferric-enterobactin	*P. aeruginosa* PAO	Dean and Poole, 1993a
	pupA	Ferripseudobactin 358 receptor	*P. putida* WCS358	Bitter *et al.*, 1991, 1994
	pupB	Ferric-pseudobactin receptor	*P. putida* WCS358	Koster *et al.*, 1993
	exbB, exbD, tonB	Membrane translocation	*P. putida* WCS358	Bitter *et al.*, 1993
Regulatory genes	*fur*	Negative regulator of biosynthetic genes	*P. aeruginosa* PA103	Prince *et al.*, 1993
			P. putida WCS358	Venturi *et al.*, 1995b
	pchR	Regulator of pyochelin synthesis and ferripyochelin receptor genes	*P. aeruginosa* PAO	Heinrichs and Poole, 1993, 1996
	pvdS	Positive regulator of biosynthetic genes	*P. aeruginosa* PAO	Cunliffe *et al.*, 1995; Miyazaki *et al.*, 1995; Leoni *et al.*, 1996
	ptxR	Regulator of siderophore production	*P. aeruginosa* PAO	Colmer and Hamood, 1995
	pfrA	Positive regulator of biosynthetic genes	*P. putida* WCS358	Venturi *et al.*, 1993
	pbrA	Transcriptional activating factor	*P. fluorescens* M114	Sexton *et al.*, 1995
	pfrI	Transcriptional activating factor	*P. putida* WCS358	Venturi *et al.*, 1995b
	sss	Member of the lambda integrase family of site specific recombinases	*P. aeruginosa* 7NSK2	Höfte *et al.*, 1994
	pupI, pupR	Regulatory system activating *pupB* expression	*P. putida* WCS358	Koster *et al.*, 1994
	pfeS, pfeR	Regulatory system activating *pfeA* expression	*P. aeruginosa* PAO	Dean and Poole, 1993b
	groES, groEL	Proteins increasing siderophore gene promoter activity	*P. putida* WCS358	Venturi *et al.*, 1994
Biosynthetic genes	*pbsC*	Enzyme acting as ATP-dependent binding of AMP to its substrate	*Pseudomonas* M114	Adams *et al.*, 1994
Excretion	*pvdA*	L-Ornithine N^5-oxygenase	*P. aeruginosa* PAO	Visca *et al.*, 1994
	pvdD	Putative peptide synthetase	*P. aeruginosa* PAO	Merriman *et al.*, 1995
	pchA, pchB	Salicylate biosynthetic genes	*P. aeruginosa* PAO	Serino *et al.*, 1995
	mexA, mexB, oprK	Excretion of pyoverdine	*P. aeruginosa* PAO	Poole *et al.*, 1993a,b
	pvdE	ABC (ATP-binding-cassette) transporter	*P. aeruginosa* PAO	Lamont, 1994, 1995

from this region which should contain some other unidentified pyoverdine-related genes (Tsuda *et al.*, 1995).

Indeed, in general, extensive similarity should exist at the level of the pyoverdine genes among the pyoverdine producer strains, as suggested by DNA cross-hybridization studies (Rombel and Lamont, 1992). However, when comparing in detail the three (ferri)pyoverdine receptor genes already sequenced from various strains which produce different pyoverdines, namely, *pupA, pupB* and *fpvA* (Bitter *et al.*, 1991; Koster *et al.*, 1993 and Poole *et al.*, 1993c, respectively), marked variability was noted for some parts of the proteins, and some other parts are well conserved (Poole *et al.*, 1993c). Indeed, an important step for reaching a better understanding of the usually strict specificity, which characterizes a (ferri)pyoverdine-mediated iron transport, is to determine the active sites involved in the recognition, binding, and translocation of the (ferri)siderophore molecules. Surprisingly, although the differences in structure which exist between pyochelin and pyoverdines, the (ferri)pyochelin-receptor gene (*fptA*) of *P. aeruginosa* PAO has significant homology with the hydroxamate (pyoverdine)-receptor genes (Ankenbauer, 1992; Ankenbauer and Quan, 1994). The hypothesis of a common set of ancestor genes for all of the receptor and biosynthetic genes developed in the different strains has been proposed (Ankenbauer and Quan, 1994; Rombel and Lamont, 1992). Another common feature which has been shown for all strains so far analyzed is that the (ferri)siderophore receptor gene is located on the bacterial chromosome within biosynthetic genes, suggesting, indeed, a clustered organization (Bitter *et al.*, 1991; Magazin *et al.*, 1986; Marugg *et al.*, 1989; Merriman *et al.*, 1995; Morris *et al.*, 1992; Tsuda *et al.*, 1995).

The reader is referred to the literature cited for detailed information about the other genes already cloned and sequenced and listed in Table III. Most are regulatory genes. It should be emphasized that a negative regulation system based on *fur* (Prince *et al.*, 1991), as for *E. coli* and many other bacteria (Bagg and Neilands, 1987a,b), is strengthened in fluorescent *Pseudomonas* by positive regulation mechanisms (Leoni *et al.*, 1996; O'Sullivan and O'Gara, 1990, 1991). Details of the topics are given in a recent review (Venturi *et al.*, 1995a).

The genetic organization related to siderophores other than pyoverdines has not been intensively investigated so far. Besides the receptor gene for (ferri)pyochelin (*fptA*, Ankenbauer and Quan, 1994) and a regulatory gene (*pchR*, Heinrichs and Poole, 1993, 1996), two genes (*pchA* and *pchB*) involved in the biosynthesis of salicylic acid, a precursor of pyochelin (Ankenbauer and Cox, 1988), have recently been described in *P. aeruginosa* PAO (Serino et al., 1995).

4. SIDEROPHORE-MEDIATED IRON UPTAKE SYSTEMS IN NONFLUORESCENT *PSEUDOMONAS* AND RELATED STRAINS

4.1. *Pseudomonas stutzeri*

P. stutzeri, which is well distributed in soil and water, has remarkable features among pseudomonads, i.e., nutritional versatility (Palleroni *et al.*, 1970, denitrification (Cuypers and Zumft, 1992), nitrogen fixation (Krotzky and Werner, 1987), natural transformation (Carlson *et al.*, 1983; Stewart and Sinigalliano, 1989), and microbial antagonism (Lim *et al.*, 1991). Although not considered a human pathogen, *P. stutzeri* has been identified in clinical material (Holmes, 1986a) and recognized as the causative agent of bacteremia associated with infection during hemodialysis (Goetz *et al.*, 1983).

The iron metabolism of *P. stutzeri* is not well documented so far. The few studies, mainly relevant to the search and description of potent siderophores, involve some collection strains and a few natural isolates. The type strain, *P. stutzeri* ATCC 17588, grown under iron starvation produces large amounts (c.a. 50–100 mg/l) of desferriferrioxamine E (dffE, Meyer and Abdallah, 1980) which is a cyclic trimer of N-succinyl-N-hydroxyl-1,5-diaminopentane (Fig. 2j). Known also under the name of nocardamin since it was first described in *Nocardia* (Stoll *et al.*, 1951), this compound belongs to the well-documented ferrioxamine family (Winkelmann, 1991) formed by at least 11 different cyclic or linear compounds, the most common is desferriferrioxamine B [Fig. 2(i)] which is commercially available in its mesylate form as a powerful iron and metal removing drug (Desferal®, Ciba-Geigy, Switzerland) used against iron overload in thalassemia. Among siderophores, desferriferrioxamines have the originality to be produced by gram (+) bacteria, e.g., *Nocardia* (Stoll *et al.*, 1951) or *Streptomyces* (Bickel *et al.*, 1960) and by gram (−) bacteria, e.g., *P. stutzeri* and also *Erwinia herbicola* (*Enterobacter agglomerans*) (Berner *et al.*, 1988), *Erwinia amylovora* (Kachadourian *et al.*, 1996), and *Hafnia alvei* (Reissbrodt *et al.*, 1990).

Together with dffE, *P. stutzeri* ATCC 17588 also produces desferriferrioxamine D_2 (dffD$_2$; Azelvandre, 1993), an analog of dffE with one of the three N-succinyl-N-hydroxyl-1,5-diamino pentyl residues replaced by a N-succinyl-N-hydroxyl-1,5-diamino butyl residue. Interestingly, the production of these two compounds can be increased (two- to three-fold) and their ratio modulated by supplementing the basal growth medium with lysine or cadaverine, which favor dffE production, or with ornithine or putrescine which favor dffD$_2$ production. Both compounds have also been detected in the other strains so far analyzed in our laboratory, namely, *P. stutzeri* ATCC 11607, *P. stutzeri* ATCC 14405 (strain ZoBell, Döhler *et al.*, 1987), and one clinical isolate. More strains should be investigated, however, to conclude that desferriferrioxamines are characteristic siderophores for *P. stutzeri* strains. The production of a catecholate siderophore (dihydroxybenzoyl-arginine) by a *P. stutzeri* natural isolate has been described (Chakraborty *et al.*, 1990), whereas another strain was described as a nonsiderophoric producer (Lim *et al.*, 1991).

P. stutzeri ATCC 17588 iron uptake is well mediated by the desferriferrioxamines E, D_2, and B and also by citrate and other siderophores of foreign origin, e.g., cepabactin of *B. cepacia* and desferriferricrocin, desferriferrirubin, and desferriferrichrysin of fungal origin. Pyochelin and pyoverdines from various fluorescent pseudomonads, however, were inefficient (Azelvandre, 1993). In agreement with these multiple potentialities of iron uptake, several iron-regulated outer membrane proteins in the 80–90 kDa range have been characterized in iron-starved cells of *P. stutzeri* (Azelvandre, 1993; Aznar and Alcaide, 1992).

4.2. *Burkholderia* (Formerly *Pseudomonas*) *cepacia* and Related Strains

B. cepacia was considered a plant pathogen since it was first isolated as the causal agent of onion soft rot (Burkholder, 1950). Widely distributed also in soil and water (Palleroni, 1984), interest in this bacterium strongly increased in the last few years as it became evident that *P. cepacia* was regularly encountered in human clinical isolates (Holmes, 1986b) and finally should be considered a potent human pathogen involved in septicaemia (Lacy *et al.*, 1993) and cystic fibrosis (Gessner and Mortensen, 1990; Govan *et al.*, 1993, Taylor and Kalamatianos, 1994). On the other hand, natural isolates from various plant rhizospheres and identified as *P. cepacia* have been recognized as powerful biocontrol agents because they antagonize soil-borne plant pathogens (Hebbar *et al.*, 1992; Parke *et al.*, 1991; Tabacchioni *et al.*, 1993). Thus, as well observed by Tabacchioni *et al.* (1995), it is of primary interest to assess the risk associated with the potential use of such strains as field inoculants by defining more precisely the degree of genetic diversity among clinical and rhizosphere *P. cepacia* strains.

The new so-called *Burkholderia cepacia* species, recently transferred as the type species of the new genus *Burkholderia*, which includes new species (Gillis *et al.*, 1995; Urakami *et al.*, 1994) and mainly formerly

Pseudomonas species (Urakami *et al.*, 1994; Yabuuchi *et al.*, 1992), has already been investigated for siderophore production by several groups. The data strongly suggest great diversity among strains. Of the 43 clinical isolates analyzed by Sokol (1986), 49% were recognized as pyochelin producers. In another study (Sokol *et al.*, 1992), 88% of the strains also produced azurechelin, a compound identified later as salicylic acid (Visca *et al.*, 1993). A few strains produced only salicylic acid. Some others produced none of these compounds. Investigating the type strain *B. cepacia* ATCC 25416, we confirmed the production of pyochelin and found another type of siderophore, cepabactin (Meyer *et al.*, 1989), already described as an iron and aluminum chelator (Winkler *et al.*, 1986) which displays antibiotic properties (Itoh *et al.*, 1979) and is produced by some *B. cepacia* natural isolates and by a *P. alcaligenes* strain (Barker *et al.*, 1979). By its structure this compound, which is 1-hydroxy-5-methoxy-6-methylpyrid-2-one [Fig. 2(c)], is closely related to the synthetic hydroxypyridinones tested as orally active drugs for iron removal (Kontoghiorghes, 1987). Moreover, a fourth category of siderophores, the ornibactins, was found later on in the *B. cepacia* ATCC 25416 strain (Meyer *et al.*, 1995). Thus, it turns out that under iron deficiency *B. cepacia* ATCC 25416 effectively produces at least three siderophores. The most important quantitatively are the ornibactins (109 mg per liter of medium), followed by cepabactin (46 mg/l) and pyochelin (11 mg/l) (Meyer *et al.*, 1995).

Ornibactins are composed of the conserved tetrapeptide chain ∂ N-OH-Orn-βOH-Asp-Ser-∂N-OH-Orn which harbors a putrescine residue and an acyl chain of 3-hydroxybutanoic acid, 3-hydroxyhexanoic acid, or 3-hydroxyoctanoic acid [Stephan *et al.*, 1993a,b; Fig. 2(d)]. The microheterogeneity in the acyl chain results in three different molecular species, the C_4-, C_6- and C_8-ornibactins, which together are excreted by iron-starved cells of *B. cepacia* ATCC 25416 in the ratio of 5, 29, and 66%, respectively. The different forms of ornibactins, however, show identical iron uptake capacity (Meyer *et al.*, 1995). As for pyoverdines (Georges and Meyer, 1995), it is postulated that the high molecular mass, iron-repressed, cytoplasmic proteins detected in *B. cepacia* ATCC 25416 and in other ornibactins producers, i.e., *B. cepacia* ATCC 17759 and *B. vietnamiensis* strains (Gillis *et al.*, 1995), should act as peptide synthetases involved in synthesizing the peptide part of the molecule (Georges and Meyer, unpublished results). Unlike pyoverdines, however, the same peptide has been found for all of the ornibactins of different bacterial origin analyzed so far (Meyer *et al.*, 1995).

Combining several procedures which allow detecting siderophores in the spent medium without extraction, we recently analyzed the siderophore production of clinical (91 strains) or plant-related (17 strains) isolates. A majority of the strains (44.5%) produced ornibactins as a unique siderophore, 30% produced ornibactins and pyochelin, 13% produced ornibactins, pyochelin, and cepabactin, whereas 7% produced only pyochelin. One strain produced ornibactins and cepabactin, and five apparently produced none of these compounds. Salicylic acid was not detected. Thus, it appears that ornibactin is the most common siderophore found in these strains because 88% of the strains were ornibactin producers, 50% pyochelin producers, and 14% cepabactin producers. No special features were distinguished at the siderophore level in clinical strains from plant-related bacteria, contrary to previous assumptions (Bevivino *et al.*, 1994). Indeed, IROMPs are present in *B. cepacia* (Anwar *et al.*, 1983). Specific receptors for several potent siderophores remain, however, to be determined.

4.3. Other Strains

Numerous nonfluorescent *Pseudomonas* or formerly *Pseudomonas* (*Burkholderia*) species remain to be investigated for their iron requirements and iron transport capacities. The few published investigations which focus on such strains also highlight the tremendous diversity found in *Pseudomonas* siderophore structures. Aerobactin [Fig. 1(f)] and schizokinen (Fig. 1(g)] are well known as siderophores produced by enterobacteria for aerobactin or *Bacillus megaterium* and the cyanobacteria *Anabaena* for schizokinen (see Winkelmann, 1991). Aerobactin has been identified as the siderophore produced by a *Pseudomonas* isolate of marine origin (Buyer *et al.*, 1991), whereas schizokinen has recently been detected in low-iron cultures of *Burkholderia solanacearum* (Münziger, 1995). For the latter, however, and as for tropolone (Fig. 1(e)] isolated from iron-deficient *Burkholderia plantarii* (Azegami *et al.*, 1988), iron uptake studies are required before they can be considered true siderophores.

Finally, some other nonfluorescent *Pseudomonas* species, e.g., *P. fragi*, *P. mendocina*, and *P. diminuta*, revealed an apparent lack of siderophore production (Meyer *et al.*, 1987). As demonstrated at least for *P. fragi* strains whose natural environment is meat and milk products, these bacteria sustain their iron requirement in opportunistic ways by using, when available, siderophores of foreign origin, i.e. some pyoverdines, desferriferrioxamines, or enterobactin for some strains, and also by efficiently removing iron from transferrin, lactoferrin, and haemoglobin by mechanisms so far unknown (Champomier-Vergès *et al.*, 1996).

5. IRON, SIDEROPHORES, AND BIOTECHNOLOGY

5.1. Iron, Siderophores, and Plant-Related Biocontrol

The publications from Schroth's laboratory (Kloepper *et al.*, 1980a,b) should be considered milestone reports in the *Pseudomonas* siderophore story because they first demonstrated the potential role of iron and siderophores in plant-growth promotion and plant protection by *Pseudomonas* and also and mainly because they generated a large number of studies devoted to *Pseudomonas* siderophores, which resulted in many developments in plant-related biotechnology and in fundamental research. According to these authors, the beneficial effect of fluorescent *Pseudomonas* developing at the rhizosphere level of plants can be explained as the result of competition for iron among the rhizosphere microbial population, including plant-deleterious and plant-pathogenic microorganisms. This competition usually favors the *Pseudomonas* strains because of the high affinity of their siderophores for iron, mainly pyoverdines.

During the last 15 years, many reports have demonstrated, however, that the iron/siderophore mechanisms is not the only one that explains the beneficial effects of *Pseudomonas*. Among other benefits are the production of antibiotics, nutrients and other secondary metabolites, such as cyanides. Root colonization or induced plant resistance are also presently well established. The reader is referred to specialized reviews on these topics (Cook *et al.*, 1995; Défago and Haas, 1990; de Weger *et al.*, 1995; Glick, 1995; Leong, 1986; Misaghi *et al.*, 1988; O'Sullivan and O'Gara, 1992). Developments in siderophore-related plant biocontrol utilizing *Pseudomonas* are still in progress with attempts to control and improve bacterial effectiveness by genetic engineering (Raaijmakers *et al.*, 1994, 1995). It can also be noted that there is an increasing interest in strains belonging to species other than the first and well-studied *P. fluorescens* or *P. putida* species, for example, *Burkholderia* (formerly *Pseudomonas*) *cepacia* and related strains (Hebbar *et al.*, 1992; Jayaswal *et al.*, 1993; Tabbachioni *et al.*, 1993).

5.2. Iron, Siderophores, and Human Pathogenicity

The ability to fulfill its nutritional requirements in situ is a prerequisite for a bacterium to multiply and to develop its virulence factors in an animal or a human. For instance, to meet the need for iron, bacteria must compete with the host for an element which is tightly bound to host proteins, e.g., ferritin, transferrin, lactoferrin, and hemoglobin.

Competition for iron is an extreme challenge because bacteria developing in situ apparently remain iron-starved (Brown *et al.*, 1984; Haas *et al.*, 1991; Shand *et al.*, 1985) and have accordingly developed iron-regulated virulent factors (Bjorn *et al.*, 1979). Among *Pseudomonas*, only a few species are considered severe pathogens, i.e., *Pseudomonas mallei* and *Pseudomonas pseudomallei* (these two species have recently been transferred to the new *Burkholderia* genus; Yabuuchi *et al.*, 1992). Information available in the literature about iron metabolism in these species is limited. An iron-chelating compound, malleobactin, has been described as the siderophore of *B. pseudomallei*. It belongs to the hydroxamate family of siderophores, but its structure remains presently unknown (Yang *et al.*, 1991). *P. aeruginosa*, although considered an opportunistic human pathogen, has a much greater medical impact compared with *B. mallei* and *B. pseudomallei* because it is involved in a major way in severe and often fatal infections in patients with cystic fibrosis (Döring, 1993), burns (Holder, 1993), ocular diseases (Berk, 1993; Fleiszig *et al.*, 1994), pneumonia (Pennington *et al.*, 1973), and various immunosuppresive diseases (Bodey *et al.*, 1983). Although there is some evidence that extracellular *P. aeruginosa* proteases hydrolyze transferrin and lactoferrin in vitro (Döring *et al.*, 1988) and in vivo (Britigan *et al.*, 1993) and, thus, that proteases may allow transfer of iron from iron proteins to the cells via siderophore systems (Döring *et al.*, 1988; Wolz *et al.*, 1994), it is now well established that *P. aeruginosa* siderophores, especially pyoverdine, can compete directly for iron, in vitro, with transferrin (Ankenbauer *et al.*, 1985; Meyer *et al.*, 1996; Sriyosachati and Cox, 1986). That the same reaction occurs in vivo is strongly supported by the following observations: (1) unlike the pyoverdine producer strains, mutants affected in pyoverdine biosynthesis are avirulent in the burned mouse model of Stieritz and Holder (1975) and (2) injection of pure pyoverdine into the infected burn site significantly restores the virulence of these mutants (Meyer *et al.*, 1996). Moreover, pyoverdine has been detected in sputa of cystic fibrosis patients (Haas *et al.*, 1991). Thus, novel ways of treatment against *P. aeruginosa* infections through iron metabolism by blocking pyoverdine biosynthesis or transport could, therefore, be postulated.

Pyochelin, the second siderophore produced by *P. aeruginosa*, also has some effect on bacterial virulence (Cox, 1982). Interestingly, some IROMPs, among them the (ferri)pyochelin receptor, facilitate the penetration of catechol-substituted cephalosporins (Gensberg *et al.*, 1994; Yamano *et al.*, 1994). This could be another means of defense against *P. aeruginosa*.

5.3. Siderotyping and Searching for New Siderophores: Back to the Bench

For many reasons it is of interest to determine the structure of the siderophore produced by a new isolate. For instance, because of the high specificity of pyoverdine-mediated iron transport, rapid recognition of the type of pyoverdine produced by a *P. aeruginosa* clinical isolate would be required if a pyoverdine-related treatment becomes available. Such "siderotyping" is presently possible by combining rapid methods of investigation like cross-feeding (Cornelis *et al.,* 1989), isoelectrofocusing (Koedam *et al.,* 1994), or iron uptake measurement (Hohnadel and Meyer, 1988). Such recent investigations, which were done on 88 *P. aeruginosa* strains, led to the conclusions that among the pyoverdine-producing strains (94%), 42% produce a pyoverdine which is identical to the one produced by *P. aeruginosa* ATCC 15692 (Group I), 42% have the same pyoverdine as *P. aeruginosa* ATCC 27853 (group II), and 16% have the same pyoverdine as strain *P. aeruginosa* Pa6 (group III). Another interesting conclusion is that no new pyoverdine species was detected among these strains (Meyer *et al.,* 1993). A new pyoverdine species, however, was recognized in another study where a set of five natural isolates belonging to *P. fluorescens* and *P. putida* species was analyzed. It was concluded from the siderotyping results that one set of two isolates is similar to one of the collection strains used as controls, whereas another set of two isolates and the fifth isolate each had original features corresponding to none of the controls. Structural determination of the pyoverdines produced by the five isolates fully confirmed the siderotyping results. Two pyoverdines were identical to the pyoverdine of *P. fluorescens* ATCC 13525, two others were assigned to the pyoverdine of *Pseudomonas* strain 244 (Hankock and Reeder, 1993), and the fifth presented an amino acid composition never described before (Voss and Budzikiewicz, personal communication). Thus, siderotyping is a powerful method of pyoverdine discrimination because results are obtained within one to two days instead of several months for obtaining a complete structure.

ACKNOWLEDGMENTS. The authors thank the following organizations for support: Université Louis-Pasteur, Strasbourg, France, Centre National de la Recherche Scientifique, Paris, France, Association Française de Lutte contre la Mucoviscidose, Paris, France, NATO Scientific Affairs Division, Brussels, Belgium.

REFERENCES

Adams, C., Dowling, D. N., O'Sullivan, D. J., and O'Gara, F., 1994, Isolation of a gene (*pbsC*) required for siderophore biosynthesis in fluorescent *Pseudomonas* sp. strain M114, *Mol. Gen. Genet.* **243:** 515–524.

Andriollo, N., Guarini, A., and Cassini, G., 1992, Isolation and characterization of pseudobactin B: A pseudobactin-type siderophore from *Pseudomonas* species strain PD30, *J. Agric. Food Chem.* **40:** 1245–1248.

Ankenbauer, R. G., 1992, Cloning of the outer membrane high-affinity Fe(III)-pyochelin receptor of *Pseudomonas aeruginosa*, *J. Bacteriol.* **174:** 4401–4409.

Ankenbauer, R. G., and Cox, C. D., 1988, Isolation and characterization of *Pseudomonas aeruginosa* mutants requiring salicylic acid for pyochelin biosynthesis, *J. Bacteriol.* **170:** 5364–5367.

Ankenbauer, R. G., and Quan, H. N., 1994, FptA, the Fe(III)-pyochelin receptor of *Pseudomonas aeruginosa*, a phenolate siderophore receptor homologous to hydroxamate siderophore receptors, *J. Bacteriol.* **176:** 307–319.

Ankenbauer, R., Sriyosachati, S., and Cox, C. D., 1985, Effects of siderophores on the growth of *Pseudomonas aeruginosa* in human serum and transferrin, *Infect. Immun.* **49:** 132–140.

Ankenbauer, R., Hanne, L. F., and Cox, C. D., 1986, Mapping of mutations in *Pseudomonas aeruginosa* defective in pyoverdin production, *J. Bacteriol.* **167:** 7–11.

Ankenbauer, R. G., Toyokuni, T., Staley, A., Rinehart, K. L., and Cox, C. D., 1988, Synthesis and biological activity of pyochelin, a siderophore of *Pseudomonas aeruginosa*, *J. Bacteriol.* **170:** 5344–5351.

Anwar, H., Brown, M. R., Cozens, R. M., and Lambert, P. A., 1983, Isolation of the outer and cytoplasmic membranes of *Pseudomonas cepacia*, *J. Gen. Microbiol.* **129:** 499–507.

Archibald, F., 1983, *Lactobacillus plantarum*, an organism not requiring iron, *FEMS Microbiol. Lett.* **19:** 29–32.

Arnow, L. E., 1937, Colorimetric determination of the components of 3,4-dihydroxyphenylalanine-tyrosine mixtures, *J. Biol. Chem.* **118:** 531–537.

Azegami, K., Nishiyama, K., and Kato, H., 1988, Effects of iron on "*Pseudomonas plantarii*" growth and tropolone and protein production, *Appl. Environ. Microbiol.* **54:** 844–847.

Azelvandre, P., 1993, Les deferriferrioxamines E et D2, sidérophores de *Pseudomonas stutzeri*, Thèse d'Université, Strasbourg, France.

Aznar, R., and Alcaide, E., 1992, Siderophores and related outer membrane proteins produced by pseudomonads isolated from eels and freshwater, *FEMS Microbiol. Lett.* **98:** 269–276.

Bagg, A., and Neilands, J. B., 1987a, Ferric uptake regulation protein acts as a repressor, employing iron(II) as a cofactor to bind the operator of an iron transport operon in *Escherichia coli*, *Biochemistry* **26:** 5471–5477.

Bagg, A., and Neilands, J. B., 1987b, Molecular mechanism of regulation of siderophore-mediated iron assimilation, *Microbiol. Rev.* **51:** 509–518.

Barbhaiya, H. B., and Rao, K. K., 1985, Production of pyoverdine, the fluorescent pigment of *Pseudomonas aeruginosa* PAO1, *FEMS Microbiol. Lett.* **27:** 233–235.

Barker, W. R., Callaghan, C., Hill, L., Nobel, D., Acred, P., Harper, P. B., Sowa, M. A., and Fletton, R. A., 1979, G1549, a new cyclic hydroxamic acid antibiotic, isolated from culture broth of *Pseudomonas alcaligenes*, *J. Antibiotics* **32:** 1096–1103.

Berk, R. S., 1993, Genetic regulation of the murine corneal and non-corneal response to

Pseudomonas aeruginosa, in: *Pseudomonas aeruginosa as an Opportunistic Pathogen* (M. Campa, M. Bendinelli and H. Friedman, eds.), Plenum Press, New York, pp. 183–206.

Berner, I., Konetschny–Rapp, S., Jung, G., and Winkelmann, G., 1988, Characterization of ferrioxamine E as the principal siderophore of *Erwinia herbicola* (*Enterobacter agglomerans*), *Biol. Metals* **1:** 51–56.

Bevivino, A., Tabacchioni, S., Chiarini, L., Carusi, M. V., Del Gallo, M., and Visca, P., 1994, Phenotypic comparison between rhizosphere and clinical isolates of *Burkholderia cepacia, Microbiology* **140:** 1069–1077.

Bickel, H., Booshardt, R., Gäumann, E., Reusser, P., Vischer, E., Voser, W., Wettstein, A., and Zähner, H., 1960, Stoffwechselprodukte von Actinomyceten. 26. Mitteilung. über die Isolierung und Charakterisierung der Ferrioxamine A-F, neuer Wuchstoffe der Sideramin-Gruppe, *Helv. Chim. Acta* **53:** 2118–2128.

Bitter, W., Marrug, J. D., de Weger, L. A., Tommassen, J., and Weisbeek, P. J., 1991, The ferric-pseudobactin receptor PupA of *Pseudomonas putida* WCS358: Homology to TonB-dependent *Escherichia coli* receptors and specificity of the protein, *Mol. Microbiol.* **5:** 647–655.

Bitter, W., Tommassen, J., and Weisbeek, P. J., 1993, Identification and characterization of *exbB, exbD, and tonB genes of Pseudomonas putida* WCS358: Their involvement in ferric-pseudobactin transport, *Mol. Microbiol.* **7:**117–130.

Bitter, W., van Leeuwen, I. S., de Boer, J., Zomer, H. W. M., Koster, M. C., Weisbeek, P. J., and Tommassen, J., 1994, Localization of functional domains in the *Escherichia coli* coprogen receptor FhuE and the *Pseudomonas putida* ' ferric-pseudobactin 358 receptor PupA, *Mol. Gen. Genet.* **245:** 694–703.

Bjorn, M. J., Sokol, P. A., and Iglewski, B. H., 1979, Influence of iron on yields of extracellular products in *Pseudomonas aeruginosa* cultures, *J. Bacteriol.* **138:** 193–200.

Bodey, G. P., Bolivar, R., Fainstein, V., and Jadeja, L., 1983, Infections caused by *Pseudomonas aeruginosa, Rev. Infect. Dis.* **5:** 279–313.

Braun, V., and Hantke, K., 1991, Genetics of bacterial iron transport, in: *Handbook of Microbial Chelates* (G. Winkelmann, ed.), CRC Press, Boca Raton, Florida, pp. 107–138.

Braun, V., Hantke, K., Eick-Helmerich, K., Köster, W., Pressler, U., Sauer, M., Schäffer, S., Schöffler, H., Staudenmaier, H., and Zimmermann, L., 1987, Iron transport systems in *Escherichia coli,* in: *Iron Transport in Microbes, Plants, and Animals* (G. Winkelmann, D. van der Helm and J. B. Neilands, eds.), VCH Verlagsgesellschaft, Weinheim, pp. 35–51.

Briskot, G., Taraz, K., and Budzikiewicz, H., 1989, Pyoverdin-type siderophores from *Pseudomonas aeruginosa. Liebigs Ann. Chem.* **1989:** 375–384.

Britigan, B. E., Hayek, M. B., Doebbeling, B. N., and Fick, R. B., Jr., 1993, Transferrin and lactoferrin undergo proteolytic cleavage in the *Pseudomonas aeruginosa*-infected lungs of patients with cystic fibrosis, *Infect. Immun.* **61:** 5049–5055.

Brown, M. R. W., Anwar, H., and Lambert, P. A., 1984, Evidence that mucoid *Pseudomonas aeruginosa* in the cystic fibrosis lung grows under iron-restricted conditions, *FEMS Microbiol. Lett.* **21:** 113–117.

Budzikiewicz, H., 1993, Secondary metabolites from fluorescent pseudomonads, *FEMS Microbiol. Rev.* **104:** 209–228.

Budzikiewicz, H., 1994, The biosynthesis of pyoverdins, *Pure & Appl. Chem.* **66:** 2207–2210.

Budzikiewicz, H., Schröder, H., and Taraz, K., 1992, Zur Biogenese der *Pseudomonas*-Siderophore: Der Nachweis analoger Strukturen eines Pyoverdin-Desferriferribactin Paares, *Z. Naturforsch.* **47c:** 26–32.

Bukowits, G. J., Mohr, N., and Budzikiewicz, H., 1982, 2-phenylthiazol-derivatives from *Pseudomonas cepacia*, *Z. Naturforsch.* **37b:**877–880.

Bulen, W. A., and LeComte, J. R., 1962, Isolation and properties of a yellow-green fluorescent peptide from *Azotobacter* medium, *Biochem. Biophys. Res. Commun.* **9:** 523–528.

Burkholder, W. H., 1950, Sour skin, a bacterial rot of onion bulbs, *Phytopathol.* **40:** 115–117.

Buyer, J. S., and Leong, J., 1986, Iron transport-mediated antagonism between plant growth-promoting and plant-deleterious *Pseudomonas* strains, *J. Biol. Chem.* **261:** 791–794.

Buyer, J. S., Wright, J. S., and Leong, J., 1986, Structure of pseudobactin A214, a siderophore from a bean-deleterious *Pseudomonas*, *Biochemistry* **25:** 5492–5499.

Buyer, J. S., de Lorenzo, V., and Neilands, J. B., 1991, Production of the siderophore aerobactin by a halophilic pseudomonad, *Appl. Environ. Microbiol.* **57:** 2246–2250.

Carlson, C. A., Pierson, L. S., Rosen, J. J., and Ingraham, J. L., 1983, *Pseudomonas stutzeri* and related species undergo natural transformation, *J. Bacteriol.* **153:** 93–99.

Chakraborty, R. N., Patel, H. N., and Desai, S. B., 1990, Isolation and partial characterization of catechol-type siderophore from *Pseudomonas stutzeri* RC 7, *Curr. Microbiol.* **20:** 283–286.

Champomier-Vergès, M. C., Stintzi, A., and Meyer, J. M., 1996, Acquisition of iron by the non-siderophore producing *Pseudomonas fragi*, *Microbiology* **142:** 1191–1199.

Cody, Y. S., and Gross, D. C., 1987, Characterization of pyoverdin Pss, the fluorescent siderophore produced by *Pseudomonas syringae* pv. *syringae*. *Appl. Environ. Microbiol.* **53:** 928–934.

Colmer, J. A., and Hamood, A. N., 1995, Isolation of a *Pseudomonas aeruginosa* chromosomal fragment which affects the regulation of siderophore production, *Pseudomonas News Letter* **20**(3): 8.

Cornelis, P., Moguilevsky, N., Jacques, J. F., and Masson, P. L., 1987, Study of the siderophores and receptors in different clinical isolates of *Pseudomonas aeruginosa*, *Antibiot. Chemother.* **39:** 290–306.

Cornelis, P., Hohnadel, D., and Meyer, J. M., 1989, Evidence for different pyoverdine-mediated iron uptake systems among *Pseudomonas aeruginosa* strains, *Infect. Immun.* **57:** 3491–3497.

Cook, R. J., Thomashow, L. S., Weller, D. M., Fujimoto, D., Mazzola, M., Bangera, G., and Kim, D., 1995, Molecular mechanisms of defense by rhizobacteria against root disease, *Proc. Natl. Acad. Sci. USA* **92:** 4197–4201.

Coves, J., and Fontecave, M., 1993, Reduction and mobilization of iron by a NAD(P)H:flavin oxidoreductase from *Escherichia coli*, *Eur. J. Biochem.* **211:** 635–641.

Cox, C. D., 1980, Iron uptake with ferripyochelin and ferric citrate by *Pseudomonas aeruginosa*, *J. Bacteriol.* **142:** 581–587.

Cox, C. D., 1982, Effect of pyochelin on the virulence of *Pseudomonas aeruginosa*, *Infect. Immun.* **36:** 17–23.

Cox, C. D., and Adams, P., 1985, Siderophore activity of pyoverdin for *Pseudomonas aeruginosa*, *Infect. Immun.* **48:** 130–138.

Cox, C. D., and Graham, R., 1979, Isolation of an iron-binding compound from *Pseudomonas aeruginosa*, *J. Bacteriol.* **137:** 357–364.

Cox, C. D., Rinehart, K. L., Moore, M. L., and Cook, J. C., 1981, Pyochelin: Novel structure of an iron-chelating growth promoter for *Pseudomonas aeruginosa*, *Proc. Natl. Acad. Sci. USA* **78:** 4256–4260.

Cunliffe, H. E., Merriman, T. R., and Lamont, I., 1995, Cloning and characterization of *pvdS*, a gene required for pyoverdine synthesis in *Pseudomonas aeruginosa*: PvdS is probably an alternative sigma factor, *J. Bacteriol.* **177:** 2744–2750.

Cuypers, H., and Zumft, W. G., 1992, Regulatory components of the denitrification gene cluster of *Pseudomonas stutzeri,* in: *Pseudomonas: Molecular Biology and Biotechnology* (E. Galli, S. Silver, and B. Witholt, eds.), American Society for Microbiology, Washington, D.C., pp. 188–197.

Csaky, T. Z., 1948, On the estimation of bound hydroxylamine in biological materials, *Acta Chem. Scand.* **2:** 450–454.

Dean, C. R., and Poole, K., 1993a, Cloning and characterization of the ferric enterobactin receptor gene (*pfeA*) of *Pseudomonas aeruginosa, J. Bacteriol.* **175:** 317–324.

Dean, C. R., and Poole, K., 1993b, Expression of the ferric enterobactin receptor (PfeA) of *Pseudomonas aeruginosa:* Involvement of a two-component regulatory system, *Mol. Microbiol.* **8:** 1095–1103.

Défago, G., and Haas, D., 1990, Pseudomonas as antagonists of soilborne plant pathogens: Modes of action and genetic analysis, in: *Soil Biochemistry* (J–M. Bollag and G. Stotzky, eds.), Marcel Dekker, New York and Basel, Vol. 6, pp. 249–291.

Demange, P., Bateman, A., Mertz, C., Dell, A., Piemont, Y., and Abdallah, M. A., 1990, Bacterial siderophores: Structures of pyoverdins Pt, siderophores of *Pseudomonas tolaasii* NCPPB 2192, and pyoverdins Pf, siderophores of *Pseudomonas fluorescens* CCM 2798. Identification of an unusual amino acid, *Biochemistry* **29:** 11041–11051.

de Weger, L. A., van Boxtel, R., van der Burg, B., Gruters, R. A., Geels, F. P., Schippers, B., and Lugtenberg, B., 1986, Siderophores and outer membrane proteins of anatagonistic, plant-growth-stimulating, root-colonizing *Pseudomonas* spp., *J. Bacteriol.* **165:** 585–594.

de Weger, L. A., von Arendonk, J. C. H. M., Recourt, K., van der Hofstad, G. A. J. M., Weisbeek, P. J., and Lugtenberg, B., 1988, Siderophore-mediated iron uptake in the plant-stimulating *Pseudomonas putida* WCS358 and other rhizosphere microorganisms, *J. Bacteriol.* **170:** 4693–4698.

de Weger, L. A., van der Bij, A. J., Dekkers, L. C., Simons, M., Wijffelman, A., and Lugtenberg, B. J. J., 1995, Colonization of the rhizosphere of crop plants by plant-beneficial pseudomonads, *FEMS Microbiol. Ecol.* **17:** 221–228.

Döhler, K., Huss, V. A. R., and Zumft, W. G., 1987, Transfer of *Pseudomonas perfectomarina* (Baumann, Bowditch, and Beaman 1983) to *Pseudomonas stutzeri* (Lehmann and Neumann 1896; Sijderius 1946), *Int. J. Syst. Bacteriol.* **37:** 1–3.

Döring, G., 1993, *Pseudomonas aeruginosa* lung infection in cystic fibrosis patients, in: *Pseudomonas aeruginosa as an Opportunistic Pathogen* (M. Campa, M. Bendinelli, and H. Friedman, eds.), Plenum Press, New York, pp. 245–273.

Döring, G., Pfestorf, M., Botzenhart, K., and Abdallah, M. A., 1988, Impact of proteases on iron uptake of *Pseudomonas aeruginosa* pyoverdin from transferrin and lactoferrin, *Infect. Immun.* **56:** 291–293.

Elliot, R. P., 1958, Some properties of pyoverdine, the water-soluble pigment of the *Pseudomonas, Appl. Microbiol.* **6:** 241–246.

Fleiszig, S. M. J., Zaidi, T. S., Fletcher, E. L., Preston, M. J., and Pier, G. B., 1994, *Pseudomonas aeruginosa* invades corneal epithelial cells during experimental infection, *Infect. Immun.* **62:** 3485–3493.

Fukasawa, K., and Goto, M., 1973, Biosynthesis of a heterocycle formed by iron-deficient *Azotobacter vinelandii* strain O, *Biochim. Biophys. Acta* **320:** 545–548.

Garibaldi, J. A., 1971, Influence of temperature on the iron metabolism of a fluorescent pseudomonad, *J. Bacteriol.* **105:** 1036–1038.

Garibaldi, J. A., 1972, Influence of temperature on the biosynthesis of iron transport compounds by *Salmonella typhimurium, J. Bacteriol.* **110:** 262–265.

Geisen, K., Taraz, K., and Budzikiewicz, H., 1992, Neue Siderophore des Pyoverdin-typs aus *Pseudomonas fluorescens, Monatsh. Chem.* **123:** 151–178.

Gensberg, K., Hughes, K., and Smith, A. W., 1992, Siderophore-specific induction of iron uptake in *Pseudomonas aeruginosa, J. Gen. Microbiol.* **138:** 2381–2387.

Gensberg, K., Doyle, E. J., Perry, D. J., and Smith, A. W., 1994, Uptake of BRL 41897A, a C(7) α-formamido substituted cephalosporin, via the ferri-pyochelin transport system of *Pseudomonas aeruginosa, J. Antimicrob. Chemother. 34:* 697–705.

Georges, C., and Meyer, J. M., 1995, High-molecular-mass, iron-repressed cytoplasmic proteins in fluorescent *Pseudomonas:* Potential peptide-synthetases for pyoverdine biosynthesis, *FEMS Microbiol. Lett.* **132:** 9–17.

Gessner, A. R., and Mortensen, J. E., 1990, Pathogenic factors of *Pseudomonas cepacia* isolates from patients with cystic fibrosis, *J. Med. Microbiol.* **33:** 115–120.

Gillam, A. H., Lewis, A. G., and Andersen, R. J., 1981, Quantitative determination of hydroxamic acids, *Anal. Chem.* **53:** 841–844.

Gillis, M., Trân Van, V., Fernandez, M. P., Goor, M., Hebbar, P., Willems, A., Segers, P., Kersters, K., Heulin, T., and Bardin, R., 1995, Polyphasic taxonomy in *Burkholderia* leading to an amended description of the genus and proposition of *Burkholderia vietnamiensis* sp. nov. for N$_2$-fixing isolates from rice in Vietnam. *Int. J. Syst. Bacteriol.* **45:** 274–289.

Gipp, S., Hahn, J., Taraz, K., and Budzikiewicz, H., 1991, Zwei Pyoverdine aus *Pseudomonas aeruginosa* R, *Z. Naturforsch.* **46c:** 534–541.

Glick, B. R., 1995, The enhancement of plant growth by free-living bacteria, *Can. J. Microbiol.* **41:** 109–117.

Goetz, A., Yu, V. L., Hanchett, J. E., and Rihs, J. D., 1983, *Pseudomonas stutzeri* bacteremia associated with hemodialysis, *Arch. Intern. Med.* **143:** 1909–1912.

Govan, J. R., Brown, P. H., Maddison, J., Doherty, C. J., Nelson, J. W., Dodd, M., Greening, A. P., and Webb, A. K., 1993, Evidence for transmission of *Pseudomonas cepacia* by social contact in cystic fibrosis, *Lancet* **342:** 15–19.

Gwose, I., and Taraz, K., 1992, Pyoverdine aus *Pseudomonas putida, Z. Naturforsch.* **47c:** 487–502.

Haas, B., Kraut, J., Marks, J., Zanker, S. C., and Castignetti, D., 1991, Siderophore presence in sputa of cystic fibrosis patients, *Infect. Immun.* **59:** 3997–4000.

Hallé, F., and Meyer, J. M., 1992a, Ferrisiderophore reductases of *Pseudomonas.* Purification, properties and location of the *Pseudomonas aeruginosa* ferripyoverdine reductase, *Eur. J. Biochem.* **209:** 613–620.

Hallé, F., and Meyer, J. M., 1992b, Iron release from ferrisiderophores. A multi-step mechanism involving a NADH/FMH oxidoreductase and a chemical reduction by FMNH$_2$, *Eur. J. Biochem.* **209:** 621–627.

Hancock, D. K., and Reeder, D. J., 1993, Analysis and configuration assignments of the amino acids in a pyoverdine-type siderophore by reversed-phase high-performance liquid chromatography, *J. Chromatogr.* **646:** 335–343.

Harding, R. A., and Royt, P., 1990, Acquisition of iron from citrate by *Pseudomonas aeruginosa, J. Gen. Microbiol.* **136:** 1859–1867.

Hebbar, K. P., Davey, A. G., Merrin, J., and Dart, P. J., 1992, *Pseudomonas cepacia,* a potential suppressor of maize soil-borne diseases - seed inoculation and maize root colonization, *Soil. Biol. Biochem.* **24:** 999–1007.

Heinrichs, D. E., and Poole, K., 1993, Cloning and sequence analysis of a gene (*pchR*) encoding an Ara C family activator of pyochelin and ferripyochelin receptor synthesis in *Pseudomonas aeruginosa, J. Bacteriol.* **175:** 5882–5889.

Heinrichs, D. E., and Poole, K., 1996, PchR, a regulator of ferripyochelin receptor gene (*fptA*) expression in *Pseudomonas aeruginosa,* functions both as an activator and as a repressor, *J. Bacteriol.* **178:** 2586–2592.

Heinrichs, D. E., Young, L., and Poole, K., 1991, Pyochelin-mediated iron transport in *Pseudomonas aeruginosa:* Involvement of a high-molecular-mass outer membrane protein, *Infect. Immun.* **59:** 3680–3684.

Höfte, M., Buysens, S., Koedam, N., and Cornelis, P., 1993, Zinc affects siderophore-mediated high affinity iron uptake systems in the rhizosphere *Pseudomonas aeruginosa* 7NSK2, *BioMetals* **6:** 85–91.

Höfte, M., Dong, Q., Kourambas, S., Krishnapillai, V., Sherratt, D., and Mergeay, M., 1994, The *sss* gene product, which affects pyoverdin production in *Pseudomonas aeruginosa* 7NSK2 is a site-specific recombinase, *Mol. Microbiol.* **14:** 1011–1020.

Hohlneicher, U., Hartmann, R., Taraz, K., and Budzikiewicz, H., 1992, The structure of ferribactin from *Pseudomonas fluorescens* ATCC 13525, *Z. Naturforsch.* **47b:** 1633–1638.

Hohlneicher, U., Hartmann, R., Taraz, K., and Budzikiewicz, H., 1995, Pyoverdin, ferribactin, azotobactin - a new triade of siderophores from *Pseudomonas chlororaphis* ATCC 9446 and its relation to *Pseudomonas fluorescens* ATCC 13525, *Z. Naturforsch.* **50c:** 337–344.

Hohnadel, D., and Meyer, J. M., 1988, Specificity of pyoverdine-mediated iron uptake among fluorescent *Pseudomonas* strains, *J. Bacteriol.* **170:** 4865–4873.

Hohnadel, D., Haas, D., and Meyer, J. M., 1986, Mapping of mutations affecting pyoverdine production in *Pseudomonas aeruginosa, FEMS Microbiol. Lett.* **36:** 195–199.

Holder, I. A., 1993, *Pseudomonas aeruginosa* burn infections: Pathogenesis and treatment, in: *Pseudomonas aeruginosa as an Opportunistic Pathogen* (M. Campa, M. Bendinelli, and H. Friedman, eds.), Plenum Press, New York, pp. 275–295.

Holloway, B. W., Römling, U., and Tümmler, B., 1994, Genomic mapping of *Pseudomonas aeruginosa* PAO, *Microbiology* **140:** 2907–2929.

Holmes, B., 1986a, Identification and distribution of *Pseudomonas stutzeri* in clinical material, *J. Appl. Bacteriol.* **60:** 401–411.

Holmes, B., 1986b, The identification of *Pseudomonas cepacia* and its occurrence in clinical material, *J. Appl. Bacteriol.* **61:** 299–314.

Itoh, J., Miyadoh, S., Takahasi, S., Amano, S., Ezaki, N., and Yamada, Y., 1979, Studies on antibiotics BN-227 and BN227-F, new antibiotics. I. Taxonomy, isolation, and characterization, *J. Antibiotics* **32:** 1089–1095.

Jacques, P., Gwose, I., Seinsche, D., Taraz, K., Budzikiewicz, H., Schröder, H., Ongena, M., Thonart, P., 1993, Isopyoverdin Pp BTP1, a biogenetically interesting novel siderophore from *Pseudomonas putida, Nat. Prod. Lett.* **3:** 213–218.

Jacques, P., Ongena, M., Gwose, I., Seinsche, D., Schröder, H., Delfosse, P., Thonart, P., Taraz, K., and Budzikiewicz, H., 1995, Structure and characterization of isopyoverdin from *Pseudomonas putida* BTP1 and its relation to the biogenetic pathway leading to pyoverdins, *Z. Naturforsch.* **50c:** 622–629.

Jayaswal, R. K., Fernandez, M., Upadhyay, R. S., Visintin, L., Kurz, M., Webb, J., and Rinehart, K., 1993, Antagonism of *Pseudomonas cepacia* against phytopathogenic fungi, *Curr. Microbiol.* **26:** 17–22.

Jurkevitch, E., Hadar, Y., Chen, Y., Libman, J., and Shanzer, A., 1992, Iron uptake and molecular recognition in *Pseudomonas putida:* Receptor mapping with ferrichrome and its biomimetic analogs, *J. Bacteriol.* **174:** 78–83.

Kachadourian, R., Dellagi, A., Laurent, J., Bricart, L., Kunesch, G., and Expert, D., 1996, Desferrioxamine-dependent iron transport in *Erwinia amylovora* CFBP1430: Cloning of the gene encoding the ferrioxamine receptor FoxR, *BioMetals* **9:** 143–150.

King, E. O., Ward, M. K., and Raney, D. F., 1954, Two simple media for the demonstration of pyocyanin and fluorescein, *J. Lab. Clin. Med.* **44:** 301–307.

Kisaalita, W. S., Slininger, P. J., and Bothast, R. J., 1993, Defined media for optimal pyoverdine production by *Pseudomonas fluorescens* 2-79, *Appl. Microbiol. Biotechnol.* **39:** 750–755.

Kleinkauf, H., and von Döhren, H., 1987, Biosynthesis of peptide antibiotics, *Annu. Rev. Microbiol.* **41:** 259–289.

Kleinkauf, H., and von Döhren, H., 1990, Nonribosomal biosynthesis of peptide antibiotics, *Eur. J. Biochem.* **192:** 1–15.

Kloepper, J. W., Leong, J., Teintze, M., and Schroth, M. N., 1980a, Enhanced plant growth by siderophores produced by plant growth promoting rhizobacteria, *Nature* **286:** 885–886.

Kloepper, J. W., Leong, J., Teintze, M., and Schroth, M. N., 1980b, *Pseudomonas* siderophores: A mechanisms explaining disease suppressive soils, *Curr. Microbiol.* **4:** 317–320.

Koedam, N., Wittouck, E., Gaballa, A., Gillis, A., Höfte, M., and P. Cornelis, 1994, Detection and differentiation of microbial siderophores by isoelectric focusing and chrome azurol S overlay, *BioMetals* **7:** 287–291.

Kolasa, T., and Miller, M. J., 1990, Synthesis of the chromophore of pseudobactin, a fluorescent siderophore from *Pseudomonas, J. Org. Chem.* **55:** 4246–4255.

Kontoghiorghes, G. J., 1987, Structure/iron binding activity of 1-hydroxypyrid-2-one chelators intended for clinical use, *Inorg. Chim. Acta* **135:** 145–150.

Korth, H., Brüsewitz, G., and Pulverer, G., 1982, Isolation of an antibacterial active tropolone from a *Pseudomonas cepacia* strain, *Zbl. Bakt. Hyg. I,* **252:** 83–86.

Koster, M., van de Vossenberg, J., Leong, J., and Weisbeek, P. J., 1993, Identification and characterization of the *pupB* gene encoding an inducible ferric-pseudobactin receptor of *Pseudomonas putida* WCS358, *Mol. Microbiol.* **8:** 591–601.

Koster, M., van Klompenburg, W., Bitter, W., Leong, J., and Weisbeek, P., 1994, Role for the outer membrane ferric siderophore receptor PupB in signal transduction across the bacterial cell envelope, *EMBO J.* **13:** 2805–2813.

Krotzky, A., and Werner, D., 1987, Nitrogen fixation in *Pseudomonas stutzeri, Arch. Microbiol.* **147:** 48–57.

Lacy, D. E., Spencer, D. A., Weller, P. H., and Darbyshire, P., 1993, Chronic granulomatous disease presenting in childhood with *Pseudomonas cepacia* septicaemia, *J. Infect.* **27:** 301–304.

Lamont, I. A., 1994, *Pseudomonas aeruginosa* OT11 pyoverdine synthetase D (*pvdD*), ferripyoverdine receptor (*fpvA*), and pyoverdine synthetase E (*pvdE*) genes, complete cds. GenBank database release 82.0, accession number U07359.

Lamont, I. A., 1995, personal communication.

Leong, J., 1986, Siderophores: Their biochemistry and possible role in the biocontrol of plant pathogens. *Annu. Rev. Phytopathol.* **24:** 187–209.

Leoni, L., Ciervo, A., Orsi, N., and Visca, P., 1996, Iron-regulated transcription of the *pvdA* gene in *Pseudomonas aeruginosa:* Effect of Fur and PvdS on promoter activity, *J. Bacteriol.* **178:** 2299–2313.

Lim, H. S., Kim, Y. S., and Kim, S. D., 1991, *Pseudomonas stutzeri* YPL-1 genetic transformation and antifungal mechanism against *Fusarium solani,* an agent of plant root rot, *Appl. Environ. Microbiol.* **57:** 510–516.

Lindberg, G. D., Larkin, J. M., and Whaley, H. A., 1980, Production of tropolone by a *Pseudomonas, J. Nat. Prod.* **43:** 592–594.

Linget, C., Azadi, P., MacLeod, J. K., Dell, A., and Abdallah, M., 1992, Bacterial siderophores: The structure of the pyoverdins of *Pseudomonas fluorescens* ATCC 13525, *Tetrahedron Lett.* **33:** 1737–1740.

Liu, P. V., and Shokrani, F., 1978, Biological activities of pyochelins: Iron chelating agents of *Pseudomonas aeruginosa, Infect. Immun.* **22:** 878–890.

Longerich, I., Taraz, K., Budzikiewicz, H., Tsai, L., and Meyer, J.–M., (1993), Pseudover-din, a compound related to the pyoverdine chromophore from a *Pseudomonas aeruginosa* strain incapable to produce pyoverdins. *Z. Naturforsch.* **48c:** 425–429.

Loper, J. E., Orser, C. S., Panopoulos, N. J., and Schroth, M. N., 1984, Genetic analysis of fluorescent pigment production in *Pseudomonas syringae* pv. syringae, J. Gen. Microbiol. **130:** 1507–1515.

Magazin, M. D., Moores, J. C., and Leong, J., 1986, Cloning of the gene coding for the outer membrane receptor protein for ferric pseudobactin, a siderophore from a plant growth-promoting *Pseudomonas* strain, *J. Biol. Chem.* **261:**795–799.

Maksimova, N. P., Blazhevich, O. V., and Fomichev, Y. K., 1992, Role of phenylalanine in the biosynthesis of the fluorescent pigment of *Pseudomonas putida,* **61:** 818–823.

Maksimova, N. P., Blazhevich, O. V., and Fomichev, Y. K., 1993, The role of pyrimidines in the biosynthesis of the fluorescent pigment pyoverdin Pm in *Pseudomonas putida* M bacteria, *Molekularn. Gen. Microbiol. Virusol.* **5:** 22–26.

Marugg, J. D., van Spanje, M., Hoekstra, W. P. M., Schippers, B., and Weisbeek, P. J., 1985, Isolation and analysis of genes involved in siderophore biosynthesis in plant-growth-stimulating *Pseudomonas putida* WCS358, *J. Bacteriol.* **164:** 563–570.

Marugg, J. D., Nielander, H. B., Horrevoets, A. J. G., van Megen, I., van Genderen, I., and Weisbeek, P. J., 1988, Genetic organization and transcriptional analysis of a major gene cluster involved in siderophore biosynthesis in *Pseudomonas putida* WCS358, *J. Bacteriol.* **170:** 1812–1819.

Marugg, J. D., de Weger, L. A., Nielander, H. B., Oorthuizen, M., Recourt, K., Lugten-berg, B., van der Hofstad, G. A. J. M., and Weisbeek, P. J., 1989, Cloning and charac-terization of a gene encoding an outer membrane protein required for siderophore-mediated uptake of Fe^{3+} in *Pseudomonas putida* WCS358, *J. Bacteriol.* **171:** 2819–2826.

Maurer, B., Muller, A., Keller–Schierlein, W., and Zähner, H., 1968, Ferribactin, ein siderochrom aus *Pseudomonas fluorescens* Migula, *Arch. Mikrobiol.* **60:** 326–339.

Menhart, N., and Viswanatha, T., 1990, Precursor activation in a pyoverdine biosynthesis, *Biochem. Biophys. Acta* **1038:** 47–51.

Menhart, N., Thariath, A., and Viswanatha, T., 1991, Characterization of the pyoverdines of *Azotobacter vinelandii* ATCC12837 with regard to heterogeneity, *Biol. Metals* **4:** 223–232.

Merriman, T. R., Merriman, M. E., and Lamont, I. L., 1995, Nucleotide sequence of *pvdD,* a pyoverdine biosynthetic gene from *Pseudomonas aeruginosa:* PvdD has similarity to peptide synthetases, *J. Bacteriol.* **177:** 252–258.

Meyer, J. M., 1992, Exogenous siderophore-mediated iron uptake in *Pseudomonas aeruginosa:* Possible involvement of porin OprF in iron translocation, *J. Gen. Microbiol.* **138:** 951–958.

Meyer, J. M., and Abdallah, M. A., 1978, The fluorescent pigment of *Pseudomonas fluo-rescens:* Biosynthesis, purification, and physicochemical properties. *J. Gen. Microbiol.* **107:** 319–328.

Meyer, J. M., and Abdallah, M. A., 1980, The siderochromes of non-fluorescent pseu-domonads: Production of nocardamine by *Pseudomonas stutzeri, J. Gen. Microbiol.* **118:** 125–129.

Meyer, J. M., and Hohnadel, D., 1992, use of nitrilotriacetic acid (NTA) by *Pseudomonas* species through iron metabolism, *Appl. Microbiol. Biotechnol.* **37:** 114–118.

Meyer, J. M., and Hornsperger, J.-M., 1978, Role of pyoverdine$_{Pf}$, the iron binding fluores-cent pigment of *Pseudomonas fluorescens* in iron transport, *J. Gen. Microbiol.* **107:** 329–331.

Meyer, J. M., Mock, M., and Abdallah, M. A., 1979, Effect of iron on the protein composition of the outer membrane of fluorescent pseudomonads, *FEMS Microbiol. Lett.* **5:** 395–398.

Meyer, J. M., Hallé, F., Hohnadel, D., Lemanceau, P., and Ratefiarivelo, H., 1987, Siderophores of *Pseudomonas* - biological properties, in: *Iron Transport in Microbes, Plants and Animals* (G. Winkelmann, D. van der Helm, and J. B. Neilands, eds.), VCH Verlagsgesellschaft, Weinheim, pp. 188–205.

Meyer, J. M., Hallé, F., and Hohnadel, D., 1989, Cepabactin from *Pseudomonas cepacia*, a new type of siderophores. *J. Gen. Microbiol.* **135:** 1479–1487.

Meyer, J. M., Hohnadel, D., Kahn, A., and Cornelis, P., 1990, Pyoverdine-facilitated iron uptake in *Pseudomonas aeruginosa:* Immunological characterization of the ferripyoverdine receptor. *Mol. Microbiol.* **4:** 1401–1405.

Meyer, J. M., Azelvandre, P., and Georges, C., 1992, Iron metabolism in *Pseudomonas:* Salicylic acid, a siderophore of *Pseudomonas fluorescens* CHA0. *Biofactors* **4:** 23–27.

Meyer, J. M., Tappe, R., Taraz, K., Budzikiewicz, H., de Vos, D., and Cornelis, P., 1993, The three pyoverdine species of *Pseudomonas aeruginosa* strains, Abstract P13, *Conference on Iron and Microbial Iron Chelates*, Brugge, Belgium, November 5–6, 1993.

Meyer, J. M., Trân Van, V., Stintzi, A., Berge, O., and Winkelmann, G., 1995, Ornibactin production and transport properties in strains of *Burkholderia vietnamiensis* and *Burkholderia cepacia* (formerly *Pseudomonas cepacia*). *BioMetals* **8:** 309–317.

Meyer, J. M., Neely, A., Stintzi, A., Georges, C., and Holder, I. A., 1996, Pyoverdine is essential for virulence of *Pseudomonas aeruginosa*, *Infect. Immun.* **64:** 518–523.

Michels, J., Benoni, H., Briskot, G., Lex, J., Schmickler, H., Taraz, K., Budzikiewicz, H., Korth, H., and Pulverer, G., 1991, Isolation and spectroscopic characterization of the pyoverdin chromophore and of its 5-hydroxy analogue. *Z. Naturforsch.* **46c:** 993–1000.

Mielczarek, E. V., Andrews, S. C., and Bauminger, R., 1992, Mössbauer spectroscopy and electron paramagnetic resonance studies of iron metabolites in *Pseudomonas aeruginosa:* Fe^{2+} and Fe^{3+} ferritin in ^{57}ferripyoverdine incubated cells and ^{57}ferric citrate fed cells, *Biol. Metals* **5:** 87–93.

Misaghi, I. J., Olsen, M. W., Cotty, P. J., and Donndelinger, C. R., 1988, Fluorescent siderophore-mediated iron deprivation - a contingent biological mechanism, *Soil Biol. Biochem.* **20:** 573–574.

Miyazaki, H., Kato, H., Nakazawa, T., and Tsuda, M., 1995, A positive regulatory gene, *pvdS*, for expression of pyoverdin biosynthetic genes in *Pseudomonas aeruginosa* PAO, *Mol. Gen. Genet.* **248:** 17–24.

Mohn, G., Taraz, K., and Budzikiewicz, H., 1990, New pyoverdin-type siderophores from *Pseudomonas fluorescens, Z. Naturforsch.* **45b:** 1437–1450.

Moore, G. R., Mann, S., and Bannister, J. V., 1986, Isolation and properties of the complex nonheam-iron-containing cytochrome *B-557* (bacterioferritin) from *Pseudomonas aeruginosa, J. Inorg. Biochem.* **28:** 329–336.

Moores, J. C., Magazin, M., Ditta, G. S., and Leong, J., 1984, Cloning of genes involved in the biosynthesis of pseudobactin, a high-affinity iron transport agent of a plant growth-promoting *Pseudomonas* strain, *J. Bacteriol.* **157:** 53–58.

Morgan, M. K., and Chatterjee, A. K., 1988, Genetic organization and regulation of proteins associated with production of syringotoxin by *Pseudomonas syringae* pv. *syringae, J. Bacteriol.* **170:** 5689–5697.

Morris, J., O'Sullivan, D. J., Koster, M., Leong, J., Weisbeek, P. J., and O'Gara, F., 1992, Characterization of fluorescent siderophore-mediated iron uptake in *Pseudomonas* sp. strain M114: Evidence for the existence of an additional ferric siderophore receptor, *Appl. Environ. Microbiol.* **58:** 630–635.

Münziger, M., 1995, Siderophore aus *Pseudomonas solanacearum* ATCC 11696, Diplomarbeit, Köln Universität, Köln, Germany.

Neilands, J. B., 1957, Some aspects of microbial iron metabolism, *Bacteriol. Rev.* **21:** 101–105.

Neilands, J. B., 1981, Microbial iron compounds, *Annu. Rev. Biochem.* **50:** 715–731.

Neilands, J. B., 1982, Microbial envelope proteins related to iron, *Annu. Rev. Microbiol.* **36:** 285–309.

Neilands, J. B., and Nakamura, K., 1991, Detection, determination, isolation, characterization, and regulation of microbial iron chelates, in: *Handbook of Microbial Iron Chelates* (G. Winkelmann, ed.), CRC Press, Boca Raton, Florida, USA, pp. 1–14.

Neilands, J. B., Konopka, K., Schwyn, B., Coy, M., Francis, R. T., Paw, B. H., and Bagg, A., 1987, Comparative biochemistry of microbial iron assimilation, in: *Iron Transport in Microbes, Plants, and Animals* (G. Winkelmann, D. van der Helm, and J. B. Neilands, eds.), VCH Verlaggesellschaft, Weinheim, Germany, pp. 3–33.

Newkirk, J. D., and Hulcher, F. H., 1969, Isolation and properties of a fluorescent pigment from *Pseudomonas mildenbergii*, *Arch. Biochem. Biophys.* **134:** 395–400.

Nowak–Thomson, B., and Gould, S. J., 1994a, A simple assay for fluorescent siderophores produced by *Pseudomonas* species and an efficient isolation of pseudobactin, *BioMetals* **7:** 20–24.

Nowak-Thomson, B., and Gould, S. J., 1994b, Biosynthesis of the pseudobactin chromophore from tyrosine, *Tetrahedron* **50:** 9865–9872.

O'Hoy, K., and Krishnapillai, V., 1987, Recalibration of the *Pseudomonas aeruginosa* PAO strain chromosome map in time units using high-frequency-of-recombination donors, *Genetics* **115:** 611–618.

Okonya, J. F., Kolsa, T., and Miller, M. J., 1995, Synthesis of the peptide fragment of pseudobactin, *J. Org. Chem.* **60:** 1932–1935.

O'Sullivan, D. J., and O'Gara, F., 1990, Iron regulation of ferric iron uptake in a fluorescent pseudomonad: Cloning of a regulatory gene, *Mol. Plant–Microbe Interact.* **3:** 86–93.

O'Sullivan, D. J., and O'Gara, F., 1991, Regulation of iron assimilation: Nucleotide sequence analysis of an iron-regulated promoter from a fluorescent pseudomonad, *Mol. Gen. Genet.* **228:** 1–8.

O'Sullivan, D. J., and O'Gara, F., 1992, Traits of fluorescent *Pseudomonas* spp. involved in suppression of plant root pathogens. *Microbiol. Rev.* **56:** 662–676.

O'Sullivan, D. J., Morris, J., and O'Gara, F., 1990, Identification of an additional ferric-siderophore uptake gene clustered with receptor, biosynthesis, and *fur*-like regulatory genes in fluorescent *Pseudomonas* sp. strain M114, *Appl. Environ. Microbiol.* **56:** 2056–2064.

Palleroni, N. J., 1984, Pseudomonas, in: *Bergey's Manual of Systematic Bateriology*, Volume 1 (N. R. Krieg, ed.), Williams and Wilkins, Baltimore, pp. 141–199.

Palleroni, N. J., Doudoroff, M., Stanier, R. Y., Solanes, R. E., and Mandel, M., 1970, Taxonomy of the aerobic pseudomonads: The properties of the *Pseudomonas stutzeri* group, *J. Gen. Microbiol.* **60:** 215–231.

Pandey, A., Bringel, F., and Meyer, J. M., 1994, Iron requirement and search for siderophores in lactic acid bacteria, *Appl. Microbiol. Biotechnol.* **40:** 735–739.

Parke, J. L., Rand, R. E., Joy, A. E., and King, E. B., 1991, Biological control of Pythium damping-off and Aphanomyces root rot of peas by application of *Pseudomonas cepacia* or *P. fluorescens* to seed, *Plant Dis.* **75:** 987–992.

Pennington, J. E., Reynolds, H. Y., and Carbone, P. P., 1973, *Pseudomonas* pneumonia: A retrospective study of 36 cases, *Am J. Med.* **55:** 155–160.

Persmark, M., Frejd, T., and Mattiasson, B., 1990, Purification, characterization, and structure of pseudobactin 589A, a siderophore from a plant growth promoting *Pseudomonas*, Biochemistry **29**: 7348–7356.

Poole, K., Young, L., and Neshat, S., 1990, Enterobactin-mediated iron transport in *Pseudomonas aeruginosa*, *J. Bacteriol.* **172**: 6991–6996.

Poole, K., Neshat, S., and Heinrichs, D., 1991, Pyoverdine-mediated iron transport in *Pseudomonas aeruginosa:* Involvement of a high-molecular-mass outer membrane protein, *FEMS Microbiol. Lett.* **78**: 1–6.

Poole, K., Heinrichs, D. E., and Neshat, S., 1993a, Cloning and sequence analysis of an EnvCD homologue in *Pseudomonas aeruginosa:* Regulation by iron and possible involvement in the secretion of the siderophore pyoverdine, *Mol. Microbiol.* **10**: 529–544.

Poole, K., Krebes, K., McNally, C., and Neshat, S., 1993b, Multiple antibiotic resistance in *Pseudomonas aeruginosa:* Evidence for involvement of an efflux operon, *J. Bacteriol.* **175**: 7363–7372.

Poole, K., Neshat, S., Krebes, K., and Heinrichs, D. E., 1993c, Cloning and nucleotide sequence analysis of the ferripyoverdine receptor gene *fpvA* of *Pseudomonas aeruginosa*, *J. Bacteriol.* **175**: 4597–4604.

Poppe, K., Taraz, K., and Budzikiewicz, H., 1987, Pyoverdine-type siderophores from *Pseudomonas fluorescens*, Tetrahedron **43**: 2261–2272.

Prince, R. W., Storey, D. G., Vasil, A. I., and Vasil, M. L., 1991, Regulation of *toxA* and *regA* by the *Escherichia coli fur* gene and identification of a Fur homologue in *Pseudomonas aeruginosa* PA103 and PAO1, *Mol. Microbiol.* **5**: 2823–2831.

Prince, R. W., Cox, C. D., and Vasil, M. L., 1993, Coordinate regulation of siderophore and exotoxin A production: Molecular cloning and sequencing of the *Pseudomonas aeruginosa fur* gene. *J. Bacteriol.* **175**: 2589–2598.

Raaijmakers, J. M., Bitter, W., Punte, H. L. M., Bakker, P. A. H. M., Weisbeek, P. J., and Schippers, B., 1994, Siderophore receptor PupA as a marker to monitor wild-type *Pseudomonas putida* WCS358 in natural environments, *Appl. Environ. Microbiol.* **60**: 1184–1190.

Raaijmakers, J. M., van der Sluis, I., Koster, M., Bakker, P. A. H. M., Weisbeek, P. J., and Schippers, B., 1995, Utilization of heterologous siderophores and rhizosphere competence of fluorescent *Pseudomonas* spp., *Can. J. Microbiol.* **41**: 126–135.

Reissbrodt, R., Rabsch, W., Chapeaurouge, A., Jung, G., and Winkelmann, G., 1990, Isolation and identification of ferrioxamine G and E in *Hafnia alvei*, *Biol. Metals* **3**: 54–60.

Rombel, L. T., and Lamont, I., 1992, DNA homology between siderophore genes from fluorescent pseudomonads, *J. Gen. Microbiol.* **138**: 181–187.

Royt, P., 1988, Isolation of a membrane-associated iron chelator from *Pseudomonas aeruginosa*, *Biochim. Biophys. Acta* **939**: 493–502.

Schröder, H., Adam, J., Taraz, K., and Budzikiewicz, H., 1995, Dihydropyoverdin sulfonic acids - Intermediates in the biogenesis?, *Z. Naturforsch.* **50c**: 616–621.

Schroeter, S., 1870, Über durch Bakterien gebildete Pigmente, *Cohn's Beitr. Biol. Pflanzen* **1**: 109–126.

Schwyn, B., and Neilands, J. B., 1987, Universal chemical assay for the detection and determination of siderophores, *Anal. Biochem.* **160**: 47–56.

Screen, J., Moya, E., Blagbrough, I. S., and Smith, A. W., 1995, Iron uptake in *Pseudomonas aeruginosa* mediated by *N*-(2,3-dihydroxybenzoyl)-L-serine and 2,3-dihydroxybenzoic acid, *FEMS Microbiol. Lett.* **127**: 145–149.

Seinsche, D., Taraz, K., and Budzikiewicz, H., 1993, Neue pyoverdin-siderophore aus *Pseudomonas putida* C, *J. Prakt. Chem.* **335**: 157–168.

Serino, L., Reimmann, C., Baur, H., Beyeler, M., Visca, P., and Haas, D., 1995, Structural genes for salicylate biosynthesis from chorismate in *Pseudomonas aeruginosa, Mol. Gen. Genet.* **249:** 217–228.

Sexton, R., Gill, P. R., Callanan, M. J., O'Sullivan, D. J., Dowling, D. N., and O'Gara, F., 1995, Iron-responsive gene expression in *Pseudomonas fluorescens* M114: Cloning and characterization of a transcription-activating factor, PbrA, *Mol. Microbiol.* **15:** 297–306.

Shand, G. H., Anwar, H., Kadurugamuwa, J., Brown, M. R., Silverman, S. H., and Melling, J., 1985, In vivo evidence that bacteria in urinary tract infection grow under iron-restricted conditions, *Infect. Immun.* **48:** 35–39.

Smith, A. W., Hirst, P. H., Hughes, K., Gensberg, K., and Govan, J. R. W., 1992, The pyocin Sa receptor of *Pseudomonas aeruginosa* is associated with ferripyoverdin uptake, *J. Bacteriol.* **174:** 4847–4849.

Smith, A. W., Poyner, D. R., Hughes, H. K., and Lambert, P. A., 1994, Siderophore activity of *myo*-inositol hexakisphosphate in *Pseudomonas aeruginosa, J. Bacteriol.* **176:** 3455–3459.

Sokol, P. A., 1986, Production and utilization of pyochelin by clinical isolates of *Pseudomonas cepacia, J. Clin. Microbiol.* **23:** 560–562.

Sokol, P. A., and Woods, D. E., 1983, Demonstration of an iron-siderophore-binding protein in the outer membrane of *Pseudomonas aeruginosa, Infect. Immun.* **40:** 665–669.

Sokol, P. A., Lewis, C. J., and Dennis, J. J., 1992, Isolation of a novel siderophore from *Pseudomonas cepacia, J. Med. Microbiol.* **36:** 184–189.

Sriyosachati, S., and Cox, C. D., 1986, Siderophore-mediated iron acquisition from transferrin by *Pseudomonas aeruginosa, Infect. Immun.* **52:** 885–891.

Stephan, H., Freund, S., Beck, W., Jung, G., Meyer, J. M., and Winkelmann, G., 1993a, Ornibactins - a new family of siderophores from *Pseudomonas, Biometals* **6:** 93–100.

Stephan, H., Freund, S., Meyer, J. M., Winkelmann, G., and Jung, G., 1993b, Structure elucidation of the gallium–ornibactin complex by 2D-NMR spectroscopy, *Liebigs Ann. Chem.* **1993:** 43–48.

Stewart, G. J., and Sinigalliano, C. D., 1989, Detection and characterization of natural transformation in the marine bacterium *Pseudomonas stutzeri* strain ZoBell, *Arch. Microbiol.* **152:** 520–526.

Stieritz, D. D., and Holder, I. A., 1975, Experimental studies of the pathogenesis of infections due to *Pseudomonas aeruginosa:* Description of a burned mouse model, *J. Infect. Dis.* **131:** 668–691.

Stintzi, A., 1993, Etude de la voie de biosynthèse de la pyoverdine de *Pseudomonas aeruginosa* PAO1, Diplôme d'Etudes Approfondies, Université de Strasbourg, France.

Stintzi, A., Cornelis, P., Hohnadel, D., Meyer, J. M., Dean, C., Poole, K., Kourambas, S., and Krishnapillai, V., 1996, Novel pyoverdine biosynthesis gene(s) of *Pseudomonas aeruginosa, Microbiology* **142:** 1181–1190.

Stoll, A., Brack, A., and Renz, J., 1951, Nocardamin, ein neues antibioticum aus einer *Nocardia. Schweiz. Z. Path. Bakteriol.* **14:** 225–233.

Tabacchioni, S., Bevivino, A., Chiarini, L., Visca, P., and Del Gallo, M., 1993, Characteristic of two rhizosphere isolates of *Pseudomonas cepacia* and their potential plant-growth-promoting activity, *Microb. Releases* **2:** 161–168.

Tabacchioni, S., Visca, P., Chiarini, L., Bevivino, A., Di Serio, C., Fancelli, S., and Fani, R., 1995, Molecular characterization of rhizosphere and clinical isolates of *Burkholderia cepacia, Res. Microbiol.* **146:** 531–542.

Tappe, R., Taraz, K., Budzikiewicz, H., Meyer, J. M., and Lefevre, J. F., 1993, Structure elucidation of a pyoverdin produced by *Pseudomonas aeruginosa* ATCC 27853, *J. Prakt. Chem.* **335**: 83–87.

Taraz, K. Seinsche, D., and Budzikiewicz, H., 1991a, Pseudobactin- and pseudobactin A-varianten: Neue peptidsidorephore vom pyoverdin-typ aus *Pseudomonas fluorescens* "E2", *Z. Naturforsch.* **46c**: 522–526.

Taraz, K., Tappe, R., Schröder, H., Hohlneicher, U., Gwose, I., Budzikiewicz, H., Mohn, G., and Lefèvre, J. F., 1991b, Ferribactins - the biogenetic precursors of pyoverdins, *Z. Naturforsch.* **46c**: 527–533.

Taylor, P. C., and Kalamatianos, C. C., 1994, *Pseudomonas cepacia* in the sputum of cystic fibrosis patients, *Pathology* **26**: 315–317.

Teintze, M., and Leong, J., 1981, Structure of pseudobactin A, a second siderophore from plant growth promoting *Pseudomonas* B10, *Biochemistry* **20**: 6457–6462.

Teintze, M., Hossain, M. B., Barnes, C. L., Leong, J., and van der Helm, D., 1981, Structure of ferric pseudobactin, a siderophore from a plant growth promoting *Pseudomonas*, *Biochemistry* **20**: 6446–6457.

Torres, L., Perez–Ortin, J. E., Tordera, V., and Beltran, J. P., 1986, Isolation and characterization of an (FeIII)-chelating compound produced by *Pseudomonas syringae*, *Appl. Environ. Microbiol.* **52**: 157–160.

Tsuda, M., Miyazaki, H., and Nakazawa, T., 1995, Genetic and physical mapping of genes involved in pyoverdin production in *Pseudomonas aeruginosa* PAO, *J. Bacteriol.* **177**: 423–431.

Turfitt, G. E., 1936, Bacteriological and biochemical relationships in the pyocyaneus-fluorescens group. I. The chromogenic function in relation to classification, *Biochem. J.* **30**: 1323–1328.

Turfitt, G. E., 1937, Bacteriological and biochemical relationships in the pyocyaneus-fluorescens group. II. Investigation on the green fluorescent pigment, *Biochem. J.* **31**: 212–218.

Turfreijer, A., Wibaut, J. P., and Boltjes, T. Y. K., 1938, The green fluorescent pigment of *Pseudomonas fluorescens*, *Rec. Trav. Chim. Pays Bas* **57**: 1397–1404.

Urakami, T., Ito-Yoshida, C., Araki, H., Kijima, T., Suruki, K.–I., and Komagata, K., 1994, Transfer of *Pseudomonas plantarii* and *Pseudomonas glumae* to *Burkholderia* spp. and description of *Burkholderia vandii* sp. nov., *Int. J. Syst. Bacteriol.* **44**: 235–245.

van der Hofstad, G. A. J. M., Marrug, J. D., Verjans, G. M. G. M., and Weisbeek, P. J., 1986, Characterization and structural analysis of the siderophore produced by the PGPR *Pseudomonas putida* strain WCS358, in: *Iron, Siderophores, and Plant Diseases* (T. R. Swinburne, ed.), Plenum Press, New York, pp. 71–75.

Venturi, V., Ottevanger, C., Leong, J., and Weisbeek, P. J., 1993, Identification and characterization of a siderophore regulatory gene (*pfrA*) of *Pseudomonas putida* WCS358: Homology to the alginate regulatory gene *algQ* of *Pseudomonas aeruginosa*, *Mol. Microbiol.* **10**: 63–73.

Venturi, V., Wolfs, K., Leong, J., and Weisbeek, P. J., 1994, Amplification of the *groESL* operon in *Pseudomonas putida* increases siderophore gene promoter activity, *Mol. Gen. Genet.* **245**: 126–132.

Venturi, V., Weisbeek, P. J., and Koster, M., 1995a, Gene regulation of siderophore-mediated iron acquisition in *Pseudomonas:* Not only the Fur repressor, *Mol. Microbiol.* **17**: 603–610.

Venturi, V., Ottevanger, C., Bracke, M., and Weisbeek, P. J., 1995b, Iron regulation of siderophore biosynthesis and transport in *Pseudomonas putida* WCS358: Involvement of a transcriptional activator and of the Fur protein, *Mol. Microbiol.* **15**: 1081–1093.

Visca, P., Serino, P., and Orsi, N., 1992a, Isolation and characterization of *Pseudomonas aeruginosa* mutants blocked in the synthesis of pyoverdin, *J. Bacteriol.* **174:** 5727–5731.

Visca, P., Colotti, G., Serino, L., Verzili, D., Orsi, N., and Chiancone, E., 1992b, Metal regulation of siderophore synthesis in *Pseudomonas aeruginosa* and functional effects of siderophore-metal complexes, *Appl. Environ. Microbiol.* **58:** 2886–2893.

Visca, P., Ciervo, A., Sanfilippo, V., and Orsi, N., 1993, Iron-regulated salicylate synthesis by *Pseudomonas* spp, *J. Gen. Microbiol.* **139:** 1995–2001.

Visca, P., Ciervo, A., and Orsi, N., 1994, Cloning and nucleotide sequence of the *pvdA* gene encoding the pyoverdin biosynthetic enzyme L-ornithine N^5-oxygenase in *Pseudomonas aeruginosa*, *J. Bacteriol.* **176:** 1128–1140.

Waring, W. S., and Werkman, C. H., 1942, Growth of bacteria in an iron-free medium, *Arch. Biochem.* **1:** 303–310.

Weinberg, E. D., 1978, Iron and infection, *Microbiol. Rev.* **42:** 45–66.

Weinberg, E. D., 1993, The iron-withholding defense system, *ASM News* **59:** 559–562.

Wiebe, C., and Winkelmann, G., 1975, Kinetics studies on the specificity of chelate iron uptake in *Aspergillus*, *J. Bacteriol.* **123:** 837–842.

Williams, P., Morton, D. I., Towner, K. J., Stevenson, P., and Griffiths, E., 1990, Utilization of enterobactin and other exogenous iron sources by *H. parainfluenzae*, and *H. paraprophilus*, *J. Gen. Microbiol.* **136:** 2343–2350.

Winkler, S., Ockels, W., Budzikiewicz, H., Korth, H., and Pulverer, G., 1986, 2-hydroxy-4-methoxy-5-methylpyridin-N-oxid, ein Al^{3+} bindender metabolit von *Pseudomonas cepacia*, *Z. Naturforsch.* **41c:** 807–808.

Winkelmann, G., 1991, *Handbook of Microbial Iron Chelates*, CRC Press, Boca Raton, Florida, USA.

Wolz, C., Hohloch, K., Ocaktan, A., Poole, K., Evans, R. W., Rochel, N., Albrecht–Gary, A. M., Abdallah, M. A., and Döring, G., 1994, Iron release from transferrin by pyoverdin and elastase from *Pseudomonas aeruginosa*, *Infect. Immun.* **62:** 4021–4027.

Xiao, R., and Kisaalita, W. S., 1995, Purification of pyoverdines of *Pseudomonas fluorescens* 2-79 by copper-chelate chromatography, *Appl. Environ. Microbiol.* **61:** 3769–3774.

Xu, G.-W., and Gross, D. C., 1988, Physical and functional analyses of the *syrA* and *syrB* genes involved in syringomycin production by *Pseudomonas syringae* pv. *syringae*, *J. Bacteriol.* **170:** 5680–5688.

Yabuuchi, E., Kosako, Y., Oyaizu, H., Yano, L., Hotta, H., Hashimoto, Y., Ezaki, T., and Arakawa, M., 1992, Proposal of *Burkholderia* gen. nov. and transfer of seven species of the genus *Pseudomonas* homology group II to the new genus, with the type species *Burkholderia cepacia* (Palleroni and Holmes, 1981) comb. nov., *Microbiol. Immunol.* **36:** 1251–1275.

Yamada, Y., Seki, N., Kitahara, T., Takahashi, M., and Matsui, M., 1970, Structure and synthesis of aeruginoic acid[2-(o-hydroxypheyl)-4-thiazolecarboxylic acid], *Agric. Biol. Chem.* **34:** 780–783.

Yamano, Y., Nishikawa, T., and Komatsu, Y., 1994, Ferric iron transport system of *Pseudomonas aeruginosa* PAO1 that functions as the uptake pathway of a novel catechol-substituted cephalosporin, S-9096, *Appl. Microbiol. Biotechnol.* **40:** 892–897.

Yang, C.-C., and Leong, J., 1984, Structure of pseudobactin 7SR1, a siderophore from a plant-deleterious *Pseudomonas*. *Biochemistry* **23:** 3534–3540.

Yang, H., Chaowagul, W., and Sokol, P. A., 1991, Siderophore production by *Pseudomonas pseudomallaei*, *Infect. Immun.* **59:** 776–780.

The Flagellum

8

THOMAS C. MONTIE

1. INTRODUCTION

The classic pseudomonads are characterized by polar flagella, ranging from a single polar flagellum, *Pseudomonas aeruginosa*, to several polar flagella in *Pseudomonas putida*, *Pseudomonas fluorescens*, and *Pseudomonas syringae*. The only exception is *Burkholderia mallei*, which is permanently immotile and interestingly requires a living host for survival. It is the causal agent of the disease, glanders, in horses. Several of the multipolar flagellar types have been removed from the *Pseudomonas* genus based on r-RNA probe comparisons (Holloway, 1996), most recently to the *Burkholderia* spp. Some of these groups are discussed separately for comparison. The bulk of this chapter, however, focuses on *P. aeruginosa* because most studies of flagella have centered on this organism, primarily because of its clinical importance, but also because of its own metabolic uniqueness and versatility.

2. BIOCHEMISTRY

2.1. General Morphology

The *P. aeruginosa* flagellum filament extends outward from a single pole greater than several lengths of the entire cell body. It is connected via a hook structure to a basal body resembling that of *E. coli* or *Salmonella*, as judged from excellent electron micrographs (EM) of these structures by Suzuki and Iino (1980). Several genes associated with the basal body components, such as *P. aeruginosa fliE* and *fliF*, have also been identified (Arora *et al.*, 1996). Although the possibility of a sheath sur-

THOMAS C. MONTIE • Department of Microbiology, The University of Tennessee, Knoxville, Tennessee 37996-0845

Pseudomonas, edited by Montie. Plenum Press, New York, 1998.

rounding the filament has not been ruled out, no evidence suggests the presence of any additional layered structure. The binding of anti-flagellin monoclonal antibodies (mAbs) to the filament surfaces, as judged by EM analysis, indicates that flagellin protein is available on the outside of the filament structure (Landsperger *et al.*, 1993).

2.2. Purification

Initial studies designed to purify flagella filaments consisted of shearing the filaments, differential centrifugation to remove whole cells, followed by ultracentrifugation to obtain the polymerized protein fila-ment pellet (stage 1) (Montie *et al.*, 1982). To reduce the level of LPS in the filament preparation, the pellets were further purified by Sephacryl 300 column separations in the presence of EDTA and sodium deoxycho-late (for details, see Kelly-Wintenberg *et al.*, 1990). This process reduced LPS to very low levels, especially in the b-type flagellins (stage 2), and was initiated mainly to purify flagella antigen for vaccine use (Kelly-Winten-berg *et al.*, 1990; Montie *et al.*, patent). LPS levels have been monitored by chemical and gel silver stain methods and by in vivo testing using the pyrogenic response in rabbits. The purity of flagellin was assayed by SDS-PAGE Coomassie blue and silver staining.

Recently, another method was introduced with emphasis on increas-ing protein purity (Cha *et al.*, 1996). Flagella protein obtained at stage 1 is loaded onto a SDS-PAGE, preparative gel (BIORAD). This method purified flagellin by continuous-elution electrophoresis from a poly-acrylamide column. As above, fractions were analyzed by running them in a mini-gel fast electrophoresis system (Pharmacia) to detect flagellin. Combined flagella fractions were dialyzed overnight and then inten-sively for 9 h with water changes every 3 h. After lyophilization, the pooled fractions were passed through an extracti-gel (Flex-column; Kontes) equilibrated with 0.05m $NAHCO_3$ to remove excess SDS. Anti-genicity of the flagellin was retained in all purification methods. Storage in the lyophilized condition at -70 °C, is recommended.

2.3. Molecular Weights of Flagellins

The molecular weight flagellin of *P. aeruginosa* was first reported to be approximately 50,000. (Ansorg and Schmitt, 1980). However, subse-quent analysis indicated heterogeneity in flagellin sizes (Allison *et al.*, 1985), which correlated with the major antigenic type grouping of a or b

(Ansorg, 1978; Lanyi, 1970) (see below). The b-types are homologous antigenically with a molecular weight of 53,000. The heterologous group (a-type) has flagellins of 43,000 to 52,000 M_r in SDS-PAGE (Table I). Inclusion in the a-type group is based on the presence of a common cross-reacting antigen designated a_0 and usually some accompanying subantigen (any combination of a_1, a_2, a_3, a_4). Classification according to antigen cross-reactivity and M_r is based on assessments using laboratory and clinical strains, including isolates from patients with cystic fibrosis and folliculitis.

Table I. Antigenic Components and Molecular Weight of Flagella of Representative Antigen Characterized (Standard) and Unknown *Pseudomonas aeruginosa* Strains

Strain[a]	Source[b]	Antigenic component[c]	Flagellin M_r
PA01	Shriners	b	53,000
M2	Shriners	b	53,000
PJ108	Tsuda/Iino	b	53,000
19660	Wayne St.	b	53,000
1244	San Antonio, McManus	b	53,000
170001	Ansorg/Lanyi	b	53,000
SBI-H	Shriners/Holder	b	53,000
1071	Shriners/Holder	b	53,000
5939	Ansorg/Lanyi	a_0a_3	52,000
5933	Ansorg/Lanyi	$a_0a_1a_2$	51,000
7191	Ansorg/Lanyi	$a_0a_1a_2$	51,000
5940	Ansorg/Lanyi	a_0a_2	47,000
170018	Ansorg/Lanyi	$a_0a_3a_4$	45,000
1210	Shriners	$a_0a_1a_2$	51,000
GNB-1	Cinn. Gen. Hosp.	a-type	49,000
3598(F)	CDC	a-type	48,000
3614(F)	CDC	a-type	48,000
2993	Shriners/Holder	a-type	ND[d]
1071	Shriners/Holder	a-type	ND
SBIN	Shriners/Holder	a-type	ND
1998	Shriners/Holder	a-type	ND
86-F(CF)	Rainbow-Clev.	a-type	48,000
572b(CF)	Rainbow-Clev.	a-type	48,000

[a](F) Refers to folliculitis (hot tub or swimming pool infectious isolates); (CF) refers to cystic fibrosis sputum isolates.
[b]Shriners indicates Shriners Burns Institute, Cincinnati, Ohio.
[c]Strain 1210 subantigens deduced from M_r; a_0, dominant subantigen of b-types confirmed by ELISA.
[d]ND, not done.

2.4. Amino Acid Composition

The amino acid composition for flagellins has been compared in several studies (Kelly-Wintenberg, K., unpublished studies; Montie et al., patent; Montie, 1992; Rotering and Dorner, 1990) (Table II). Preparations purified by the Sephacryl 300 column-detergent method were analyzed. New values were obtained by an improved sensitive technique. For comparison b-type data (M-2) obtained previously (Rotering and Dorner, 1990) are compared with more recent results (Montie and Kelly-Wintenberg, unpublished data). The recent data reveal several significant differences from the prior data (Rotering and Dorner, 1990). Our recent data show three to seven proline residues and methionine in all flagella types tested, except strain 12993. The presence of proline and methionine and the lack of tryptophan residues is consistent with the nucleotide-derived amino acid composition (Brimer et al., 1997; Brickman et al., 1997; Spangenberg et al., 1996; Totten and Lory, 1990). (See also section 4.) The derived amino acids obtained for the PAK a-type

Table II. Amino Acid Analyses of *P. aeruginosa* Flagellins

Amino acid	Strain and flagellin type						
	M-2[a]	M-2	PJ108	PA103 PK[b]	170018	5940	2993
	(b)	(b)	(b)	(b)	(a₀ a₃ a₄)	(a₀ a₂)	(a-type)
ASP	74	70	75	73	63	67	80
GLU	49	47	48	49	45	47	50
SER	48	53	57	56	41	47	40
GLY	51	47	51	54	47	50	40
HIS	0	0.6	1.2	0.8	3.6	1.6	0
ARG	18	21	20	19	19	16	19
THR	48	52	47	47	35	40	40
ALA	91	87	87	87	64	69	80
PRO	0	8	7	7	10	7.8	0
TYR	4	7	7	6.4	6.7	4.9	3
VAL	38	35	32	32	26	28	30
MET	0	1.5	1.5	1.8	4.2	3.5	3
ILE	30	27	27	27	20	23	30
LEU	43	41	41	41	35	35	35
PHE	13	10	12	12	11	11	10
LYS	18	20	17	18	16	15	20
kDa	53	53	53	53	45	46.5	50

[a]Analysis in a Beckman analyzer using ninhydrin reagent.
[b]Flagella isolated from PA103 Fla+ after cloning the PAO1 fla gene into a PA103 Fla-mutant.

flagellin nucleotide sequences (Totten and Lory 1990) include two tyrosines, two methionines, and two prolines. The sequence for b-types, compared to a-types, shows increased hydroxy amino acids, especially tyrosine, in the central variable region. The absence of histidine, cysteine, and tryptophan is also consistent with all of the nucleotide-derived amino sequence analyses for both a- and b-types (see section 4). All the analyses show that flagella protein has a low aromatic amino acid content. Thus the *P. aeruginosa* amino acid composition is typical of most true bacterial flagellins where distinct regions of the primary structure are critical for the biosynthesis, export, assembly, and function of this surface organelle which must have a rigid wave and internal, tight, helical structure.

The amino acid composition (Table II) shows that the number of tyrosine residues increases from one in a-types to between four to seven depending on the strain in b-types in general agreement with derived sequences (see section 4). Interestingly, the average number of methionine residues increases in the b-types versus a-types from approximately one to two residues to three to four residues, respectively. Because the molecular weights are lower in a-types and the number of methionines residues doubles, this variation represents a notable change, perhaps with important molecular implications for classification. As mentioned, the lower number of hydroxy amino acids in a-types compared to be-types may be reflected in the relative decrease in specific activity in a-types following phosphate radiolabeling of flagella (Kelly-Wintenberg et al., 1993) (see section 3.1.). The decrease in threonines and serines by approximately 13 and 17 residues accordingly may relate to the requirement for glycoprotein linkages at specific sites in the variable region. The decrease in tyrosine residues in a-types to two residues would certainly limit potential tyrosine phosphorylations in a-types.

Generally, the molecular weights determined by total amino acid composition agree with the M_r data obtained in gel electrophoresis determinations. However, the SDS-PAGE calculated molecular weights for a-types were too high because the sequence-derived size of all a-types is approximately 40,000 (Brimer et al., 1997). The high M_r values are caused in part by posttranslational modifications with glycans and probably phosphate moieties (Brimer et al., 1997).

Direct analyses of nine N-terminal amino acids was reported for a number of both a- and b-type flagellines examined and those were, Ala-Leu-Thr-Val-Asn-Apn-Ile-Ala (Montie et al., U.S. Patent; Rotering and Dorner, 1990). These findings were later substantiated independently, and six additional amino acids were identified (#10–15), Ser-Leu-Asn-Thr-Gln-Arg-(Montie et al., unpublished data) (Table 2). These sequences agree with the derived amino acid sequences of a-type and b-type flagellins (Totten and Lory, 1990; Spangenberg et al., 1996). These data together indicate that the a- and b-type N-terminal sequences are essentially the same and also show commonality compared with several other gram-negative and gram-positive organisms (Table II). Totten and Lory (1990) matched PAK a-type flagellin gene-derived 150 N-terminal amino acids and 80 C-terminal with five diverse gram-negative and gram-positive bacteria. The overall similarity it as follows: R. cecicola, 77%; S. marcescens, 77%; B. subtilis, 75%; E. coli, 72%; Salmonella muenchen, 61%; Borrelia burgdorferi, 63%; Campylobacteria coli, 64%; and the 28-KDa flagellin of C. crescentus, 53%.

3. POSTTRANSLATIONAL MODIFICATIONS OF FLAGELLINS

3.1. Phosphorylation

There was considerable doubt for a long time that protein phosphorylation occurs in prokaryotes, although it was identified early on as an important facet of the regulation of eukaryotes and has since proved to be the key biochemical factor in central metabolic regulation of these cells. However, protein phosphorylation in bacteria was recognized in the late 1970s (for a recent review, see Kennedy and Potts, 1996). The first phosphorylations identified involved serine/threonine phosphorylations on central metabolic enzymes (Garnak and Reeves, 1978; Wang and Koshland, 1978). Subsequently during the 1980s sensor histidine kinases, which autophosphorylate and then transfer the phosphate to response regulatory proteins at aspartate, were viewed as the bacterial counterpart to phosphorylation cascade systems in eukaryotes. These two-component systems are associated with chemotaxis and environmental sensor systems and are dominant facets of bacterial transport and general metabolic and transcriptional controls. For many years it was believed that these systems were peculiar to bacteria, but recently histidine kinase two-component regulators similar to those in bacteria have been found in yeast, mammalian mitochondria, and Archaea (Kennedy and Potts, 1996). On the other hand, ser/thr kinases and protein tyrosine phosphorylations, the latter believed to be strictly a eukaryotic characteristic, are emerging as important features of bacteria. The recent report of tyrosine phosphorylations in three different Archaea (Smith et al., 1997) has indicated the primal origin of tyrosine kinase which was previously thought to be solely in the eukaryotic domain (Kennedy and Potts, 1996). Tyrosine and ser/thr phosphorylated proteins have been identified in cyanobacteria (McCartney *et al.*, 1997), *Acinetobacter, Streptomyces, Yersinia, Pseudomonas,* and *Escherichia coli.* In multicellular prokaryotes, such as *Myxococcus, Streptomyces,* and cyanobacteria, eukaryotic-like ser/thr and tyrosine kinases based on sequence similarity have been identified (see the recent review by Zhang, 1996). It is believed these kinases function in cell-to-cell signaling similar to eukaryotic systems, but little is actually known about their function. It remains to be seen whether any kinases in unicellular prokaryotes are related to the Archaea and eukaryotic kinases. Nevertheless, the importance of protein phosphorylations in prokaryotes at multiple O-linked sites is becoming more apparent.

In *Pseudomonas* the presence of phosphorylation on flagellin proteins has been reported in at least three different species, *P. aeruginosa, P. fluorescens* (Kelly-Wintenberg *et al.*, 1990; South *et al.*, 1994) and a putative membrane kinase and possibly flagellin *Burkholderia solonacearum* (Atkinson *et al.*, 1992), and a fourth, *P. syringae,* exhibited tyrosine phosphorylation of a cytoplasmic protein (Ray *et al.*, 1994). Results with *P. aeruginosa* a-type and b-type flagellins showed that they are phosphorylated. However, the degree of phosphorylation seemed to be higher in b-types. Recently, more extensive testing using a more stringent two-dimensional electorphoresis has not verified the presence of tyrosine phosphate (Cha *et al.*, 1996) although anti-phosphotyrosine antibodies remain positive in some assays. Recent results using the highly specific mAb described by McCartney *et al.* (1997), have not confirmed these findings (Brimer *et al.*, 1997).

Unfortunately, with *P. aeruginosa* flagellin removal of phosphate coincided with partial hydrolysis to obtain single amino acids. A small peptide exhibited almost the same mobility as P-Tyr making it difficult to distinguish the amino acid. No other phosphorylated amino acids were identified in these preparations. A putative 42-kDa membrane kinase, however, phosphorylated poly Glu-Tyr and other peptides containing only tyrosine in vitro (Cha *et al.*, 1996; South *et al.*, 1994). Of interest was the finding that flagellin threonine and serine were also phosphorylated in vitro presumably by the same 42-kDa kinase which itself is also autophosphorylated at threonine and serine (Cha *et al.*, 1996). Phosphorylated serine was identified only in flagellin following partial enzymatic digestion before acid hydrolysis. Specific phosphatases for Ser/Thr-P, but not P-Tyr, removed labeled phosphate from in vitro labeled flagellin. It may be possible to obtain a confirmatory identification of phosphotyrosine by combining enzymatic and acidic or alkaline digestion. The fact that flagellin is prelabeled in vivo with orthophosphate, but does not reveal any -O-linked amino acids in 2-dimensional electrophoresis assays indicates that there is a particular sensitivity of these groups to either acid or alkaline which results in their removal during short term hydrolysis (45 min to 1 hr). In light of the phosphorylation of *P. solanacearum* putative flagellin at tyrosines (Allen, personal communication) and phosphorylation of *Campylobacter* flagellin at serine, these observations underscore the significance of these results with *P. aeruginosa*. It is obvious that flagellin is phosphorylated in vivo and in vitro. The indication of phosphorylation at both Thr/Ser and possibly Tyr suggests that the enzyme is a dual specific kinase (Cha *et al.*, 1996). Pertinent to the *P. aeruginosa* results are the reports of a eukaryotic-type membrane dual specific kinase found in *Streptomyces coelicolor* (Matsumoto *et al.*, 1994) and a dual specific phosphatase from the cyanobacterium *Nostoc commune* (McCartney *et al.*, 1997). One can speculate that a dual-functional enzyme may have evolved early in evolution as an efficient means of providing multiple recognition signals.

3.2. Glycosylation

A very recent and novel finding has been the detection of glycan residues on a-type *P. aeruginosa* flagellins (Brimer *et al.*, 1997). A major discrepancy between the sequence-derived molecular weights of 40,000 for strains 5933 and 5939 and the apparent M_rs of 51,000 to 52,000 posed the question of the presence of glycan residues in addition to phosphates which would alter electrophoretic mobility. Other a-type flagellins, notably PAK and 170018, also gave a sequence-derived mass of approximately 40 kDa, but an apparent of M_r of 45 kDa. Glycan analysis of flagellin blots showed that 5933 and 5939 flagellins are glycosylated, whereas 170018 gives a weak reaction and b-types were not glycosylated (Brimer *et al.*, 1997). Treatment of the flagellins with trifluoromethane sulfonic acid to deglycosylate the proteins decreased the M_r by 5 to 6 kDa from 52 and 51 kDa to 45 kDa for 5933 and 5939 and by a slight shift of approximately 1 kDa in flagellins M-2 (b-types) and 170018. Preliminary assays of these samples for glycan moieties demonstrated that the flagellin bands showing the sharply decreased M_rs (5933 and 5939) were deglycosylated. It is likely that a-type flagellins giving apparent M_rs in the higher ranges >45 kDa are more heavily glycosylated. The presence of phosphorylations on the flagellins may have contributed in part to an increased molecular weight shift in gels from 40 kDa to 45 kDa. Preliminary experiments indicated that the glycosylation is not involved in antigenicity to a human mAb (Brimer and Montie, 1997) as recently reported for *Campylobacter* flagella (Doig *et al.*, 1996). In *Campylobacter*, the removal of sialic acid from the flagellin results in loss of antigenicity.

An intriguing question is, why are there glycan moieties on some *P. aeruginosa* flagellins and not others or why are they present at all? Although one cannot rule out the possibility of a minute amount of glycosylation on low molecular weight a-types, such as 170018, it does not appear to be glycosylated. Gel patterns of a-type flagellins indicate that the amount of glycosylation is variable. Furthermore, b-types do not seem to contain glycan groups. If antigenicity is not a factor, then possibly the glycan group protects flagellin against environmental proteases and/or against environmental denaturants in general. The variation among similar groups in this experiment may have originated in the *Archaea*, which are also soil–water organisms and where glycosylation occurs in approximately 50% of the species. Similar to *Pseudomonas*, it is puzzling that in many cases, closely related species of *Archaea* differ with respect to glycosylation. In *Halobacterium halobium*, it has been suggested that the glycosyl groups may allow the flagellin filaments to slip past one another when the flagella switches rotational direction, thus allowing the filaments to remain in a bundle without separating (Alam and Oesterhelt, 1984). Glycan additions have also been identified recently on the polar flagellum of *Azospirillum brasilense* (Moens *et al.*, 1995) and the pili of *Neisseria miningitidis* (Virjii *et al.*, 1993). In *Azospirillum* this may contribute to their attachment to plants.

As observed with the a-type flagellins, the *Archaea* flagellins have higher M_rs than would be predicted by their primary sequence. In *Methanococcus hungatei* and *Halobacterium halobium*, a reduction of 10 kDa occurred following chemical deglycosylation (Jarrell *et al.*, 1996; Lechner and Wieland, 1989; Southam *et al.*, 1990). Of importance is the finding that at least some of the *Archaea* flagellins contain a number of the carbohydrate linkage consensus sequences, Asn-X-Ser/Thr, which represent N-linkage sites in *Methanococcus spp.* These are quite common, and one such linkage has been associated with an isolated glycosylated hexapeptide (Lechner *et al.*, 1985). A number of these sequences occur in the a-type flagellins. However, we believe that an O-linkage is more likely because this is apparently the case in the eubacterium *Campylobacter* (Doig *et al.*, 1996). The shift from alanine to threonine or serine in 170018 flagellins to 5933 and 5939 flagellins further suggests the importance of O-linkages in *Pseudomonas* flagellins. The *Archaea* may require glycosylation for proper flagellar filament assembly (Jarrel *et al.* 1996). Evidence indicates that this posttranslational modification occurs at the cell surface, even outside the cell membrane. If this is the important feature for such organelles, then it may be a special feature of export where phosphorylation is also coupled to this process. This seems more likely to be the case in *P. aeruginosa*. Also, the fact that *P. aeruginosa* is an opportunistic pathogen suggests a possible role for these modifications in some strains in virulence, such as concealment from antibody recognition, enzyme resistance, or attachment.

4. COMPARATIVE ANALYSIS OF THE *fliC* GENES

The first nucleotide sequence of a flagellin gene was reported by Totten and Lory (1990) using strain PAK, an a-type strain. As discussed above, the N- and C-terminal regions are conserved, and the central region is variable, as is common in other flagellins. One would predict that this difference would persist between the a- and b-type flagellins because the central region contains the antigenic site in enterics (Xiao-Song *et al.*, 1994).

Recently the *P. aeruginosa* flagellin genes of a number of strains have been sequenced and compared (Brickman *et al.*, 1997; Brimer *et al.*, 1997). In every case a- or b-antigen types were confirmed by colony blotting with type-specific Mabs. The b-genes were first obtained by the PCR technique modified from Winstanley *et al.* (1996) by C. Brickman (Brickman *et al.*, 1997) and followed by amplification and isolation of these a-type genes (Brimer *et al.*, 1997). This technique amplified 85 to 90% of the gene including all but a small amount of the highly conserved N- and C-terminal regions. In agreement with Winstanley's results, PCR products of the standard strains give the b-type flagellins of 1.25 Kb and the a-type of 1.02 Kb (Winstanley *et al.*, 1996; Winstanley *et al.*, 1994). The presence of a *fliC* gene fragment was also confirmed using PAK a-type flagellin gene (Totten and Lory, 1990) as the probe in Southern blots. The PCR products were cloned into *E. coli* for sequencing.

4.1. Flagellin b-Type Genes

The b-type flagellin genes of PAO1 code for proteins 488 amino acids long (derived molecular weight of 49.3 kDa) (Baker, GenBank accession no. 454775) whereas the a-type flagellin genes code for 391 amino acids (Totten and Lory, 1990). The PCR-amplified gene fragments cloned represented 90% of the flagellin protein. The missing regions included 46 N-terminal amino acids and 8 C-terminal amino acids. The protein sequence of strain M-2 was compared to the sequence of stains PAO1 and three Fla⁻ strains AK 1153, 96E, and 448bb (Brickman *et al.*, 1997). Strains M-2 and PAO1 are wild-types that synthesize b-type flagellins. Within the 434 amino acid sequence of the amplified M-2 flagellin fragment, a difference was found only in a single conserved amino acid. The glycine at amino acid 377 in PAO1 is changed to an alanine in M-2. This change was also noted in five other b-type strains. This finding definitively demonstrated and confirmed the homologous nature of these flagellins. Spangenberg *et al.* (1996) recently reported similar results for several putative b-type flagellins indicating the homogeneity of b-type flagellin gene amino acid sequences.

The high level of conserved sequences in b-types was utilized to delineate readily potential mutations in flagellins of Fla⁻ isolates (Brickman *et al.*, 1997), found in abnormally high numbers in CF *P. aeruginosa* isolates (Luzar *et al.*, 1985; Mahenthiralingam *et al.*, 1994). The advantage of the rapidity of the PCR technique is that it might be applied to confirm *P. aeruginosa* infectious levels in CF patients if it were validated for Fla⁻ strains. It was shown that the PCR technique is a reliable approach for detecting Fla⁻ flagellin gene fragments (Brickman *et al.*, 1997). The Fla⁻ strains were readily classified as b- or a-types based on gene fragment sizing of the amplified DNA. The three Fla⁻ gene fragments were compared with wild-types M-2 and PAO1 (Fig. 1). Protein sequence analysis of isolate 448bb illustrated that there is a stop codon at amino acid 84 (T-A-G nucleotides 230–232) which is apparently responsible for the Fla⁻ phenotype of this particular CF isolate. Despite the stop codon in the 448bb flagellin sequence, these proteins share 99% identity. Interestingly, the protein sequence of 96e (CF Fla⁻ isolate) shares 100% identity with the wild-type flagellin from PAO1. The Fla⁻ b-type strain AK1153 (Fla⁻) exhibits 99% homology with both PAO1 and M2 flagellins (data not shown). Although there may be a potential mutation in the flagellin regions not obtained by PCR, the central variable region and the major portion of the N- and C-terminals were well conserved. The complete N-terminal region would be particularly important to sequence because it is mainly responsible for the export of flagellin in the absence of a signal sequence. However, no evidence of precursor flagellins in the cytoplasm was detected for any Fla⁻ strains (Brickman *et al.*, 1997). Assuming that the N- and C-terminal regions not sequenced were not mutated, one or more regulatory mutations are possible explanations for the Fla⁻ phenotype of AK1153 and 96e. A lack of requirement for glutamine suggested they are not *rpoN* mutants. It was concluded that the PCR technique for amplifying flagellin DNA not only facilitated rapid isolation of the bulk of the flagellin gene for comparative studies, but also would afford rapid detection of both Fla⁺ and Fla⁻ strains in the sputum of CF patients.

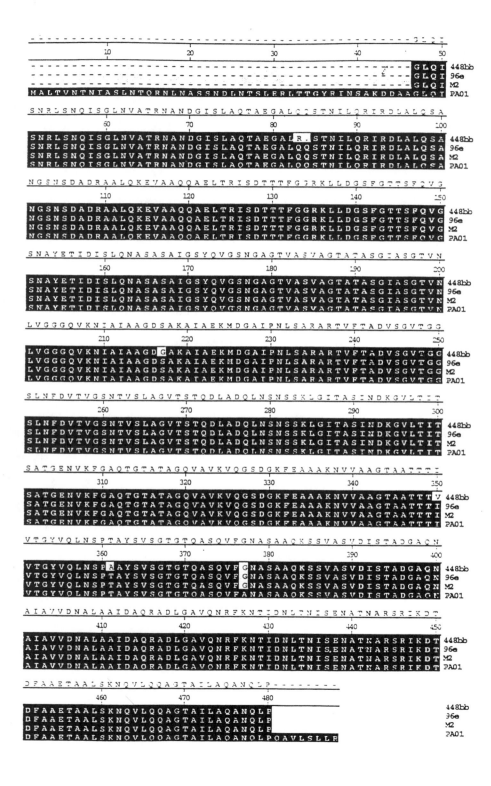

4.2. Flagellin a-Type Genes

A study of a-type flagellin genes showed increased heterogenicity compare to b-types (Brimer *et al.*, 1997; Spangenberg *et al.*, 1996). PCR products were confirmed as a-genes by size and specific Mab reactivity (Brimer *et al.*, 1997). PCR products were cloned and sequenced for strain 170018, 5933, and 5939, and these were compared with the PAK flagellin (Totten and Lory, 1990). A high degree of similarity was revealed among these a-type strains although some distinct differences were present (Fig. 2) (Brimer *et al.*, 1997). Amino acid sequence comparisons of the a-type strains 170018, 5933, 5939, and PAK (Totten and Lory, 1990) revealed high similarity among the a-type strains although there were some distinct differences (Fig. 2). There were only two amino acid residues that differ between the sequences of 170018 and 5939, whereas there were thirty-one residues that vary between 170018 and 5933. An amino acid residue change from an alanine to a threonine occurred at position 279 in the 5939 sequence compared to the other a-type strains. A seven amino acid difference was seen from residue 139 to 145 only in the 5933 sequence. Additionally, the amino acid sequence of strain 5933 showed variation in sequence not found in the other a-type strains. Another important occurrence in the 5933 sequence was that it exhibited single amino acid changes which involved changes from alanine to threonine at residues 265, 282, 284, and 286. Of the a-type strains compared, strain 5933 flagellin was the most divergent in amino acid sequence. A specific variation in the amino acid sequence of PAK was seen between residues 302–312 and 320–325 compared with the sequences of 170018, 5933, and 5939. Divergences from the PAK sequence were also found in other putative a-types (Spangenberg *et al.*, 1996). Certain deletion patterns of these variable amino acids may afford a means of forming subgroups of strains.

When comparing the sequence of 170018, 5933, and 5939 with the sequence of PAK (Totten and Lory, 1990), the number of N- and C-terminal residues missing is apparent because these residues are highly conserved among flagellin proteins (Figs. 1 and 2). Approximately 45 amino acids from the N-terminus and eight amino acids from the C-terminus were missing from the deduced amino acid sequence of the 1.02-kg fragments of *fli*C. When the amino acid sequence of PAK was used

Figure 1. Protein sequence comparison between nonmotile flagellin b-type strains CF448bb and CF96e (GenBank accession #AF016230) and wild-type b-type strains M2 (GenBank accession #AF016229) and PA01 (GenBank accession #U54775). Alignment was performed by the clustal method using the computer program DNA*.

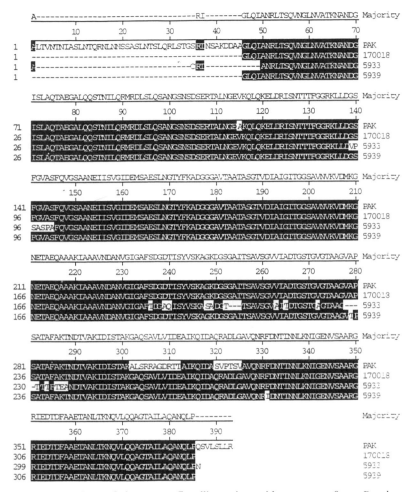

Figure 2. Comparison of the a-type flagellin amino acid sequences from *Pseudomonas aeruginosa* strains PAK (Totten and Lory, 1990), 170018 (GenBank accession #U76543), 5933 (GenBank accession #AF003906) and 5939 (GenBank accession #AF003905). Alignment was performed by the clustal method using the computer program DNA*.

for substitution of the missing highly conserved N- and C-terminal residues, the molecular mass of these missing amino acids was 5.6 kDa. Thus 5.6 kDa can be added to the molecular weight deduced which gives an adjusted molecular weight of approximately 40 kDa for all of these a-type strains. The adjusted molecular weights do not correspond to the apparent M_r which is 5 kDa higher for 170018 and PAK and 11–12 kDa higher for 5933 and 5939 flagellins. This discrepancy is explained in part by posttranslational modification resulting in the higher apparent M_r through the addition of glycosyl groups (see section 3.2).

Comparison of the amino acid sequences of an a-type (170018) and a b-type (PAO1) strain revealed significantly more divergence than comparing a-types (Fig. 3) (Brimer *et al.*, 1997). The most obvious difference between the a- and b-type flagellin sequences was that the 170018 sequence shows missing stretches of sequence in the central region. Specif-

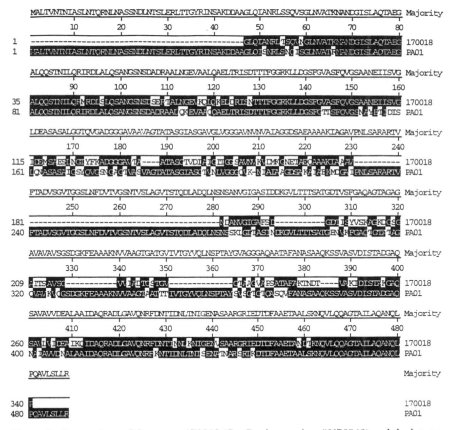

Figure 3. Comparison of the a-type 170018 (GenBank accession #U76543) and the b-type PA01 (GenBank accession #U54775) flagellin amino acid sequences. Alignment was performed by the clustal method using the computer program DNA*.

ically, these stretches of residues are 231–281, 294–304, 329–339, 351–364, and 384–387. This would account for the molecular weight differences between the two flagellin types. Interestingly, many single amino acid changes are from alanine to threonine/serine and the reverse is true in some cases. When the four a-types and PAO1 were compared, the N-terminal 106 amino acids and the C-terminal 68 amino acids were conserved, defining a central variable region.

5. FLAGELLA AND CHEMOTAXIS GENES AND REGULATION

5.1. Regulation of Flagella Biosynthesis and Assembly

Tsuda and Iino (1983a,b) first studied the flagellin genes which were mapped in regions I and II of the chromosome. They showed that region I contains one motility and five flagella cistrons. Region II contains ten flagella and two chemotaxis genes.

Initial observations that σ^{54} (rpoN) is required to express the genes of nitrogen assimilation, and especially glutamine synthestase, were followed by the discovery of a wide range of genes that are also transcribed by the σ^{54} − RNA polymerase, such as bacterial adhesins and enzymes for amino acid biosynthesis (Gussin *et al.*, 1986; Jin *et al.*, 1994; Thony and Hennecke, 1989). It has been shown that in *P. aeruginosa* the rpoN gene product is required to express pilin and also flagellin (Ishimoto and Lory, 1989; Totten and Lory, 1990; Totten *et al.*, 1990). When Totten and Lory (1990) cloned and sequenced the *fliC* gene, they identified several potential *rpoN*-dependent promoters and the upstream sequence resembling the recognition site for σ^{28}. Subsequently, it was found that the flagellin gene promoter is not of the σ^{54} class but is under the direct control of a σ^{28}-like factor (*fliA/rpoF*) (Starnbach and Lory, 1992). These authors proposed that the dependence on *rpoN* for transcription of flagellin results from its direct or indirect control of *fliA* expression or its control of a gene encoding an uncharacterized transcriptional activation of *fliC* (Starnbach and Lory, 1992). Interestingly, *fliA* also probably controls a Pseudomonal analogue of one component of the chemotaxis two-component regulatory system CheY, because it is just downstream of the σ^{28} (*fliA*) regulatory site. More recently, additional flagellar genes have been identified analogous to *Escherichia coli* and *Salmonella typhimurium* flagellar genes. These genes were identified primarily by transposon mutagenesis and complementation (Simpson *et al.*, 1995).

A putative *Salmonella* FliO homologue of apparently small molecular weight (14,842) was observed as a probable membrane protein because of its high hydrophobicity and potential membrane-spanning regions. Upstream from *fliO* is *fliN* (predicted MW, 16,608) and a putative *fliM* gene. In the same region with the putative *fliO* gene, *fliP* apparently contains a signal sequence. Interest in these genes arose from the observation by the Lory group that defects in them led to the absence of flagellin synthesis and nonpiliated adherence to mucin (Simpson *et al.*, 1995). Thus the location of *fliMNOP* in *Pseudomonas* is homologous to *E. coli*. *fliO*, which also showed analogy to mop3 of *Erwinia cartovora* has an unknown function in *E. coli* but is involved in *Pseudomonas* flagellin biosynthesis (Malakooti *et al.*, 1994). Loss of *fliO* monospecifically in *P. aeruginosa* results in loss of mucin adhesion. Its probable membrane location and association with the switch protein operon suggested that it may function to assemble the flagellin at this site. Unfortunately, it is apparently too unstable to isolate. *fliN* on *fliM* are interesting proteins because they are not only switch components, but also they are homologous to virulence determinants in *Yensinia* and *Salmonella* which are involved in the secretion apparatus.

Another gene group is involved in mucin adhesin and motility (Arora *et al.*, 1996). These are the *P. aeruginosa fliEFG* genes which are most homologous to genes coding for basal body proteins in *E. coli* and *Salmonella*. The *fliF* gene product is a part of the MS ring inserted into the cytoplasmic membrane and is a platform utilized for exporting various proteins during assembly of the flagellum (Kuboni *et al.*, 1992). Insertions into the *fliF* gene caused a loss of motility and adhesion. Arora *et al.* (1996) concluded that the *fliO* and *fliF* genes, which are involved in the assembly and secretion of flagellin proteins, are also associated with mucin adhesion. It did not seem that flagella filament (*fliC*) or its regulatory apparatus (RpoF) are directly involved because these gene functions can be separated from the nonpilus adhesion by mutational segmentation and still retain the adhesion (Arora *et al.*, 1996). Altogether, this remains a puzzling phenomenon because the basal body genes cannot themselves code for adhesion because these proteins are not exposed at the cell surface. Thus, an association between the flagellin biosynthesis and regulation and/or flagellar export pathway seems probable.

It appears that control of a motility component and adhesion occurs at the level of transcription. A mutation in *rpoN* affected both motility and adhesion. Recently, Ritchings *et al.*, (1995) identified a flagellar two-component regulatory system (FleS/FleR) with the closest homology of FleR to the *Caulobacter crescentus* FlbD regulatory component. Mutations in this system eliminated motility and the nonpilus adhesion in *P. aeruginosa*. The evidence suggests that this two-component system is controlled by RpoN. Interestingly, in *Caulobacter* the primary function of the flbD product is in ordering the expression of flagellar genes by a temporal control mechanism (Ranakrishran *et al.*, 1994).

5.2. Chemotaxis Genes

The chemotaxis system of *P. aeruginosa* was first characterized by Moulton and Montie (1979). The properties include response to a wide variety of amino acids. Both glucose and citrate taxis require induction by their respective carbon sources. Response to amino acids is controlled by the richness of the nitrogen source (Craven and Montie, 1985) with the best response observed for basic amino acids and peptides (Craven and Montie, 1985; Kelly-Wintenberg and Montie, 1994). Growth on organic nitrogen represses the taxis response to amino acids whereas growth on nitrate gives maximum response. A methylated chemotaxic receptor protein (MCP) of approximately 75 kDa was identified (Craven and Montie, 1983). A gene for a similar MCP has recently been cloned and sequenced (Ohtake *et al.* 1996). It has a derived molecular weight of 68,042, and show 30% identity to *E. coli* Tsr MCP. A mutant in the above gene (*pct1*) is deficient in taxis to glycine, serine, threonine, and valine. A cluster of chemotaxis genes has been located, *cheY*, *cheZ*, *cheA*, *cheB*, and *cheJ* (Ohtake *et al.*, 1996; Starnbach and Lory, 1992). The first four genes resemble those of *E. coli*, whereas a *cheT* gene shows no homology to any known sequences.

Chemotaxis to inorganic phosphate (Pi) has also been investigated (Ohtake *et al.*, 1996). An operon independent of *phoB* and *phoR*, but involving *phoU*, was detected. PhoU negatively regulates Pi taxis. This operon is located just downstream from the phosphate transport genes, and this system may utilize the Pi periplasmic-binding protein required for Pi transport under limiting Pi conditions (Poole and Hancock, 1984).

6. ANTIGENICITY AND IMMUNOGENICITY

6.1. Antigenicity

Lanyi (1970) first recognized two major groups of *P. aeruginosa* flagella using agglutination assays. One homologous group was designated type 1 and a second heterologous group type 2. Ansorg (1978) developed and expanded the typing system using indirect immunofluorescence. He designated the homologous group b-type and the heterologous a-type. A common subantigen (designated a_0) was found in a-types in various combinations with a_1, a_2, a_3, or a_4 subantigens.

An association was found between the antigen type and the M_r of a given flagellin (Allison *et al.*, 1985). Homologous b-type flagellins had a M_r of 53 kDa, whereas a-types ranged from 45 to 52 kDa because of their heterogeneity. All isolates whether clinical or environmental can be classified as a- or b-type. Direct cross-reactivity among a-types and among b-types, but not between a- and b-types, has been demonstrated in ELISA (Montie and Anderson, 1988) protection studies in traumatized mice (Drake and Montie, 1987; Holder and Neely, 1989; Holder *et al.*, 1982; Montie *et al.*, 1987) and in opsonophagocytosis studies (Anderson and Montie, 1987, 1989).

The lack of cross-reactivity between a- and b-type antigens suggests that a distinct antigenic region should be present in these flagellins. Utilizing the PCR technique to isolate the number of flagellar gene fragments of both types provided a means to approximate a central variable region for flagellin. The comparison of four a-types with PAO1 showed that the 106 N-terminal amino acids and the 68 C-terminal amino acids are conserved, defining a central variable region for flagellin (Brimer *et al.*, 1997). At least five hydrophilic peptide regions are observed in the four a-types in this region (Brimer *et al.*, 1997). These represent potential antigenic sites and could be further tested using analogous linear peptides. Because denatured and/or deglycosylated a-type flagellins do not lose their antigenicity, this suggests the presence of linear epitope(s). Similarly, three potential hydrophilic sites have been identified in the b-type variable region (Brickman *et al.*, 1997). Because these central variable regions of flagellins are believed to be exposed in the native filament, it is likely that these regions are significant antigenically in the intact bacterium.

6.2. Immunogenicity and Passive Protection

Flagella filament protein is an excellent antigen. This antigen is protective and generates a specific antibody in infected burned mice (Holder *et al.*, 1982; Montie *et al.*, 1987). Polyclonal rabbit antiflagellar antibody protects challenged burned mice and mice immunosuppressed with cyclophosphamide (Drake and Montie, 1988). Mouse monoclonal antiflagellar antibodies also protect in a burned mouse model (Rosok *et al.*, 1990). A correlation between the reactivity of a given mouse mAb in antimotility plate assays and its potential as a protective antibody was demonstrated (Montie *et al.*, 1987). Antibody made to strain 170018, one of the a-type flagellins with an M_r of 45 kDa, reacted best with its own strain and with strain 5940 flagella with a similar M_r. Human MAbs made to flagellins of M_r of approximately 49 kDa showed better cross-reactivity with high M_r flagellins 5933 and 5939 (Brimer *et al.*, 1997; Landsperger *et al.*, 1994). Although glycan groups on flagellins are not directly responsible for antigenicity, glycan groups (Brimer *et al.*, 1997) could influence the degree of cross-reactivity, or there could be several cross-reactive peptide sites involved in determining the common a_0 site in the a-type flagellin variable region.

Several human mAbs have been generated for both a- and b-type flagellins. These give good protection in neutropenic mouse models (Oishi, 1993), a burned mouse model (Landsperger *et al.*, 1993), and attenuate *P. aeruginosa* in induced pneumonia in an immunocompetent rat (Landsperger *et al.*, 1994).

In studies with CF patients, sera ELISA assays have demonstrated high levels of both a- and b-type antibodies associated with colonized patients (Anderson *et al.*, 1989). A shift in the type of flagellar antibody produced in CF patients was also seen from a distribution of IgG1, IgG2, and IgG3 in patients with a good pulmonary status to a high level of IgG1 and particularly IgG3 in patients with poor pulmonary status (Lagace *et al.*, 1995). This shift was suggested as a contributing factor to inflammation in the lung. It is clear that flagellated *P. aeruginosa* must have been present either initially and/or in a selected, spreading, resurgent population during outbreaks of expansion in the chronically infected CF lung. This is somewhat inconsistent with the unusual finding that a large percentage of CF isolates are Fla⁻ (Luzar *et al.*, 1985; Mahenthiralingam *et al.*, 1994). Although flagella and pili seem to be required for colonization, environmental and phagocytic pressures contribute to selection for a large number of Fla⁻ strains, where microcolony formation (e.g., mucoidy) replaces the need for invasive virulence factors in some strains. Flagella seem to be required for nonopsonic phagocytosis by murine macrophages (Mahenthiralingam *et al.*, 1994).

7. VIRULENCE

In 1980 McManus first showed that nonmotile mutants are avirulent in a scalded rat model (McManus *et al.*, 1980). At about the same time studies were completed in challenged burned mice showing a correlation between motility or chemotaxis and lack of virulence in several laboratory strains (Craven and Montie, 1981; Holder *et al.*, 1982). A more definitive study included results from a burned mouse study which showed that a Fla⁻ (nonmotile) isogenic mutant of the highly virulent *P. aeruginosa* strain M-2 lost its virulence (Montie *et al.*, 1982).

Importantly, a spontaneous reversion of the mutant to a Fla$^+$ (motile) form completely restored virulence. Fla$^-$ isolates from CF patients were also avirulent in the burned mouse model (Luzar and Montie, 1985). When virulence levels were titrated in the burned mouse model, it was shown that Fla$^-$ strains are 10^4 to 10^6 less virulent than parent or Fla$^+$ strains (Drake and Montie, 1987). A strain with a nonfunctional flagellum (Mot$^-$) also showed decreased virulence by a factor of three logs. Because the Fla$^-$ strains accumulate in the burned skin region, but do not invade the bloodstream, it was concluded that the flagellum are required for invasive virulence, as defined by this animal model. A similar infectious profile is observed when virulent strains (Fla$^+$) are neutralized with antiflagellar antibodies in the burned-mouse model (Drake and Montie, 1987).

8. OTHER PSEUDOMONADS

Compared to *P. aeruginosa*, little research has been performed with the flagella of other *Pseudomonas* species. Isolated *Burkholderia capacia* flagellins showed a pattern similar to *P. aeruginosa* in M_r distribution (Montie and Stover, 1983). On the basis of molecular weight and serology, a type I homologous group was identified, (M_r of 33,000) and a type II heterologous group with M_rs from 44,000 to 46,000. Results showed cross-reactivity among the isolates within each group. *P. (Xanthomonas) maltophilia* flagellin has an M_r of 33,000, whereas the value for *P. stutzeri* flagellin is 55,000. Values based on uncorrected sedimentation coefficients suggest masses between 38 and 42 kDa for *P. aeruginosa*, *P. putida*, *P. fluorescens*, and *P. syringae* pv. *morsprunorum* (Whiteside and Rhodes-Roberts, 1985). These values generally agree with estimated values based on methionine content from amino analysis. Although most of these molecular weight values do not agree with other published data (see section 2.3), the amino acid composition and distribution of the flagellins are generally consistent with those for most flagellins, no cysteine, little to no histidine, and small amounts of aromatic amino acids.

Winstanley *et al.*, (1994) cloned *P. putida* flagellin genes and reported a deduced molecular mass of 49.2 kDa which is similar to the SDS-PAGE value of 50 kDa for *P. putida* PR 52000. Strain PAN 8 flagellin has an unusually high SDS-PAGE value of 81 kDa compared to a predicted molecular mass of 68.2 kDa. These data suggest that some posttranslational additions occur in *P. putida* PAN 8 flagellin.

The flagellin of *P. pseudomallei* has been isolated and characterized (Brett *et al.*, 1994). Analysis by SDS-PAGE gave the same mass value of 43 kDa from several strains which was somewhat larger than the predicted value of 40.1 kDa obtained from amino acid analysis.

REFERENCES

Alam, M., and Oesterhelt, D., 1984, Morphology function and insolation of halobacteria flagella, *J. Mol. Biol.* **176:** 459–475.

Allison, J., Dawson, M., Drake, D., and Montie, T. C., 1985, Electrophoretic separation and molecular weight characterization of *Pseudomonas aeruginosa* H-antigen flagellins, *Infect. Immun.* **49:** 770–774.

Anderson, T. R., and Montie, T. C., 1989, Flagellar antibody stimulated opsono-phagocytosis of *Pseudomonas aeruginosa* associated with response to either a- or b-type flagellar antigen, *Can. J. Microbiol.* **35:** 755–763.

Anderson, T. R., and Montie, T. C., 1987, Opsonophagocytosis of *Pseudomonas aeruginosa* treated with antiflagellar serum, *Infect. Immun.* **55:** 3204–3206.

Anderson, T. R., Montie, T. C., Murphy, M. D., and McCarthy, V. P., 1989, *Pseudomonas aeruginosa* flagellar antibodies in patients with cystic fibrosis, *J. Clin. Microbio.* **27:** 2789–2793.

Ansorg, T., 1978, Flagella specific H antigenic schema of *Pseudomonas aeruginosa*, *Zentbl. Bakt. Mikrobiol. Hyg.* **242:** 228–238.

Ansorg, R., and Schmitt, W., 1980, Immunologishe and elektrophoretische charakterisierung der flagelline unterschied licher H-typen von *Pseudomonas aeruginosa. Med Microbiol. Immunol.* **163:** 217–226.

Arora, S. K., Titchings, B. W., Almira, E. C., Lory, S., and Ramphal, R., 1996, Cloning and characterization of *Pseudomonas aeruginosa fliF*, necessary for flagellar assembly and bacterial adherence to mucin, *Infect. Immun.* **64:** 2130–2136.

Atkinson, M., Allen, C., and Sequeira, L., 1992, Tyrosine phosphorylation of a membrane protein from *Pseudomonas solanacearum*, *J. Bacteriol.* **174:** 4356–4360.

Brett, P. J., Mah, D. C. W., and Woods, D. E., 1994, Isolation and characterization of *Pseudomonas pseudomallei* flagellin proteins, *Infect. Immun.* **62(5):** 1914–1919.

Brickman, C. S., Kelly-Wintenberg, K., and Montie, T. C., 1997, Comparative analysis of flagellin genes of wild-type fla+ and non-motile fla− *P. aeruginosa* strains, In: *Abstracts of the 97th General Meeting of the American Society for Microbiology*, 1997, American Society for Microbiology, Washington D.C., #D-42, p. 215.

Brimer, C. D., Kelly-Wintenberg, K., and Montie, T. C., 1997, Cloning and characterization of *Pseudomonas aeruginosa*. a-type flagellin genes, In: *Abstracts of the 97th General Meeting of the American Society for Microbiology*, 1997, American Society for Microbiology, Washington, D.C., #D-43, p. 215.

Cha, H., R. Nichols, and Montie, T. C., 1996, Posttranslational phosphorylation of *Pseudomonas aeruginosa* by an envelope kinase in vitro, In: *Abstracts of the 96th General Meeting of the American Society for Microbiology*, 1996. American Society for Microbiology, Washington, D.C., D-73, p. 254.

Craven, R. C., and Montie, T. C., 1985, Effect of nitrogen source on the chemotaxis of *Pseudomonas aeruginosa* toward amino acids, *J. Bact.* **164:** 544–549.

Craven, R. C., and Montie, T. C., 1981, Motility and chemotaxis of three strains of *Pseudomonas aeruginosa* used for virulence studies, *Can. J. Microbiol.* **25:** 458–460.

Craven, R. C., and Montie, T. C., 1985, Effect of nitrogen source on the chemotaxis of *Pseudomonas aeruginosa* toward amino acids. *J. Bacteriol.* **164:** 544–549.

Doig, P., Kinsella, N., Guerry, P., and Trust, T. J., 1996, Characterization of post-transnational modification of *Campylobacter* flagellin: Identification of a sero-specific glycosylation moiety. *Mol. Microbiol.* **19**(2): 379–387.

Drake, D., and Montie, T. C., 1988, Flagella, motility, and invasive virulence of *Pseudomonas aeruginosa. J. Gen. Microbiol.* **134**: 43–52.

Drake, D., and Montie, T. C., 1987, Protection against *Pseudomonas aeruginosa* infection by passive transfer of anti-flagellar serum, *Can. J. Microbiol.* **33**: 755–763.

Garnak, M., and Reeves, H. C., 1978, Phosphorylation of isocitrate dehydrogenase of *Escherichia coli, Science.* **203**: 1111–1112.

Gussin, G. N., Ronson, C. W., and Ausubel, F. M., 1986, Regulation of nitrogen fixation genes, *Annu. Rev. Genet.* **20**: 567–591.

Holder, I. A., and Neely, A. N., 1989, Combined host and specific anti-Pseudomonas-directed therapy for *Pseudomonas aeruginosa* infections in burned mice: Experimental results and theoretic considerations, *J. Burn Care Rehabil.* **10**: 131–137.

Holder, I. A., Wheeler, R., and Montie, T. C., 1982, Flagellar preparations from *Pseudomonas aeruginosa* animal protection studies, *Infect.* Immun. **35**: 276–280.

Holloway, B. W., 1996, *Pseudomonas* genetics and taxonomy, in: *Molecular Biology of Pseudomonas.*, (T. Nakazawa, K. Furakawa, D. Haas, and S. Silver, eds.), ASM Press, Washington, D.C., pp. 22–32.

Ishimoto, K. S., and Lory, S., 1989, Formation of pilin in *Pseudomonas aeruginosa* requires the alternative sigma factor (RpoN) of RNA polymerase, *Proc. Natl. Acad. Sci. USA.* **86**: 1954–1957.

Jarrell, K. F., Bayley, D. P., and Kostyukova, A. S., 1996, The archeal flagellum: A unique motility structure, *J. Bacteriol.* **178**(17): 5057–5064.

Jin, S., Ishimoto, K., and Lory, S., 1994, Nucleotide sequence of the *rpoN* gene and characterization of two downstream open reading frames in *Pseudomonas aeruginosa*, *J. Bacteriol.* **176**: 1313–1322.

Kelly-Wintenberg, K., Anderson, T., and Montie, T. C., 1990, Phosphorylated tyrosine in the flagellum filament of *Pseudomonas aeruginosa*, *J. Bacteriol.* **172**: 5135–5139.

Kelly-Wintenberg, K., South, S., and Montie, T. C., 1993, Tyrosine phosphate in a- and b-type flagellins of *Pseudomonas aeruginosa*, *J. Bacteriol.* **175**: 2458–2461.

Kelly-Wintenberg, K., and Montie, T. C., 1994, Chemotaxis to oligopeptides by *Pseudomonas aeruginosa*, *Appl. Environ. Microbiol.* **60**: 363–367.

Kennedy, P. J., and Potts, M., 1996, Fancy meeting you here! A fresh look at "prokaryotic" protein phosphorylation, *J. Bacteriol.* **178**: 4759–4764.

Kuboni, T., Shimamoto, N., Yanaguchi, S., Yanaguchi, K., and Aizawa, S., 1992, Morphological pathway of flagellar assembly in *Salmonella typhimurium*, *J. Mol. Biol.* **226**: 433–446.

Landsperger, W. J., Kelly-Wintenberg, K., Montie, T. C., Knight, L. S., Hansen, M. B., Huntenburg, C. C., and Schneidkraut, M. J., 1994, Inhibition of bacterial motility with human anti-flagellar monoclonal antibodies attenuates *Pseudomonas aeruginosa*-induced pneumonia in the immunocompetent rat, *Infect. Immun.* **62**: 4825–4830.

Landsperger, W. J., South, S. L., Kelly-Wintenberg, K., Montie, T. C., and Huntenberg, C. C., 1993, Immunoreactivity of human IgG1 monoclonal antibodies with flagella of *Pseudomonas aeruginosa*, *Abstracts ASM*, **E55**: 152.

Lanyi, B., 1970, Serological properties of *Pseudomonas aeruginosa*. II. Type-specific thermolabile (flagellar) antigens, *Acta Microbiol. Acad. Sci. Hung.* **17**: 35–48.

Lechner, J. F., and Wieland, F., 1989, Structure and biosynthesis of prokaryotic glycoproteins, *Ann. Rev. Biochem.* **58**: 173–194.

Lechner, J. F., Weiland, F., and Sumper, M., 1985, Transient methylation of dilichol oligosacharides as an obligatory step in halobacterial sulfated glycoprotein biosynthesis, *J. Biol. Chem.* **260:** 8984–8989.

Legace, J., Pelaquin, L., Kermani, P., and Montie, T. C., 1995, IgG subclass responses to *Pseudomonas aeruginosa* a- and b-type flagellins in patients with cystic fibrosis: A prospective study. *J. Med. Microbiol.* **43:** 270–276.

Luzar, M. A., and Montie, T. C., 1985, A virulence and altered physiological properties of cystic fibrosis strains of *Pseudomonas aeruginosa, Infect. Immun.* **50:** 572–576.

Luzar, M. A., Thomassen, M. J., and Montie, T. C., 1985, Flagella and motility alterations in *Pseudomonas aeruginosa* strains from patients with cystic fibrosis: Relationship to patient clinical condition, *Infect. Immun.* **50:** 577–582.

Mahenthiralingam, E., Campbell, M. E., and Speert, D. P., 1994, Nonmotility and phagocytic resistance of *Pseudomonas aeruginosa* isolates from chronically colonized patients with cystic fibrosis, *Infect. Immun.* **62:** 596–605.

Malakooti, J., Ely, B., and Matsumura, P., 1994, Molecular characterization, nucleotide sequence, and expression of the *fliO, fliP, fliQ,* and *fliR* genes of *Escherichia coli, J. Bacteriol.* **176:** 189–197.

Matsumoto, A., Hong, S. K., Ishizuka, H., Horinouchi, S., and Beppu, T., 1994, Phosphorylation of the AFSR protein involved in secondary metabolism in streptomyces species by a eucaryotic-type protein kinase. *Gene* **146:** 47–56.

McCartney, B., Howell, L., Kennelly, P. J., and Potts, M., 1997, Protein tyrosine phosphorylation in the cyanobacterium *Anabaena* sp. strain PCC 7120, *J. Bacteriol.* **179**(7): 2314–2318.

McManus, A. T., Moody, E. E., and Mason, A. D., 1980, Bacterial motility: A component in experimental *Pseudomonas aeruginosa* burn wound sepsis, *Burns* 6: 235–239.

Moens, S., Michiels, K., and Vanderleyden, J., 1995, Glycosylation of the flagellin of the polar flagellum of *Azospirillum brasilense,* a gram-negative nitrogen-fixing bacterium, *Microbiology* **141:** 2651–2657.

Montie, T. C., and Anderson, T. C., 1988, Enzyme linked immunosorbent assay for detection of *Pseudomonas aeruginosa* H (flagellar) antigen, *Eur. J. Clin. Microbiol. Infect. Dis.* **7:** 256–260.

Montie, T. C., Dorner, F., McDonel, J. C., and Mitterer, A., U.S. Patent 4,831,121, May 1989.

Montie, T. C., Doyle-Huntzinger, D., Craven, R. C., and Holder, F. A., 1982, Loss of virulence associated with absence of flagellum in an isogenic mutant of *Pseudomonas aeruginosa* in the burned-mouse model, *Infect. Immun.* **38:** 1296–1298.

Montie, T. C., Drake, D., Sellin, H., Slater, O., and Edmonds, S., 1987, Motility, virulence, and protection with a flagella vaccine against *Pseudomonas aeruginosa* infection, In: *Antibiotics and Chemotherapy* (D. Doring, I. A. Holder, and K. Botzenhart, eds.), Karger, Basel, pp. 233–248.

Montie, T. C., Philips, D., and Landsperger, W., 1997, Characterization of monoclonal antibodies to *Pseudomonas aeruginosa* type-A flagellar antigen, *Behring Inst. Mitt.* **98:** 424–433.

Montie, T. C., and Stover, G. B., 1983, Flagellar preparations from *Pseudomonas* species: Isolation and characterization by molecular weight, *J. Clin. Microbiol.* **18:** 452–456.

Moulton, R. C., and Montie, T. C., 1979, Chemotaxis by *Pseudomonas aeruginosa. J. Bacteriol.* **137:** 274–280.

Ochi, H., Ohtsuka, H., Yokota, S., Uezumi, I., Terashima, M., Irie, K., and Noguchi, H., 1991, Inhibitory activity on bacterial motility and *in vivo* protective activity of human monoclonal antibodies against flagella of *Pseudomonas aeruginosa, Infect Immun.* **59**(2): 550–554.

Ohtake, H., Kato, J., Kuroda, A., Taguchi, K., and Sakai, Y., 1996, Chemotactic signal transduction network in Pseudomonas aeruginosa, in: *Molecular Biology of Pseudomonads* (T. Nadazawa, K. Furakawa, D. Haas, S. Silver, eds.), ASM Press, Washington, D.C., pp. 188–194.

Oishi, K., Sonoda, F., Iwagaki, A., Penglertnapagorn, P., Watanabe, K., Nagatake, T., Siadiak, A., Pollack, M., and Matsomoto, K., 1993, Therapeutic effects of human antiflagella monoclonal antibody in a neutropenic murine model of *Pseudomonas aeruginosa* pneumonia, *Antimicrob. Agents Chemother.* **37:** 164–170.

Poole, K., and Hancock, R. E. W., 1983, Secretion of alkaline phosphatase and phospholipase C is specific and does not involve an increase in outer membrane permeability. *FEMS Microbiol. Lett.* **16:** 25–29.

Ranakrishran, F., Zhao, J-L., and Newton, A., 1994, Multiple stimulational proteins are required for both transcriptional activation and flagellar genes, *J. Bacteriol.* **176:** 7587–7600.

Ray, M. K., Kumar, G. S., and Shivaji, S., 1994, Tyrosine phosphorylation of a cytoplasmic protein from the Antarctic psychrotrophic bacterium *Pseudomonas syringae*, *FEMS Microbiol. Lett.* **122:** 49–54.

Ritchings, B. W., Almira, E. C., Lory, S., and Ramphal, R., 1995, Cloning and phenotypic characterization of *fleS* and *fleR*, new response regulators of *Pseudomonas aeruginosar which regulate motility and adhesion to mucin*, *Infect. Immun.* **63:** 4868–4876.

Rosok, M. J., Stebbins, M. R., Connelly, K., Lostrom, M. E., and Sidiak, A. W., 1990, Generation and characterization of murine antiflagellum monoclonal antibodies that are protective against lethal challenge with *Pseudomonas aeruginosa*, *Infect. Immun.* **58:** 3819–3828.

Rotering, H., and Dorner, F., 1989, Studies on a *Pseudomonas aeruginosa* flagella vaccine, in: *Antibiotics and Chemotherapy*, (N. Hoiby, S. S. Pederson, G. H. Shanel, G. Doring, and I. A. Holder, eds.), Karger, Basel, Vol. 42, pp. 218–228.

Simpson, D. A., Ramphal, R., and Lory, S., 1995. Characterization of *Pseudomonas aeruginosa fliO*, a gene involved in flagellar biosynthesis and adherence, *Infect. Immun.* **63:** 2950–2957.

Smith, S. C., Kennelly, P. J., and Potts, M., 1997, Protein tyrosine phosphorylation in the archaea, *J. Bacteriol.* **179:** 2418–2420.

South, S., Nichols, R., and Montie, T. C., 1994, Tyrosine kinase activity in *Pseudomonas aeruginosa*, *Mol. Microbiol.* **12:** 903–910.

Southam, G., Kalmokoff, M. L., Jarrell, K. F., Koval, S. F., and Beveridge, T. J., 1990, Isolation, characterization, and cellular insertion of the flagella from two strains of the archaeabacterium *Methanospirillum hungatei*, *J. Bacteriol.* **176**(6): 3221–3228.

Spangenberg, C., Heuer, T., Burger, C., and Tummïler, B., 1996, Genetic diversity of flagellins of *Pseudomonas aeruginosa*, *FEBS Lett.* **396**(2–3): 213–217.

Starnbach, M. N., and Lory, S., 1992, The *fliA* (*rpoF*) gene of *Pseudomonas aeruginosa* encodes an alternative sigma factor required for flagellin syntheses, *Mol. Microbiol.* **6**(4): 459–469.

Suzuki, T., and Iino, T., 1980, Isolation and characterization of multiflagellate mutants of *Pseudomonas aeruginosa*, *J. Bacteriol.* **143:** 1471–1479.

Thony, B., and Hennecke, H., 1989, The -24/-12 promoter comes of age, *FEMS Microbiol.* **5:** 341–357.

Totten, P. A., Lara, J. C., and Lory, S., 1990, The *rpoN* gene product of *Pseudomonas aeruginosa* is required for expression of diverse genes, including the flagellin gene, *J. Bacteriol.* **172:** 389–396.

Totten, P. A., and Lory, S., 1990, Characterization of the type a flagellin gene from *Pseudomonas aeruginosa* PAK, *J. Bacteriol.* **172:** 7188–7199.

Tsuda, N., and Iino, T., 1983a, Ordering of the flagellar genes in *Pseudomonas aeruginosa* by insertions of mercury transposon Tn501, *J.* Bacteriol. **153:** 1008–1017.

Tsuda, N., and Iino, T., 1983b, Transductional analysis of the flagellar genes in *Pseudomonas aeruginosa*, *J. Bacteriol.* **153:** 1018–1926.

Virjii, M., Saunders, J. R., Siims, G., Makepeace, K., Maskell, D., and Fergeson, D. J., 1993, Pilus-facilitated adherence of *Neisseria meningitidis* to human epithelial and endothelial cells: Modulation of adherence phenotype occurs concurrently with changes in primary amino acid sequence and the glycosylation status of pilin, *Mol. Microbiol.* **10:** 1013–1028.

Wang, J. Y., and Koshland, D. E., Jr., 1978, Evidence for protein kinase activities in the prokaryote *Salmonella typhimurium*, *J. Biol. Chem.* **253:** 7605–7608.

Whiteside, T. M., and Rhodes-Roberts, M. E., 1985, Biochemical and serological properties of purified flagella and flagellins of some *Pseudomonas spp*, *J. Gen. Microbiol.* **131:** 873–883.

Winstanley, C., Coulson, M. A., Wepner, B., Morgan, J. A. W., and Hart, C. A., 1996, Flagellin gene and protein variation amongst clinical isolates of *Pseudomonas aeruginosa*, *Microbiology* **142:** 2145–2151.

Winstanley, C., Morgan, J. A. W., Pickup, R. W., and Saunders, J. R., 1994, Molecular cloning of two *Pseudomonas* flagellin genes and basal body structural genes, *Microbiology* **140:** 2019–2031.

Xiao-Song, H. E., Rivkina, M., Stocker, B. A. D., and Robinson, W. S., 1994, Hypervariable region IV of Salmonella *fliC* encodes a dominant surface epitope and a stabilizing factor for functional flagella, *J. Bacteriol.* **176:** 2406–2414.

Zhang, C-C., 1996, Bacterial signaling involving eukaryotic-type kinases, *Mol. Microbiol.* **20**(1): 9–15.

Selected Industrial Biotransformations

9

MARCEL G. WUBBOLTS and
BERNARD WITHOLT

1. INTRODUCTION

The development and application of biocatalysts, be it whole cell or enzymatic systems, rely on understanding the metabolism of biotic and xenobiotic substances by biological systems. The knowledge of metabolic routes has traditionally been applied to synthesize 'common' metabolites, such as citric acid, amino acids, and more complex substances like vitamins or antibiotics. The vast amount of information currently available on metabolic pathways, however, also allows us to design and construct biocatalysts that perform reactions leading to the formation of substances unknown to biological systems.

Some of the compounds now produced by microorganisms were long thought to be confined to organic synthesis because of their low solubility in aqueous systems and their toxicity to biological systems. The observation that (partially) purified enzymes and even intact cells can be efficiently utilized in mixed solvent systems or even in the presence of anhydrous solvents has caused a boom in applying biological catalysts in organic synthesis in the last two decades.

Most of the biocatalysts currently applied in industry have been developed from natural isolates of microorganisms, selected by a screening procedure. Strikingly many of these isolates belong to the genus *Pseudomonas*, which has adapted to a wide variety of compounds. The metabolic versatility of these microorganisms was already observed in

MARCEL G. WUBBOLTS • DSM Research, 6160 MD Geleen, The Netherlands.   BERNARD WITHOLT • Institute of Biotechnology, ETH Hönggerberg, HPT, CH-8093 Zürich, Switzerland.

Pseudomonas, edited by Montie. Plenum Press, New York, 1998.

the early days of *Pseudomonas* microbiology (den Dooren de Jong, 1926; Stanier *et al.*, 1966) and is probably so because *Pseudomonas* strains can share and acquire genetic information required to degrade a large number of compounds (Holloway *et al.*, 1992; van der Meer, 1994). Therefore it is no surprise that these microorganisms are widely encountered in nature and that pseudomonads are found in ecological niches ranging from contaminated soil to pathological samples from patients.

1.1. Metabolites of *Pseudomonas* in Biotechnology

A well known biotechnological production process for a natural compound on an industrial scale with the help of *Pseudomonas* strains is the production of Vitamin B_{12} by *P. denitrificans* (Florent, 1986). This process is especially efficient in media (such as beet molasses) rich in betaine (Fa *et al.*, 1984; Kusel *et al.*, 1984). Production of vitamin B_{12} by methylotrophic *Pseudomonas* sp. AM-1 on the cheap carbon source methanol (Tsuchiya and Nishio, 1980) has not been able to replace the successful original industrial process (Florent, 1986).

Microbial polysaccharides are important biotechnological products applied in the food, agricultural, and chemical industries (Baird *et al.*, 1983; Bushell, 1983; reviewed in Morris, 1992; Sutherland, 1990, 1991). Exopolysaccharide production is common to *P. fluorescens*, *P. putida*, *P. aeruginosa*, *P. cepacia*, and other *Pseudomonas* strains (Allison and Goldsbrough, 1994; Marques *et al.*, 1986; Sutherland *et al.*, 1994). These compounds are believed to serve as a defense mechanism against the host immune system and as protection against bacteriophages or dehydration. Plant pathogens may use polysaccharides as an anchoring material (Glazer and Nikaido, 1995).

The production of *Pseudomonas* exopolysaccharides, such as alginate, from *P. aeruginosa* and gellan gum from *P. elodea* have been commercialized, but these polysaccharides have not reached the application level of Xanthan gum, the principally used microbial polysaccharide from *Xanthomonas campestris* (Glazer and Nikaido, 1995). The role of alginates in cystic fibrosis has triggered numerous scientific efforts to elucidate the molecular biology of their synthesis (Deretic *et al.*, 1993; Martins and Sa-Correia, 1991; Schurr *et al.*, 1994), which may well result in more efficient ways to produce this polysaccharide frequently used to immobilize cells and enzymes in biotransformative procedures.

Other biopolymers of industrial importance are the bacterial polyesters, poly-3-hydroxyalkanoates (Anderson and Dawes, 1990; reviewed in Steinbuchel and Valentin, 1995). These biopolymers are biodegradable and provide an environmentally friendly alternative to common plastics.

The commercialization of Biopol, a poly-3-hydroxybutyrate-poly-3-hydroxyvalerate copolymer from *Alcaligenes eutrophus* by Zeneca (ICI) has been realized (Anonymous, 1985) and the development of other biodegradable plastics is underway. Poly-3-hydroxyalkanoates from *P. oleovorans* (de Smet *et al.*, 1983a), an alkane degrading microorganism (Schwartz, 1973; Schwartz and McCoy, 1973), are of particular interest because these bacterial polyesters can be produced with variable side groups, depending on the (co)substrate added to the culture (Lageveen *et al.*, 1988; Lenz *et al.*, 1992; reviewed in Steinbuchel and Valentin, 1995). Biopolymers with variable length side chains, unsaturated side chains, and aromatic pendant groups have been produced by this microorganism (Steinbuchel and Valentin, 1995), but large scale commercial application of these medium-chain-length poly-3-hydroxyalkanoates has yet to be realized.

1.2. Biocatalysis Versus Chemical Catalysis

The growing demand for optically pure drugs and agrochemicals, devoid of enantiomeric ballast, and the improvement of chemical catalysts has initiated the development of enantiospecific synthetic routes for chiral substances. For many compounds, however, chemical catalysts that introduce a chiral center or distinguish between the optical antipodes of an existing chiral center are unavailable.

By virtue of their regio- and stereoselectivity, enzymes can be utilized to complement the catalytic arsenal of organic chemists, and numerous multistep chemical routes to optically active compounds have been developed by now. The integration of one or more enzymatic steps that catalyze a regioselective reaction or introduce chirality in such pathways provides a route to products difficult to synthesize by 'classical' organic chemistry.

1.2.1. Enzyme Classes and Their Practical Application

The utilization of enzymes in biocatalytic processes, either in (partially) purified form or in whole-cell systems, is not evenly distributed over the different classes of the International Union of Biochemistry Enzyme Commission (EC) (Faber, 1995). Table 1 summarizes enzymes that have been utilized in synthesis at laboratory or industrial scale.

Biocatalysts based on the hydrolases (EC class 3) are most utilized in synthetic reactions, because these enzymes are cofactor independent, are in some cases commercially available, and because many hydrolases can be used in (the presence of) organic solvents. Oxidoreductases (EC

Table I. EC Classification of *Pseudomonas* Biocatalysts

Strain[a]	Enzyme	EC number[b]	Process or product	Reference or patent
Class 1-Oxidoreductases				
P. testosteroni	3β,17β-Hydroxysteroid dehydrogenase	1.1.1.51	Ketosteroids, muscarine	Deamici *et al.*, 1991
P. diminuta	Malate dehydrogenase	1.1.1.82	NAD(P)H regeneration, L-alanine	Suye *et al.*, 1992
P. putida M-10	Morphine dehydrogenase	1.1.1.218	Hydromorphone, hydrocodone	Hailes and Bruce, 1993
Pseudomonas sp. PED	Alcohol dehydrogenase	1.1.1.-	(*R*)-Alcohols	Bradshaw *et al.*, 1992a
P. acidovorans	Ketonic acid reductase	1.1.1.-	(*S*)-β,δ,γ-Hydroxycarboxylic acids	Kula and Peters, 1993
P. maltophilia	Ketopantoic acid reductase	1.1.1.-	D-Pantoic acid	Shimizu and Yamada, 1991
P. putida ATCC 31916	Catechol-1,2-dioxygenase	1.13.11.1	*cis,cis*-Muconic acids	Maxwell, 1988
Pseudomonas sp.	Catechol-2,3-dioxygenase	1.13.11.2	Picolinic acid and pyridine derivatives	Asano *et al.*, 1994; Hagedorn *et al.*, 1989
P. putida UV4	Benzene dioxygenase	1.14.12.3	*cis*-Dihydrodiols	Boyd *et al.*, 1993a,b
P. putida NCIMB 11767, *P. putida* NG1, *P. putida* F1, *P. putida* F39/D	Toluene dioxygenase	1.14.12.11	Sulfoxides, *cis*-dihydrodiols	Allen *et al.*, 1995; Resnick *et al.*, 1993; Brand *et al.*, 1992; Jenkins *et al.* 1987
P. putida PpG7	Naphthalene dioxygenase	1.14.12.12	Indigo, *cis*-dihydrodiols	Ensley, 1994; Resnick *et al.*, 1993
P. putida	Salicylate hydroxylase	1.14.13.1	Catechols	Suzuki *et al.*, 1991a
P. putida NCIMB 10007	Cyclopentanone monooxygenase	1.14.13.16	Baeyer–Villiger catalyst, sulfoxides	Roberts and Willetts, 1993
P. putida ATCC 17453	2,5-Diketocamphane-1,2-monooxygenase	1.14.15.2	Baeyer–Villiger catalyst	Taylor and Trudgill, 1986
P. aeruginosa, *P. oleovorans*,	Alkane monooxygenase	1.14.15.3	Optically active epoxides,	Idemitsu, 1986; Kusunose *et*

Organism	Enzyme	EC number	Product/Application	References
P. denitrificans			optically active sulfoxides, heterocyclic aromatic acids	al., 1967; May and Katopodis, 1986; van der Linden and Huybrechtse, 1967; reviewed in Wubbolts et al., 1995
P. putida mt-2	Xylene monooxygenase	1.14.15.-	Heterocyclic aromatic acids, optically active styrene oxides	Kiener, 1992; Wubbolts et al., 1995
P. aeruginosa, P. cepacia	Tridecanone monooxygenase	1.14.-.-	Baeyer–Villiger catalyst	Britton and Markovetz, 1977; Forney and Markovetz, 1969
P. putida	*p*-Cresol methylhydroxylase	1.14.-.-	Optically active benzylic alcohols	McIntire et al., 1984, 1985
P. putida JD1	*p*-Ethylphenol methylene hydroxylase	1.14.-.-	(*R*)-Benzylic alcohols	Reeve et al., 1989, 1990
P. putida UV4	Benzocycloalkane hydroxylase	1.14.-.-	(*R*)-Benzylic alcohols	Boyd et al., 1991
Class 2-Transferases				
Pseudomonas sp. MS31	Serine hydroxymethyl transferase	2.1.2.1	L-Serine	Watanabe et al., 1987a
Pseudomonas sp. NRRL B-18375	Cyclodextrin glycosyl transferase	2.4.1.19	Cyclodextrins	Allenza et al., 1991
P. riboflavina	Alanine dioxovalerate aminotransferase	2.6.1.-	δ-Aminolevulinate	Rhee et al., 1987
P. fluorescens, P. fluorescens ATCC 11250	Phenylalanine transaminase	2.6.1.58	L-Phenylalanine	Evans et al., 1987; Marquardt et al., 1988
Class 3-Hydrolases				
P. fluorescens DSM 7033, *Pseudomonas sp.*	Esterase	3.1.1.2	(*S*)-1-Phenylalkanols, (*S*)-α-cyanobenzylalcohols, (*S*)-naproxen	Chan and Salazar, 1991; Krell and Rasor, 1993; Mitsuda et al., 1991
P. aeruginosa, P. cepacia, P. fluorescens, Pseudomonas	Lipase	3.1.1.3	Enantioselective ester hydrolysis, alcoholysis and	Reviewed in Drauz and Waldmann, 1995; Inagaki

(continued)

Table I. (*Continued*)

Strain[a]	Enzyme	EC number[b]	Process or product	Reference or patent
sp., P. glumae			ester synthesis	et al., 1992; Jaeger and Wohlfarth, 1993; Kanerva et al., 1994; Soberonchavez and Palmeros, 1994
P. myxogenes	Esterase	3.1.1.-	(S)-2-Ethylhexanol	Hou, 1993
P. fragi IFO 3458	Esterase	3.1.1.-	L-Carnitine	Francalanci et al., 1993
P. fluorescens W, P. amyloderamosa	α-Glucosidase	3.2.1.20	Starch processing, α-1,4-exoglucanase	Kelly and Fogarty, 1983
P. amyloderamosa	Isoamylase	3.2.1.68	Amylopectin and glycogen hydrolysis, α-1,6-endoglucanase, amylose from starch	Amemura and Futai, 1992
Pseudomonas sp.	S-Adenosyl-L-homocysteine hydrolase	3.3.1.1	S-Adenosyl-L-homocysteine	Yamada et al., 1986
Pseudomonas sp., Pseudomonas sp. AD1	Epoxide hydrolase	3.3.2.3	Resolution of epoxides	Jacobs et al., 1991; Warhurst and Fewson, 1994
P. putida ATCC 12633, Pseudomonas sp., P. fluorescens IFO3081	Amidase	3.5.1.4	D- and L-Amino acids, α-methylphenylalanine	Kerkhoffs and Boesten, 1989; Sakashita et al., 1993; Schoemaker et al., 1992; Ube, 1982
Pseudomonas sp.	N-Acetyl-L-amino acid acylase	3.5.1.14	L-Amino acids, peptide antibiotics	Konishi et al., 1992; Verkhovskaya and Yamskov, 1991
Pseudomonas sp. NS671	L-N-Carbamoylase	3.5.1.77	L-Amino acids	Ishikawa et al., 1993

Microorganism	Enzyme	EC number	Product	Reference
Pseudomonas sp. AJ-11220, P. putida 77	D-N-Carbamoylase	3.5.1.77	D-Amino acids	Shimizu et al., 1986; Syldatk et al., 1987; Yokozeki et al., 1987b
Pseudomonas sp., P. diminuta	Glutaryl acylase	3.5.1.-	7-Aminocephalosporanic acid (7-ACA)	Fukagawa et al., 1992; Ichikawa et al., 1988
P. fluorescens DSM84, P. putida, Pseudomonas sp., Pseudomonas sp. AJ-11220, P. striata	D-Hydantoinase	3.5.2.2	D-Amino acids	Morin et al., 1990; Takahashi, 1983; Yokozeki et al., 1987a,b
Pseudomonas sp. NS671	L-Hydantoinase	3.5.2.2	L-Amino acids	Ishikawa et al., 1993; Olivieri et al., 1985
P. solanacearum	Lactamase	3.5.2.-	Resolution of lactams	Evans et al., 1992
P. desmolytica AJ-11898; P. thiazolinophilum	D,L-2-Amino-Δ^2-thiazoline-4-carboxylate hydrolase	3.5.4.-	L-Cysteine	Sano et al., 1977; Yokozeki et al., 1988
Pseudomonas sp.	Nitrilase	3.5.5.1	Glycine	Shimizu et al., 1993
P. synthaxa, P. ovalis	Nitrilase	3.5.5.1	D- and L-Lactic acid	Yamagami et al., 1993
Pseudomonas sp., Pseudomonas sp. MY-1	Nitrilase	3.5.5.1	(R)-Mandelic acid, α-hydroxycarboxylic acids	Hashimoto et al., 1994; Layh et al., 1992; Murakami, 1993
Pseudomonas sp., P. putida 109, Pseudomonas sp. 113	Haloacid dehalogenase	3.8.1.2	D-3-Chlorolactic acid, D- and L-lactic acid, resolution of haloacids	Motosugi et al., 1984; Onda et al., 1990
Pseudomonas sp	Haloalkane dehalogenase	3.8.1.5	(R)-3-Chloro-1,2-propanediol, glycidol	Suzuki et al., 1993
Class 4-Lyases P. dacunhae	L-Aspartate-4-decarboxylase	4.1.1.12	D-Aspartic acid, L-alanine	Senuma et al., 1989; Takamatsu et al., 1986
Pseudomonas sp.	Oxynitrilase	4.1.2.-	α-Hydroxycarboxylic acids	Hashimoto et al., 1994
P. pseudoalcaligenes	Maleate hydratase	4.2.1.31	(R)-Malic acid	van der Werf et al., 1995

(continued)

Table I. (*Continued*)

Strain[a]	Enzyme	Process or product	EC number[b]	Reference or patent
P. putida	2-Keto-4-pentenoate hydratase	2-Keto-4-hydroxycarboxylic acids	4.2.1.80	Kunz et al., 1981
P. chloraphis B23	Nitrile hydratase	Acrylamide	4.2.1.84	Yamada and Tani, 1987
Pseudomonas sp. B21C9	Nitrile hydratase	(S)-2-Isopropyl-4'-chloro-phenylacetic acid	4.2.1.84	Masutomo et al., 1995
Pseudomonas sp.	Carnitine hydrolyase	L-Carnitine	4.2.1.89	Hoeks, 1991; Kulla, 1991
P. fluorescens, P. putida, P. trefolii	Aspartase	L-Aspartic acid, L-phenylalanine	4.3.1.1	Marquardt et al., 1988; Takagi et al., 1984
P. putida	Aspartase	Aspartame	4.3.1.1	Tuneo et al., 1989
P. putida CR 1-1	3-Chloro-D-alanine chloride-lyase	D-Cysteine derived amino acids	4.5.1.2	Nagasawa et al., 1983a,b
Class 5-Isomerases				
P. putida, P. putida ATCC 17642, P. striata	Amino acid racemase	Racemization of amino acids, D- and L-amino acids, L-tryptophan, S-adenosyl-L-homocysteine	5.1.1.10	Ishiwata et al., 1990; Lim et al. 1993; Roise et al., 1984; Yamada, 1986
Pseudomonas sp. NS671	Hydantoin racemase	D- and L-Amino acids	5.1.1.-	Ishikawa et al., 1993
P. cepacia	Mannose isomerase	D-Fructose	5.3.1.7	Allenza, 1991
Pseudomonas sp. ST-24	D-Tagatose epimerase	D-Psicose	5.3.1.-	Itoh et al., 1995
Class 6-Ligases				
P. putida M	Phenyl acetyl CoA ligase	Penicillin analogs	6.2.1.21	Fernandez-Valverde et al., 1993; Ferrero et al., 1991

[a]Some of the strains have been reclassified: P. testosteroni is now Comomonas testosteroni; P. acidovorans has been reclassified as Comomonas acidovorans; P. maltophilia has been renamed Xanthomonas maltophilia, and P. cepacia is currently known as Burkholderia cepacia.
[b]No EC classification available.

class 1) are relatively complex enzymes that require cofactors, frequently consist of more than one protein component, and are therefore mostly used as whole-cell biocatalysts. Despite their complexity and difficulty in handling, these biocatalysts are important tools by virtue of their regio- and stereoselectivity. The enzymes that catalyze addition-elimination reactions, lyases (EC class 4), are underrepresented in biotransformations although some biocatalysts of this type are utilized in large scale industrial applications, e.g., to produce acrylamide (Yamada and Tani, 1983).

The development of biocatalysts that catalyze group transfer reactions (transferases, EC class 2) or isomerizations (isomerases, EC class 5) and ligase-based biotransformative reactions (EC class 6) is in a relatively early stage with few realized industrial applications. Enzymes of these classes generally are highly substrate-specific, which makes them less suited for a wide range of biocatalytic applications.

1.2.2. Sources of *Pseudomonas* Biocatalysts

Pseudomonads are renowned for their role in degrading environmental pollutants. The utilization of these microorganisms and improved recombinants thereof in environmental biotechnology has been well documented (Gibson and Subramanian, 1984; Harayama and Don, 1985; Ramos *et al.*, 1994; Reineke, 1986; reviewed in Timmis, 1995; Timmis *et al.*, 1994). The detailed knowledge gathered from the biodegradation field has also provided the means to develop specialized (recombinant) biocatalysts with well-defined biotransformative capacities (Favre Bulle *et al.*, 1991; Gibson *et al.*, 1995; Resnick *et al.*, 1993; Wubbolts *et al.*, 1994b; Yen *et al.*, 1992). In addition to these peripheral catabolic enzymes, catabolic and anabolic enzymes of the central metabolism of *Pseudomonas* have also been used for bioconversion reactions (Faber, 1995; Wong and Whitesides, 1994).

In the following sections we describe selected biotransformations of industrial importance grouped in the different EC classes. Realized industrial processes and processes with industrial potential are discussed.

2. BIOCATALYSTS AND THEIR PRACTICAL APPLICATION

2.1. Oxidoreductases

2.1.1. Dehydrogenases

The use of dehydrogenases in enzymatic biocatalyses has been strongly stimulated by the development of efficient in situ cofactor regeneration systems (Faber, 1995; Wandrey and Bossow, 1986), which have reduced cofactor costs to a fraction of the overall production costs of an enzymatic biotransformation.

2.1.1a. Reduction of Aldehydes and Ketones. A stereospecific alcohol dehydrogenase (EC 1.1.1.−) from *Pseudomonas* sp. PED, which transfers the hydride of NADH to the *si* face of carbonyl compounds, thus producing (R) alcohols from a range of aliphatic, cyclic and aromatic substrates, has been described by Bradshaw *et al.* (1992a). The enzyme has been used with in situ cofactor regeneration with the same enzyme by using 2-propanol to generate reducing equivalents. Using this enzyme, optically active 1-deuterohexanol, a building block for synthesizing chiral polyisocyanated liquid crystals, was produced from hexanal and deuterated NADH (Bradshaw *et al.*, 1992b). The reduction of (β,γ,δ)-ketonic acid esters to (S)- (β,γ,δ)-hydroxycarboxylic acids by an NADH-dependent ketonic acid reductase (EC 1.1.1.−) from *P. acidovorans*, which also accepts substituted aliphatic, alicyclic and aromatic ketones, ketoacetals, and aldehydes as substrates, provides a route to new optically active building blocks (Kula and Peters, 1993).

The production of D-pantothenic acid (vitamin B₃) (Fig. 1) has been performed by chemoenzymatic synthesis with reduction of ketopantoic acid to D-pantoic acid by ketopantoic acid reductase (EC 1.1.1.−) from *P. maltophilia* (Shimizu and Yamada, 1991).

Examples of highly regioselective dehydrogenation of complex molecules are the conversion of morphine and codeine to hydromorphone (analgesic drug) and hydrocodone (antitussive drug), respectively, by *P. putida* M-10 (morphine dehydrogenase; EC 1.1.1.218) and of hydroxysteroids to ketosteroids by 3β,17β-hydroxysteroid dehydrogenase (EC 1.1.1.51) from *P. testosteroni* (Deamici *et al.*, 1991 (now reclassified as *Commomonas testosteroni*). The latter enzyme, which has been cloned and sequenced (Abalain *et al.*, 1993), has been used together with β-hydroxysteroid dehydrogenase from *Streptomyces* to demonstrate the controlled chemoenzymatic synthesis of all eight stereoisomers of muscarine, a mushroom toxin with three optical centers (Deamici *et al.*, 1991).

P. diminuta IFO-13182 L-malate:NAD (P)⁺ oxidoreductase (malate dehydrogenase (NADP⁺), EC 1.1.1.82) catalyzes the oxidation of (S)-malic acid to oxaloacetic acid, which was employed to produce pure reduced coenzyme (Suye *et al.*, 1989) and provides an efficient cofactor [NAD (P)⁺] regeneration system to produce L-alanine from pyruvate by alanine dehydrogenase from *Corynebacterium flaccumfaciens* AHU-1622 (Suye *et al.*, 1992).

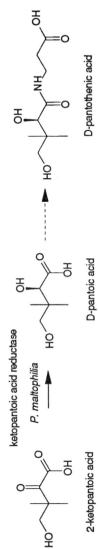

Figure 1. Production of D-pantothenic acid using ketopantoic acid reductase from *P. maltophilia* (Shimizu and Yamada, 1991).

2.1.1b. Reduction of Unsaturated Carbon–Carbon Bonds. The enantioselective reduction of prochiral C $=$ C bonds is a reaction that is difficult to perform chemically, but using whole cells of baker's yeast (*S. cerevisiae*), excellent enantioselectivities have been demonstrated for double bonds activated by an electron-withdrawing group (complied in Faber, 1995). Although pathways have been described that could contain an unsaturated carbon–carbon bond reductase (Andreoni *et al.*, 1995), we have not encountered reports of commercial application of such enzymes from *Pseudomonas* strains.

2.1.2. Oxygenases

2.1.2a. Monooxygenases

Aromatic ring monooxygenases. Aromatic ring monooxygenation reactions are widespread within the psuedomonads. Hydroxylation of the aromatic ring to yield catechol or protocatechuate derivatives is required to destabilize the aromatic system, thus enabling opening of the aromatic ring, which eventually results in complete degradation of aromatic compounds (Gibson and Subramanian, 1984; Harayama and Don, 1985). These enzymes (EC 1.14.13.−) regioselectively introduce oxygen into aromatic rings carrying a functional group, such as (substituted) phenols (Bartilson *et al.*, 1990; Kohler *et al.*, 1993; Reichlin and Kohler, 1994; Shingler *et al.*, 1989), toluenes (Whited and Gibson, 1991), nitrobenzenes (Haigler and Spain, 1991), and heteroaromatic acids (Hoeks and Venetz, 1992). Salicylate hydroxylase (EC 1.14.13.1) from *P. putida* is an exceptional aromatic ring monooxygenase that hydroxylates the ring of nitro-, amino- and halophenols and removes the ortho-substituted groups to produce catechols, i.e., it cleaves the C–N and C–X bonds of ortho-substituted phenols (Suzuki *et al.*, 1991a).

Although the aromatic ring monooxygenases have not received much industrial attention because of the absence of chirality in the product, such interest is now increasing (Hoeks and Venetz, 1992; Yen *et al.*, 1992).

Alkane and alkene monooxygenases. The degradation of alkyl groups containing substrates, such as alkanes and alkylbenzenes, by *Pseudomonas* strains proceeds via oxidation of the substrates to terminal alkanols by monooxygenases and subsequent conversion of these compounds to carboxylic acids (review: Bestetti and Galli, 1987; Fukuda *et al.*, 1989; Gibson, 1991; Kunz and Chapman, 1981; Lee and Chandler, 1941; Robinson, 1964; Worsey and Williams, 1975).

The alkane hydroxylase system (EC 1.14.15.3) from *P. oleovorans* (Lee and Chandler, 1941), a microorganism that grows on linear alkanes

(Baptist *et al.*, 1963; Schwartz, 1973); is probably the best studied non-heme iron monooxygenase. The biochemistry (Ruettinger *et al.*, 1974, 1977; McKenna and Coon, 1970; Peterson *et al.*, 1967) and genetics of the system (Kok *et al.*, 1984, 1989a,b; Eggink *et al.*, 1988; Eggink, 1987; Owen *et al.*, 1984; Chakrabarty, 1976) and its application in biocatalysis have been extensively studied. Closely related enzyme systems have been found in *P. aeruginosa* and *P. denitrificans* (Kusunose *et al.*, 1967; van der Linden and Huybrechtse, 1967).

The alkane hydroxylase enzyme from *P. oleovorans* is capable of regioselective oxidation of terminal methyl groups of linear and branched alkanes and alkylbenzenes (Baptist *et al.*, 1963; de Smet *et al.*, 1983b; van Beilen *et al.*, 1994), stereoselective epoxidation of terminal alkenes (Abbott and Hou, 1973; de Smet, 1983; de Smet *et al.*, 1981; May and Abbott, 1972; May and Schwartz, 1974) and alkadienes (May and Schwartz, 1974), oxidation of alcohols to aldehydes, and demethylation of methylethers and stereospecific sulfoxidation reactions (May and Katopodis, 1986; Katopodis *et al.*, 1984).

The catalytic potential of the alkane hydroxylase system has resulted in the filing of a number of patents (Philips *et al.*, 1986a; Philips *et al.*, 1988; Idemitsu, 1986; Kiener, 1992b,c; Phillips *et al.*, 1986; Witholt and Lageveen, 1988) and has found application in industry to synthesize optically active β-blockers (Fig. 2) (Johnstone *et al.*, 1987) and heterocyclic aromatic alcohols and carboxylic acids (Kiener, 1992a, 1995). Recombinant microorganisms of *E. coli* (Favre Bulle *et al.*, 1991) and *Pseu-*

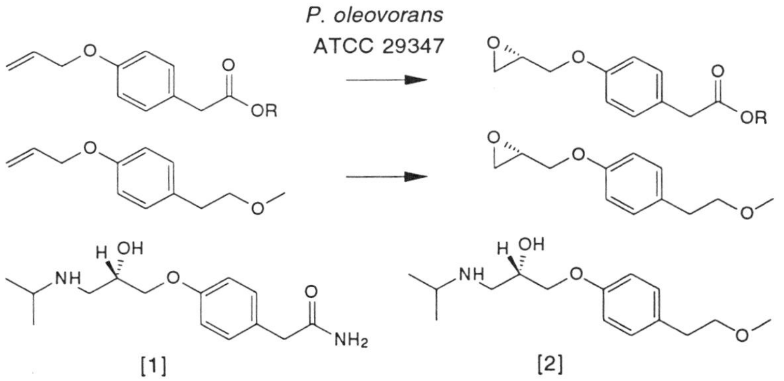

Figure 2. Oxidation of 4-allyloxyphenylacetate ester and 4-(2-methoxyethyl) phenylallyl ether to optically active epoxides by *P. oleovorans* ATCC29347. Synthesis routes to β-adrenergic blockers (*S*)-(−)-atenolol [1] and (*S*)-(−)-metoprolol [2] (Gist-Brocades, 1988; Johnstone *et al.*, 1987).

domonas (Bosetti *et al.*, 1992) containing the alkane hydroxylase system were used to produce alkanoic acids and terminal n-alkanols respectively.

Xylene oxygenase (EC 1.14.15.−) from *P. putida* mt-2 (Kunz and Chapman, 1981) is a monooxygenase that is closely related to the alkane hydroxylase system (Kok *et al.*, 1992; Suzuki *et al.*, 1991b), but the substrate ranges of both enzymes show hardly any overlap. Xylene oxygenase oxidizes pendent methyl groups on aromatic rings (Wubbolts *et al.*, 1994b) whereas alkane hydroxylase does not accept toluence derivatives as substrates (van Beilen *et al.*, 1994; reviewed in Wubbolts *et al.*, 1995). Xylene oxygenase producing recombinant strains of *Escherichia coli* have been used to produce the blue dye indigo (Mermod *et al.*, 1986) and optically active epoxides (37 to >98% e.e.) from styrene and styrene derivatives (Wubbolts *et al.*, 1994a) whereas wild-type *P. putida* mt-2 was used to produce five- and six-ring heteroaromatic acids, such as 5-methyl-pyrazine-2-carboxylic acid from 2,5-dimethylpyrazine, on the an industrial scale (Kiener, 1992a, 1995; Zimmermann *et al.*, 1992; Zimmermann *et al.*, 1993).

Pseudomonas monooxygenases which hydroxylate methylene functionalities, as opposed to the enzymes described above that primarily oxidize the terminal methyl group of alkanes, have provided synthetic routes to optically active (R)-benzylic alcohols. Such monoxygenases have been reported for *P. putida* (*p*-cresol methylhydroxylase, EC 1.14.−.−), *P. putida* JD1 (*p*-ethylphenol methylene monooxygenase, EC 1.14.−.−) and *P. putida* UV4 (benzocycloalkane hydroxylase, EC 1.14.−.−) (Boyd *et al.*, 1991; McIntire *et al.*, 1984, 1985; Reeve *et al.*, 1989, 1990). The conversion of *iso*-butyric acid to (S)-3-hydroxy-*iso*-butyric acid (>95% e.e.) by *P. putida* ATCC 21244 is an example of a branched acid oxidation which yields an important chiral synthon for the production of optically active vitamins, flavors, and antibiotics. In this case, the (R)-enantiomer is produced using the yeast *Candida rugosa* IFO 1542 (Faber, 1995).

Sulfoxidation reactions. Optically active sulfoxides are important chiral synthons in organic chemistry, but a general method to synthesize these compounds is lacking (Faber, 1995). The stereoselective oxidation of sulfur groups in thioethers to sulfoxides has been described for a number of mono- and dioxygenases from *Pseudomonas*.

Monooxygenases that produce optically active sulfoxides include the alkane hydroxylase system (EC 1.14.15.3) of *P. oleovorans* (May and Katopodis, 1986), a monooxygenase from *Pseudomonas* sp. ATCC 19286 (Idemitsu, 1986), and a monooxygenase from camphor-grown cells of *P. putida* NCIMB 10007 (Beecher *et al.*, 1994). Various *Pseudomonas* strains that degrade benzothiophenes via the corresponding chiral sulfoxides have been isolated, but the optical activity of the intermediates has not yet been resolved (Kropp *et al.*, 1994; Saftic *et al.*, 1992). Unexpectedly, the toluene dioxygenase of *P. putida* NCIMB 11767 (EC 1.14.12.11) produces optically active sulfoxides (monooxygenase reaction) at high enantiomeric excess rather than the achiral sulfones (dioxygenase reaction) from sulfides (Allen *et al.*, 1995).

Baeyer–Villiger reactions. The Baeyer–Villiger reaction, common in organic chemistry, is used to produce esters and lactones from (cyclic) ketones and (Roberts *et al.*, 1995). It can also be catalyzed by monooxygenases, in which case it proceeds stereospecifically, in contrast to the chemically catalyzed reaction (Fig. 3) (Faber, 1995). Monooxygenases (EC 1.14.15.–) catalyze Bayer–Villiger reactions of cyclic ketones forming optically active lactones, frequently with excellent enantioselectivity, but the yield of the reaction can be low because of subsequent hydrolysis of the lactones by hydrolases also present in the microorganisms. Hydrolysis can be avoided by adding hydrolase inhibitors (Faber, 1995) or by using monooxygenases in vitro with cofactor recycling systems (Grogan *et al.*, 1992; Roberts and Willetts, 1993).

Pseudomonas strains that contain Baeyer–Villiger monooxygenases include *P. aeruginosa*, *P. cepacia*, *P. putida* ATCC 17453, and *P. putida* NCIMB 10007 (Britton and Markovetz, 1977; Forney, and Markovetz, 1969; Roberts and Willetts, 1993; Taylor and Trudgill, 1986).

The NADH-dependent Baeyer–Villiger monoxygenase from *P. putida* NCIMB 10007 (EC 1.14.13.16) is more useful in vitro than the commonly used enzyme from *Acinetobacter* sp. NCIMB 9871, which is NADPH-dependent, because NADH is easily recycled using formate dehydrogenase (Roberts and Willetts, 1993; Wandrey and Bossow, 1986), whereas NADPH regeneration is tedious (Faber, 1995). The combined use in tandem of an NADH-generating alcohol dehydrogenase

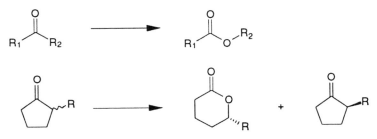

Figure 3. Baeyer–Villiger oxygenations of linear and cylic ketones.

and Baeyer–Villiger monooxygenase from *P. putida* NCIMB 10007 to convert secondary alcohols via the corresponding ketones into optically active lactones in vitro is an elegant alternative cofactor regeneration system (Faber, 1995; Roberts and Willetts, 1993). *Acinetobacter* sp. NCIMB 9871, a strain commonly used for Baeyer–Villiger reactions with whole cells (Faber, 1995), is a suspected pathogen (Lenn and Knowles, 1994) in contrast to *P. putida* strains, which also makes the latter microorganisms the preferred biocatalysts for in vivo systems.

Enantiomeric resolution of monocyclic lactones (Fig. 3) using Baeyer–Villiger monooxygenases from camphor-grown *P. putida* NCIMB 10007 and other *Pseudomonas* strains is effective and has provided routes to chiral lactones that cannot be produced by other biocatalysts (Adger *et al.*, 1995; Faber, 1995; Grogan *et al.*, 1992, 1993; Roberts *et al.*, 1995).

Whole-cell biocatalysts of this type frequently contain genetic information for multiple Baeyer–Villiger monooxygenases. Because these are inducible enzymes (Roberts *et al.*, 1995), the substrate on which cells are grown is important for the specificity and the stereoselectivity of the reaction catalyzed. Cells of *P. putida* NCIMB 10007 grown on (+)-camphor versus (−)-camphor show different stereospecificities (Wright *et al.*, 1994). In addition, different Baeyer–Villiger monooxygenase activities are observed in *P. putida* NCIMB 10007 grown on (+)-camphor as a function of the growth phase (Grogan *et al.*, 1993). These effects have to be taken into account to avoid the undesired side reactions. Otherwise the use of (partially) purified enzymes is required.

The Baeyer–Villiger oxidation of bicyclo[3.2.0]hept-2-ene-6-one (Shipston *et al.*, 1992) and derivatives (reviewed in Faber, 1995; Roberts and Willetts, 1993) by whole cells of *P. putida* NCIMB 10007 (or *Acinetobacter* sp. NCIMB 9871) and the purified monooxygenase yields two regioisomeric lactones at high enantiomeric excess (>94% e.e.), which illustrates the diversity of these in vivo biocatalysts and the specificity of these monooxygenases. The (−)-isomer of bicyclo[3.2.0]hept-2-ene-6-one is oxidized to (−)- (1R,5S)-3-oxabicyclo [3.3.0]oct-6-en-2-one, whereas the (+)-enantiomer is converted to (−)- (1S,5R)-2-oxabicyclo[3.3.0]oct-6-en-3-one by a single enzyme (Fig. 4) (Shipston *et al.*, 1992).

Baeyer–Villiger oxidations using enzymes from *P. putida* NCIMB 10007 have resulted in the development of chemoenzymatic synthetic routes to azadirachtin (Gagnon *et al.*, 1995), a complex alkaloid insect control agent, and the coenzyme (R)-α-lipoic acid (Adger *et al.*, 1995), a drug to treat liver diseases.

Although Baeyer–Villiger monooxygenase reactions are relatively new to biocatalysis and have yet to be applied on industrial scale, they

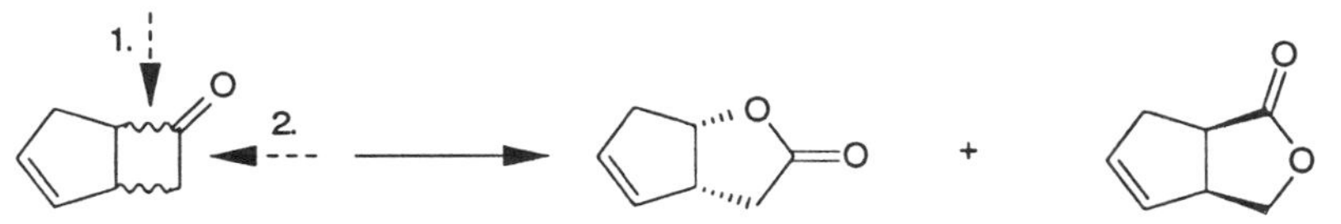

bicyclo[3.2.0]hept-2-en-6-one

(1S,5R)-2-oxabicyclo
[3.3.0]oct-6-en-3-one
(via 1.)

(1R,5S)-3-oxabicyclo
[3.3.0]oct-6-en-2-one
(via 2.)

Figure 4. Regioisomers produced by Baeyer–Villiger monooxygenase of *P. putida* NCIMB 10007 from bicyclo[3.2.0]hept-2-en-6-one (Shipston *et al.*, 1992).

hold great promise for synthesizing optically active lactone-derived synthons.

Oxidative dehalogenations. Dehalogenation of halogenated compounds can be catalyzed by hydrolytic dehalogenases, glutathione transferases, hydratases, and monooxygenases (reviewed in Janssen *et al.*, 1994). In contrast to hydrolytic dehalogenases (see section 2.3.6), monooxygenase-catalyzed dehalogenation reactions performed by pseudomonads have not proven useful for synthetic biotransformative reactions. On the other hand, these reactions are useful for biodegradation. In environmental biotechnology, aromatic ring monooxygenases have a prominent role in degrading halogenated compounds, such as trichloroethylene (TCE), a recalcitrant compound and a major pollutant, which is efficiently degraded by wild-type and recombinant *Pseudomonas* and *E. coli* strains expressing toluene monooxygenase (Aust *et al.*, 1994; Hur *et al.*, 1994; Winter *et al.*, 1989; Yen *et al.*, 1992).

2.1.2b. Dioxygenases

Aromatic Ring Dioxygenases. The dihydroxylation of aromatic compounds to optically active cis-diols by dioxygenases and the subsequent oxidation of these compounds to catechols by dihydrodiol dehydrogenases is one of the principal routes in degrading aromatic hydrocarbons by microorganisms (Gibson and Subramanian, 1984; Harayama and Don, 1985). Pseudomonads have a variety of dioxygenases with different specificities: benzene (EC 1.14.12.3) (Irie *et al.*, 1987a; Tan *et al.*, 1993), toluene (EC 1.14.12.11) (reviewed in Gibson and Subramanian, 1984), halobenzene (EC 1.14.12.–) (Bestetti *et al.*, 1992; reviewed in van der Meer, 1994), benzoate (EC 1.14.12.10) (Hansen *et al.*, 1992; Harayama *et al.*, 1986; Romanov and Hausinger, 1994), naphthalene (EC 1.14.12.12) (Connors and Barnsley, 1982; Davies and Evans, 1964; Ensley and Gibson, 1983), and biphenyl (EC 1.14.12.–) (reviewed in

Furakawa *et al.*, 1992) dioxygenases have been isolated and characterized, biochemically and genetically.

The aromatic ring dioxygenases of pseudomonads are stereospecific, are flexible in their substrate range, and oxidize various ring substituents and even heteroaromatic rings, which makes them especially suited for biocatalysis (Kieslich, 1991). The dioxygenases catalyze the formation of derivatives of *cis*-(1*S*,2*R*)-1,2-dihydroxycyclohexa-3,5-dienes (cis-diols) at excellent e.e. values.

The two chiral centers make cis-diols extremely useful as chiral synthons, but when produced with whole cell biocatalysts, these compounds are efficiently oxidized further to catechols by dihydrodiol dehydrogenases. Dihydrodiol dehydrogenase negative mutants of *Pseudomonas* strains containing aromatic ring dioxygenases have therefore been generated to permit the accumulation of cis-diols (Boyd *et al.*, 1993; Jenkins *et al.*, 1987; Resnick *et al.*, 1993; Williams *et al.*, 1990). Alternatively, recombinant microorganisms carrying only the genes encoding the dioxygenase have been successfully utilized (Brand *et al.*, 1992; Resnick *et al.*, 1993; Wubbolts and Timmis, 1990; Zeyer *et al.*, 1985).

In recombinant *E. coli* strains, which express the benzene, naphthalene or, toluene dioxygenase genes, the blue textile dye indigo has been formed (Ensley *et al.*, 1983; Irie *et al.*, 1987b; Murdock *et al.*, 1993; Stephens *et al.*, 1989). The oxidation of the *E. coli* metabolite indole to an unstable intermediate (indole-2,3-dihydrodiol), which spontaneously converts to indoxyl, followed by the dimerization of two indoxyl molecules, eventually leads to the formation of indigo (Fig. 5) (Ensley and Gibson, 1983). By metabolic engineering, an *E. coli* strain that carries the naphthalene dioxygenase genes and accumulates indole from glucose has been constructed (Ensley, 1994; Murdock *et al.*, 1993; Serdar *et al.*, 1992). This recombinant strain is used in a process that is in develop

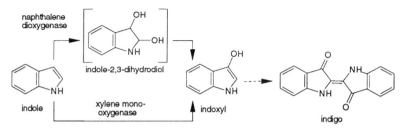

Figure 5. Production of indigo using recombinant *E. coli* strains expressing naphthalene dioxygenase (Amgen process) or xylene monooxygenase from *P. putida*. The dimerization of indoxyl to indigo occurs spontaneously (Ensley and Gibson, 1983; Mermod *et al.*, 1986).

ment at Amgen Inc. (Murdock *et al.*, 1993; Serdar *et al.*, 1992) to produce indigo as a biotechnological alternative to the chemical synthetic routes which are an environmental burden because of the use of toxic reagents (Amato, 1991).

Aromatic ring dioxygenases also catalyze reactions that are atypical to the dioxygenases, such as stereospecific sulfoxidations (Lee *et al.*, 1995) (see section 2.1.2.a), monooxygenations (Boyd *et al.*, 1993b; Brand *et al.*, 1992; Gibson *et al.*, 1995; Resnick *et al.*, 1994; Swanson, 1992) and desaturation reactions (Gibson *et al.*, 1995).

The production of cis-diols has received considerable industrial interest from Shell (Schofield, 1989), ICI (Anonymous, 1986), Minnesota Mining Co. (Mader and Tautvydas, 1990), Japan Synthetic Rubber Co. (Matsubara *et al.*, 1991), and General Electric (Johnson and Mondello, 1991) and is currently performed on a commercial scale (Kieslich, 1991; reviewed in Sheldrake, 1992).

The application of a variety of cis-diols produced by *Pseudomonas* F39/D in chemoenzymatic synthesis is illustrated by the elegant work of the group of Hudlicky, who have performed enantiocontrolled syntheses of optically active cyclopentanoids and cyclohexanoids, such as prostaglandin precursors, terpenes, different inositol stereoisomers, and other rare carbohydrates (Fig. 6) (Hudlicky *et al.*, 1988, 1994a,b). A method for converting cis-diols to trans-diols, recently developed by the same group (McKibben *et al.*, 1995), augments the applicability of these synthons even further.

In the polymer field, cis-diols of benzene give access to a biotechnological route for producing polyphenylene, a polymer that is hard to synthesize by chemical means (Ballard *et al.*, 1983), and cis-diols derived from phenylacetylene have been applied to synthesize acetylene-terminated resins (Mader and Tautvydas, 1990; Williams *et al.*, 1990).

The combined use of a dioxygenase and a cis-diol dehydrogenase (EC 1.3.1.−) results in forming catechols (1,2-dihydroxybenzenes). *P. putida* MST, an α-methylstyrene degrader, produces 3- and 4-chloro substituted catechols from chlorobenzene (Bestetti *et al.*, 1992), and *Pseudomonas* T-12 was use by researchers from Smith Kline to produce catechols with halogen and nitrile substituents (Johnston and Renganathan, 1987). Shell investigated the production of substituted catechols, carrying hydrogen, halogen, cyano, and carboxyl groups, by *P. putida* NCIB 12190, which contains benzene dioxygenase and a dihydrodiol dehydrogenase, from benzene derivatives (Geary and Haives, 1998; Geary and Pryce, 1990). Similarly, using *E. coli* recombinants carrying TOL plasmid-derived genes that encode the broad substrate range enzyme toluic acid-1,2-dioxygenase (Harayama *et al.*, 1986) and 1,2-di-

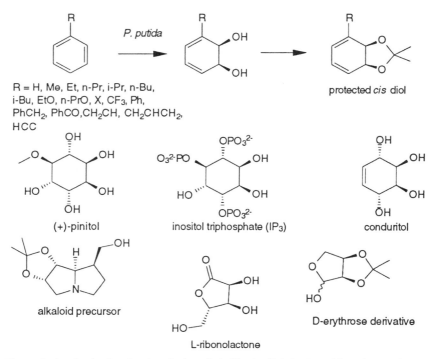

R = H, Me, Et, n-Pr, i-Pr, n-Bu,
i-Bu, EtO, n-PrO, X, CF₃, Ph,
PhCH₂, PhCO,CH₂CH, CH₂CHCH₂,
HCC

(+)-pinitol

inositol triphosphate (IP₃)

conduritol

alkaloid precursor

L-ribonolactone

D-erythrose derivative

Figure 6. Synthesis of variously substituted *cis*-dihydrodiols by *P. putida* and several examples of complex molecules made using this synthon (Faber, 1995; Hudlicky *et al.*, 1994a).

hydroxycyclohexa-3,5-diene carboxylate dehydrogenase, a biocatalyst for the synthesizing substituted catechols from benzoic acid derivatives was created (Zeyer *et al.*, 1985).

Catechol dioxygenases. Catechols were degraded by *Pseudomonas* via a ring opening reaction between (ortho cleavage) or adjacent to (meta cleavage) the hydroxyl groups, and this is catalyzed by catechol-1,2-dioxygenases (EC 1.13.11.1) and catechol-2,3-dioxygenases (EC 1.13.11.2), respectively. The products of these ring cleavage reactions are *cis,cis*-muconic acids (catechol-1,2-dioxygenases) and 2-hydroxymuconic semialdehyde derivatives (catechol-2,3-dioxygenases) (reviewed in Hirose *et al.*, 1994).

P. putida ATCC 31916, which contains a benzoate dioxygenase, 1,2-dihydro-dihydroxybenzoate dehydrogenase, and catechol-1,2-oxygenase, but lacks catechol-2,3-oxygenase activity, has been used to convert catechol and substituted catechols quantitatively to *cis,cis*-muconic

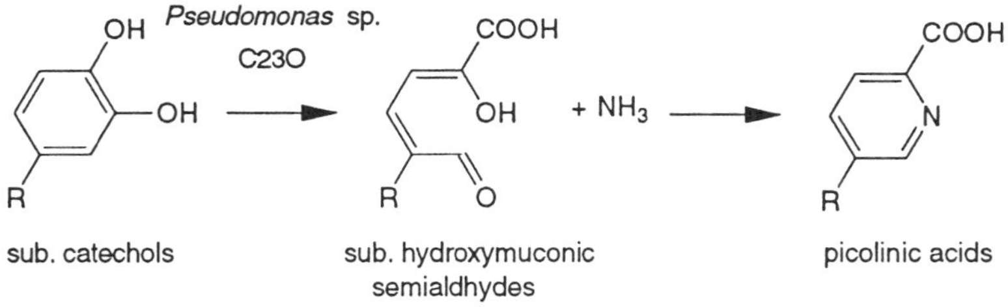

Figure 7. Synthesis of picolinic acids using catechol-2,3-dioxygenase (C230) from *Pseudomonas* sp. (Asano *et al.*, 1994).

acids by Celgene (Maxwell, 1988). The chemoenzymatic production of picolinic acid derivatives (Fig. 7) from catechols has been realized by reacting 2-hydroxymuconic semialdehyde derivatives with ammonia or a primary amine (Asano *et al.*, 1994; Hagedorn *et al.*, 1989).

2.2. Transferases

Transferases catalyze group transfer reactions. Although some acyltransferases, alkyltransferases, aminotransferases, sugar group transferases, and phosphoryltransferases are used in biocatalytic processes, the application of transferases from pseudomonads on an industrial scale is rare.

2.2.1. Acyltransferases

Acyltransferases from *Penicillum chrysogenum* are particularly useful for synthesizing new β-lactam antibiotics from 6-aminopenicillanic acid (6-APA). For this purpose an integrated two-enzyme process for producing benzylpenicillin from 6-APA and 3-furylacetic acid using a phenyl acetyl CoA ligase from *P. putida* M and the *Penicillum chrysogenum* acyl-CoA:6-APA acyltransferase was developed (Martinez-Blanco *et al.*, 1990) (see section 2.6). We did not come across acyltransferases from *Pseudomonas* that are applied in biocatalysis.

2.2.2. Alkyltransferases

Production of L-serine from the achiral amino acid glycine and methanol was realized using serine hydroxymethyl transferase (EC 2.1.2.1) from *Pseudomonas* MS31 (Fig. 8). The yield of L-serine produced

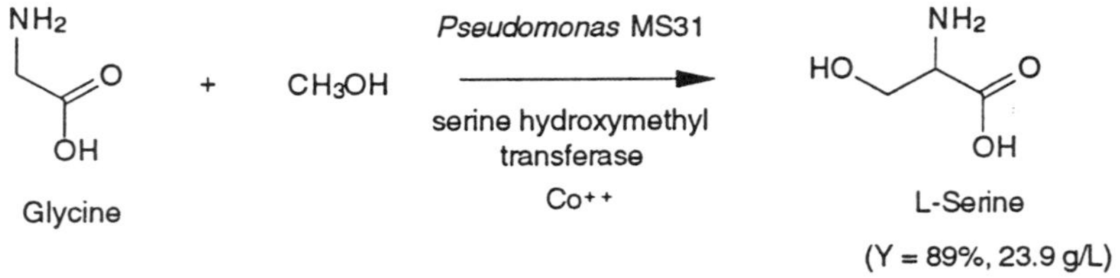

Figure 8. Biotransformation of glycine to L-serine using *Pseudomonas* (Watanabe *et al.*, 1987a,b).

with his organism was improved by inhibiting the serine degradation pathway with Co^{++} (Watanabe *et al.*, 1987a,b).

2.2.3. Aminotransferases

Alanine dioxovalerate aminotransferase (EC 2.6.1.−) from *P. riboflavina* has been used to produce δ-aminolevulinate, an insecticide, from alanine and 4,5-dioxovalerate (Fig. 9) (Rhee *et al.*, 1987). A process patented by Hoechst for producing phenylalanine uses *P. fluorescens* recombinants containing a cloned aspartase [aspartate ammonia-lyase (EC 4.3.1.1); see section 2,4.3] and a phenylalanine aminotransferase (EC 2.6.1.58). The former enzyme converts fumarate and ammonia to aspartic acid, which is subsequently used by the latter enzyme to produce L-phenylalanine from phenylpyruvate (Marquardt *et al.*, 1988).

2.2.4. Glycosyltransferases

Cyclodextrins are useful because of their stereoselective complexing characteristics. A patented procedure to isolate cyclodextrin glycosyl transferase (EC 2.4.1.19) from *Pseudomonas* sp. NRRL B-18375 which can be used to synthesize cyclodextrins was filed by Universal Oil Products (UOP Inc.) (Allenza *et al.*, 1991). The enzymatic production of maltotetraose from starch by cyclodextrin glycosyl transferase and a maltotetraose-forming enzyme (G4 amylase, EC 3.2.1.60) from *P. stutzeri* IFO-3773 has been described by Nippon Shinyaku (Maruo *et al.*, 1993).

2.2.5. Phosphotransferases

Although phosphotransferases are of central importance to bacterial metabolism and its control (reviewed in Cozzone, 1993), no industrial processes based on *Pseudomonas* phosphotransferases have yet been

Figure 9. Production of δ-aminolevulinate using *P. riboflavina* aminotransferase (Rhee *et al.*, 1987).

developed. One of the potential applications is the use of cloned enzymes that are subject to a different mode of control in metabolically engineered strains. An example is the production of lysine by introducing the cloned aspartokinase (EC 2.7.2.4) from a lysine analog resistant mutant of *P. acidovorans* into *E. coli.* In contrast to the analogous enzyme from the host strain, the cloned aspartokinase is not subject to feedback control by lysine (Alvarez-Jacobs *et al.*, 1990).

2.3. Hydrolases

Most industrial and laboratory enzyme processes are based on hydrolases. These enzymes do not utilize cofactors, usually consist of one catalytic component, and are readily available from commercial suppliers. Furthermore, the ability of hydrolases to catalyze synthesis reactions in nonaqueous environments has been thoroughly investigated and has provided paths to the synthesis of complex esters and amides.

2.3.1. Amide Hydrolysis

Resolution of racemic amides by amide hydrolysis is a principal route to producing optically active natural and unnatural L-amino acids, used as food additives, in medicine, and as compounds for chiral synthesis. Production of D-amino acids, which are rarely encountered in biological systems, is of interest for manufacturing antibiotics.

It is of principal importance for a resolution process that the undesired enantiomer is easily racemized. This can be achieved after the reaction has been completed and the product and the unreacted starting compound separated or, which is more desirable, during the biotransformative process.

2.3.1a. Amidases. Chuo Kaseihin in Japan developed a process for producing L-carnitine, a compound involved in fatty acid metabolism which is used as a food additive, from racemic carnitineamide using a *Pseudomonas* amidase (EC 3.5.1.4) (Nakayama *et al.*, 1988). A method for producing optically active 2-aryl-alkanoic acids from the corresponding racemic amides using an enantioselective amidase from a.o. *Pseudomonas* has been described by Novo Nordisk (Stieglitz *et al.*, 1992).

Amidases (EC 3.5.1.–) are most prominently used in synthesizing natural and unnatural amino acids from amino acid amides (Fig. 10). *Pseudomonas* sp. strains containing an amidase are used for this purpose by the Nitto company in Japan (Sakashita *et al.*, 1993). DSM in the Netherlands uses an isolated α-aminoacyl amidase (EC 3.5.1.4) from *P.*

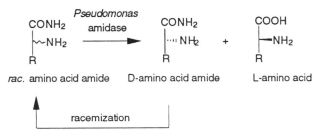

Figure 10. Production of L-amino acids using *Pseudomonas* amidases.

putida ATCC12633 (Kerkhoffs and Boesten, 1989). By in situ racemization with an amino acid amide racemase (EC 5.1.1.10) from *Pseudomonas* (see section 2.5), the latter process is led to completion, yielding L-amino acid exclusively.

2.3.1b. Hydantoinases. Hydantoinases (dihydropyrimidinases, EC 3.5.2.2) catalyze the hydrolysis of hydantoins to N-α-carbamyl amino acids and are particularly useful for producing both D- and L-α-amino acids. Hydantoinases specific for both the D- and the L-enantiomers are available, and racemization of the unreacted hydantoin is accomplished at alkaline pH. The starting compounds are racemic substituted hydantoins, readily obtained from inexpensive precursors by chemical synthesis (Faber, 1995; Roberts *et al.*, 1995).

A hydantoinase from *P. fluorescens* strain DSM 84 that specifically produces D-N-α-carbamyl amino acids (Fig. 11) has been characterized biochemically and genetically (Lapointe *et al.*, 1994; Morin *et al.*, 1986), and several other D-hydantoinases have been identified in *Pseudomonas sp.*, *Pseudomonas sp.* AJ-11220, *P. striata* and *P. putida* strains (Morin *et al.*, 1990; Takahashi, 1983; Yokozeki *et al.*, 1987a,b).

An enzymatic process for producing D-amino acids using a *P. striata* D-hydantoinase has been developed by Kanegafuchi Chemical Co. in Japan (Takahashi, 1983). Hydrolysis of racemic 5-(4-hydroxyphenyl)-hydantoin by D-hydantoinase from *P. striata* leads to the formation of the unnatural amino acid 4-hydroxy-D-phenylglycine, a building block for synthesizing the β-lactam antibiotic amoxicillin (Fig. 12). This process is carried out on a multiton scale by the Kanegafuchi company. Although the formation of amino acids from N-carbamoyl derivatives is readily catalyzed by nitrous acid, Kanegafuchi has developed an enzymatic alternative for his reaction based on a D-N-carbamoyl-α-amino acid amidohydrolase (EC 3.5.1.77) which converts D-N-carbamoyl-α-amino

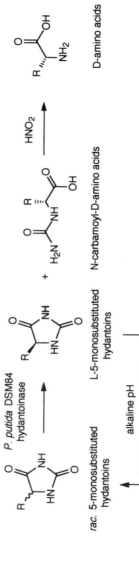

Figure 11. Production of D-amino acids using *P. fluorescens* strain DSM 84 hydantoinase.

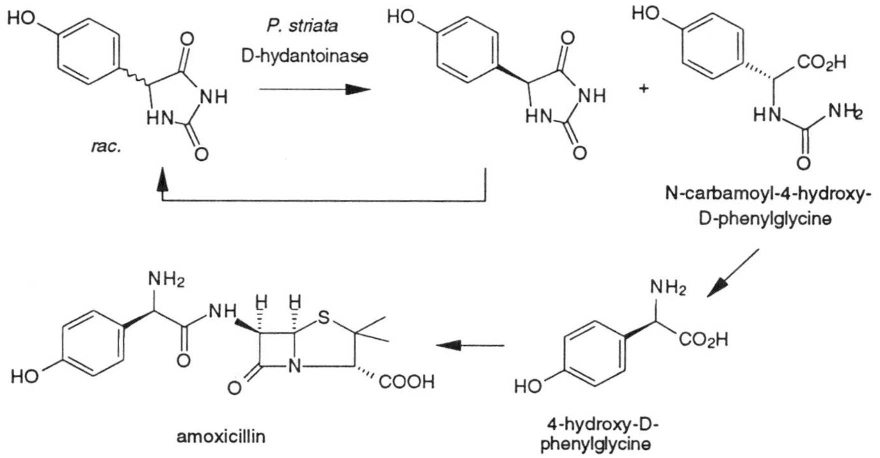

Figure 12. Amoxicillin production with a *P. striata* hydantoinase (Takahashi, 1983).

acid to D-αamino acid (Nanba *et al.*, 1992). A similar process based on *Pseudomonas sp.* AJ-11220, which contains both a D-hydantoinase and a D-N-carbamoyl-α-amino acid amidohydrolase, was developed by the Ajinomoto company (Yokozeki *et al.*, 1987a,b).

The production of L-amino acids with L-hydantoinases from *Pseudomonas* sp. strain NS671 using intact cells was described by the Nippon Soda company (Ishikawa *et al.*, 1993), and an enzymatic process for producing L-amino acids using hydantoinases from psuedomonads was developed by Sclavo in Italy (Olivieri *et al.*, 1985).

2.3.1c. Aminoacylases. The hydrolysis of N-acyl amino acid derivatives, catalyzed by N-acyl-L-amino acid acylases (EC 3.5.1.14), provides another route to produce optically pure amino acids. For this purpose aminoacylases of mostly eukaryotic origin (e.g., *Aspergillus*) are applied on an industrial scale (Faber, 1995; Verkhovskaya and Yamskov, 1991). A process for synthesizing peptide antibiotics using an aminoacylase from *Pseudomonas sp.* was developed by Bristol–Myers Squibb (Konishi *et al.*, 1992).

To produce 7-aminocephalosporanic acid, a building block for semisynthetic cephalosporins, the Fujisawa Pharmaceutical Company has developed a two-step process from cephalosporin C that involves a glutaryl-7-aminocephalosporanic acid aminoacylase (EC 3.5.1.−) from *Pseudomonas* (Fukagawa *et al.*, 1992). An analogous process was invented earlier by Asahi Kasei Kogyo using *Pseudomonas sp* SE-83 or SE-495 (Ichikawa *et al.*, 1988).

2.3.1d. Lactamases. Lactamases that stereospecifically hydrolyze β- and γ-lactams produce β- and γ-amino carboxylic acids with potentially two chiral centers, which are versatile chiral substances. A stereospecific γ-lactamase (EC 3.5.2.−) from *P. solanaceaum* was thus utilized to resolve racemic 2-azabicyclo[2.2.1]hept-5-en-3-one to optically pure (1*R*,4*S*)-2-azabicyclo[2.2.1]hept-5-3n-3-one (Evans *et al.*, 1992). The resolved product has been used to chemically synthesize the anti-HIV drug (−)-carbovir (Evans *et al.*, 1992; Faber, 1995) and 4-*cis*-aminocyclopent-2-en-1-carboxylic acid.

Lactamase-deficient biocatalysts can be used to synthesize β-lactam antibiotics. Amoxicillin can be synthesized by a *P. melanogenum* KY 3987 strain containing a penicillin acylase (EC 3.5.1.11) which is devoid of β-lactamases (EC 3.5.2.6) to avoid product hydrolysis (Kawamori *et al.*, 1983).

2.3.2. Ester Hydrolysis

2.3.2a. Esterases. From a culture collection of over 700 microorganisms screened for 2-ethylhexylbutyrate hydrolyzing activity, *P. myxogenes* was found to produce an esterase (EC 3.1.1.−) capable of asymmetrically hydrolyzing 2-ethylhexylbutyrate to (*S*)-2-ethyl-1-hexanol (80% e.e.) (Hou, 1993).

Esterases can also distinguish between both enantiomers of chiral epoxides. An esteraes from *P. fragi* IFO 3458, able to hydrolyze the S-enantiomer of racemic 3,4-epoxybutyric acid, was used by Enichem to accumulate (*R*)-3,4-epoxy-butyric acid, which was transformed to L-carnitine by hydrolysis and subsequent trimethylamide addition (Fig. 13) (Francalanci *et al.*, 1993).

P. fluorescens DSM 7033 esterase (EC 3.1.1.2) cleaves only the (S)-enantiomer of racemic 1-arylalkyl esters, which provides a synthetic route developed by Boehringer Mannheim to optically pure (*S*)-1-phenylalkanols (Krell and Rasor, 1993). An esterase from *Pseudomonas sp.* was exploited by the Sumitomo Chemical Co. to synthesize (*S*)-α-cyanobenzylalcohols, which are of use for synthesizing insecticides (Mitsuda *et al.*, 1991). The use of recombinant *E. coli* strain expressing a *P. fluorescens* esterase to produce the anti-inflammatory drug (*S*)-Naproxen by hydrolysis of racemic Naproxen esters has been described by Syntex Pharmaceuticals (Chan and Salazar, 1991).

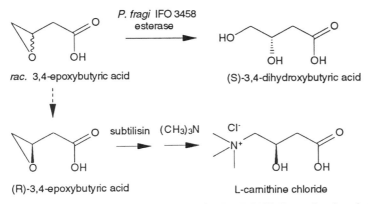

Figure 13. Production of L-carnitine using *P. fragi* IFO 3458 (Francalanci *et al.*, 1993).

2.3.2b. Proteases. Despite the fact that proteases (EC 3.4.−.−) from various sources (e.g., subtilisin, chymotrypsin, and papain) are efficient and frequently used catalysts for ester forming and breaking reactions, proteases from *Pseudomonas* have not received much attention. This is hardly surprising considering the virtual unavailability of *Pseudomonas* proteases from commercial sources.

2.3.2c. Lipases. The pseudomonads are particularly rich in catalytically useful lipases. Lipases (EC 3.1.1.3) from *P. fluorescens* (PFL), *Pseudomonas* sp. (PSL), *P. cepacia* (PCL), and *P. aeruginosa* (PAL) have been and continue to be 'major players' in enantioselective ester hydrolysis and esterification reactions with technical or commercial applications (reviews: Drauz and Waldmann, 1995; Jaeger and Wohlfarth, 1993; Theil, 1995).

Lipases are the most used hydrolases in biotransformations because these enzymes are highly enantioselective, are usually single component enzymes which require no cofactor, and are readily available from commercial suppliers. In addition, lipases are catalytically active in aqueous solutions but also in more hydrophobic environments, such as water–cosolvent mixtures and even organic solvents.

Lipases are used primarily to differentiate between enantiotopes of prochiral substrates, usually ester groups (deacylation) or hydroxy groups (acylation), which can result in producing chiral products at high e.e. from achiral substrates. Alternatively, lipases are employed to separate racemic mixtures of chiral substrates, such as esters of racemic carboxylic acids or alcohols, by kinetic resolution.

Depending on the reaction conditions, lipases can catalyze either ester hydrolysis or transesterification reactions, such as acylation or alcoholysis. To achieve the latter type of reaction, the equilibrium of the reaction is shifted toward ester synthesis by utilizing low water or anhydrous environments.

Hydrolysis. Lipase-catalyzed hydrolysis of prochiral esters is a useful route to optically pure compounds from achiral precursors. The enantiotopos discriminative hydrolysis of prochiral acyclic and cyclic diol diacetates by lipases from *Pseudomonas* has been extensively utilized to produce optically active monoacetate esters at moderate to excellent e.e. values (reviewed in Drauz and Waldmann, 1995; Faber, 1995). Similarly, lipases from *Pseudomonas* can hydrolyze racemic chiral acyclic and cyclic esters with high stereoselectivity to optically active primary and secondary alcohols (reviews: Drauz and Waldmann, 1995; Faber, 1995; Jaeger and Wohlfarth, 1993; Kieslich, 1991; Theil, 1995). Some industrial examples of *Pseudomonas* lipase-assisted syntheses that involve enantiospecific ester hydrolysis are the production of the herbicide (S)-$(-)$-fenpropimorph by Caffaro SpA, Italy (Sunjic and Gelo, 1995), (R)- and (S)-atenolol β-blockers by ICI, India (Bevinakatti and Banerji, 1992), a thromboxane antagonist drug (Patel *et al.*, 1992), and the anticancer drug paclitaxel (Patel *et al.*, 1994) both developed by Bristol-Myers Squibb, U.S.A.

Alcoholysis. When the presence of water should be avoided (e.g., when product or substrate are unstable in aqueous environments), cleavage of the ester bond by lipases can be performed by an alcohol (alcoholysis) in organic solvents. Intermolecular alcoholysis by *Pseudomonas* lipases has thus been used for resolving chiral lactones, epoxides, and anhydrides, and lipase-catalyzed intramolecular alcoholysis provides a route optically active lactones from racemic hydroxycarboxylic acids. The yield and the enantioselectivity of the reaction increases markedly compared to the water-catalyzed reaction (review: Drauz and Waldmann, 1995).

Acylation. Lipase-mediated acylation of prochiral diols and transesterification of prochiral diesters to optically active products requires the use of a suitable acyl donor and low water or anhydrous organic solvents, although a small amount of water (0.5–1%) is required for proper function of the lipase ('bound water'). Acyl donors, such as carboxylic acid anhydrides, oxime esters, or vinyl esters, are frequently employed. Vinyl acetate and isoprenyl acetate are very useful acyl donors because vinyl alcohol and isoprenyl alcohol tautomerize to acetaldehyde and acetone, respectively, which impedes the reverse reaction and forces the equilibrium in the desired direction.

Pseudomonas lipases are particularly useful for the enantioselective acylation of racemic primary, secondary, and tertiary alcohols in organic solvents, which in many cases provides a way to resolve racemic mixtures of these alcohols to optically pure compounds (Fig. 14). The enantioselectivity, on the one hand, and the broad substrate range of *Pseudomonas* lipases, on the other hand, allow the kinetic resolution of an enormous variety of racemic alcohols (reviewed in Drauz and Waldmann, 1995; Faber, 1995). Optically pure halohydrins are important building blocks for the chiral synthesis of β-blockers (Kloosterman *et al.*, 1988) and provide a general route to optically active epoxides (Faber, 1995; Kutsuki *et al.*, 1986). Optically active cyanohydrins are synthons needed to produce chiral α-amino acids, aldehydes, and amino alcohols (review: Faber, 1995). Both chiral synthons are industrially produced in a lipase-catalyzed resolution process starting from racemic halohydrins and cyanohydrins, respectively (Inagaki *et al.*, 1992; Kanerva *et al.*, 1994; Kutsuki *et al.*, 1986).

In addition to resolution of racemic alcohols by lipase-catalyzed acylation, chiral hydroperoxides have also been resolved using a *P. fluorescens* lipase and isoprenyl acetate as an acyl donor (Baba *et al.*, 1988).

2.3.3. Epoxide Hydrolases

Epoxide hydrolases (EC 3.3.2.−) provide an attractive route to optically active 1,2-diols and optically active epoxides, important chiral synthons that are notoriously difficult to synthesize chemically. Although optically active epoxides can be obtained by oxidizing vinylic bonds using monooxygenases (Furuhashi, 1992; Wubbolts *et al.*, 1995), epoxide hydrolase-catalyzed resolutions are cofactor-independent and more easily applied in enzyme systems.

Most of these biotransformations have been done with epoxide hydrolases from eukaryotic sources, such as rabbit liver microsomal and fungal epoxide hydrolases (Faber, 1995; Pedragosa-Moreau *et al.*, 1993).

Although a number of *Pseudomonas* strains that degrade epoxides via a hydrolytic mechanism have been isolated (Rustemov *et al.*, 1992; van den Wijngaard *et al.*, 1989; Warhurst and Fewson, 1994), the enzymes involved have been insufficiently characterized to develop a biocatalytic process. *Pseudomonas* sp. AD1 epoxide hydrolase (EC 3.3.2.3) has been purified and characterized, but this enzyme is not enantioselective (Jacobs *et al.*, 1991).

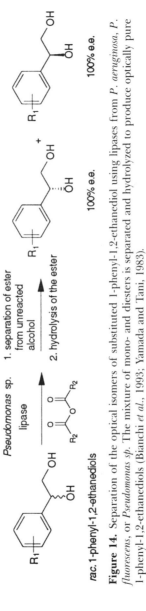

Figure 14. Separation of the optical isomers of substituted 1-phenyl-1,2-ethanediol using lipases from *P. aeruginosa*, *P. fluorescens*, or *Pseudomonas sp.* The mixture of mono- and diesters is separated and hydrolyzed to produce optically pure 1-phenyl-1,2-ethanediols (Bianchi *et al.*, 1993; Yamada and Tani, 1983).

2.3.4. Nitrile Hydrolases

Nitrile hydrolases or nitrilases (EC 3.5.5.1) catalyze the one-step hydrolysis of nitriles to carboxylic acids with the release of ammonia. The enzymatic reaction proceeds under milder conditions and produces less hazardous waste products compared to the alternative chemical reaction (Faber, 1995), which explains why the enzymatic route is chosen even when chemo-, regio-, or enantioselectivity is not required.

A synthetic route to (*R*)-mandelic acid from racemic mandelonitrile (Endo and Tamura, 1991) and of other α-hydroxycarboxylic acids containing a phenyl group from the corresponding racemic nitriles (Hashimoto *et al.*, 1994), which uses whole cells of *Pseudomonas* or other microorganisms containing a nitrilase, has been developed by Nitto in Japan. These biocatalysts also contain an oxynitrilase (EC 4.1.2.10), a lyase (see section 2.4.2 and Table 1) that allows producing the precursor nitriles from derivatives of benzaldehyde and HCN (Fig. 15). Likewise, *Pseudomonas* strains capable of enantioselective hydrolysis of racemic *o*-acetylmandelonitrile to *R* (−)-*o*-acetylmandelic acid have been isolated (Layh *et al.*, 1992).

Idemitsu Kosan invented a general process to prepare optically active α-substituted carboxylic acids, such as 2-chloropropionic acid and 2-methylbutyric acid, from racemic nitriles catalyzed by *Pseudomonas* MY-1 (Murakami, 1993). A similar process was developed by another Japanese company, Asatu Kasei (Yamaoto *et al.*, 1990).

Nitto has also applied nitrilases in an enzymatic route to glycine starting from glycinonitrile using the enzyme from a number of microorganisms including *Pseudomonas* (Shimizu *et al.*, 1993) and in a biocatalytic process to synthesize D- and L-lactic acid from racemic lactonitrile using whole cells of a.o. *P. synxantha* or *P. ovalis* (Yamagami *et al.*, 1993).

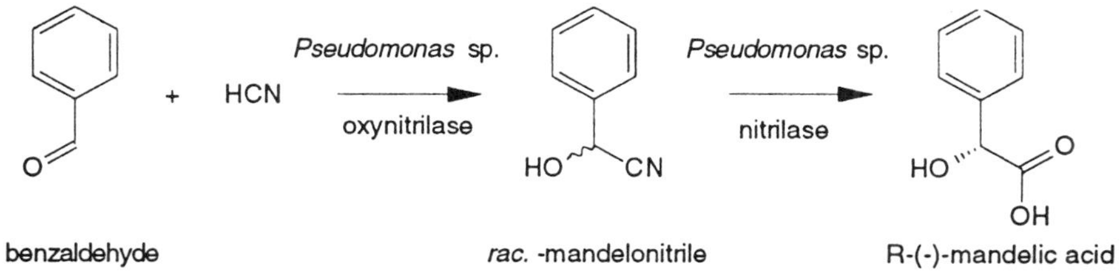

benzaldehyde   rac. -mandelonitrile   R-(-)-mandelic acid

Figure 15. The nitrilase-based Nitto process for the production of α-hydroxycarboxylic acids (Hashimoto *et al.*, 1994). The *Pseudomonas* biocatalyst contains both oxynitrilase and nitrilase activity.

2.3.5. Glycosidases

The hydrolytic enzymes most used in starch processing stem from food-grade microorganisms, such as *Bacillus licheniformis* (α-amylase, EC 3.2.1.1) and *Aspergillus niger* (glycoamylase, EC 3.2.1.3) (Drauz and Waldmann, 1995). Consequently, glucosidases from *Pseudomonas* used in starch processing are for nonfood products or they catalyze a reaction not easily performed by other enzymes.

α-Glucosidase (EC 3.2.1.20) is a glucosidase of limited industrial relevance. It is an α-1,4-exoglucanase that produces α-D-glucose from starch and can be found in many microorganisms, such as *P. fluorescens* W and *P. amyloderamosa* (Kelly and Fogarty, 1983). *P. amyloderamosa* contains another, more important glucosidase, isoamylase (EC 3.2.1.68), an endoglucanase that hydrolyzes the α-1,6-linkages of glycogen and amylopectin (Amemura and Futai, 1992). This enzyme (debranching enzyme) is seldom found in other microorganisms and provides a way to synthesize linear polysaccharides from α-1,6-branched polysaccharides (Drauz and Waldmann, 1995).

2.3.6. Dehalogenases

Dehalogenases from different sources have proven useful for synthesizing optically active molecules (review: Fetzner and Lingens, 1994). *Pseudomonas* halo-acid dehalogenases (EC 3.8.1.2) have been used for enantiomeric resolution of haloacids. The useful chiral synthon D-3-chlorolactic acid was produced from 2,3-dichloropropionic acid by *P. putida* 109 (Onda *et al.*, 1990), and by using L-2-haloacid dehalogenase from *Pseudomonas sp.*, a chemoenzymatic synthetic route from racemic 2-chloropropionic acid to the herbicide (*R*)-fluazifop was realized (Roberts *et al.*, 1995). One report (Motosugi *et al.*, 1984) describes the production of D- and L-lactic acid from racemic 2-chloropropionic acid by successively using two haloacid dehalogenases, an enantioselective L-2-chloropropionic acid dehalogenase from *P. putida* and and a DL-2-chloropropionic acid dehalogenase from *Pseudomonas* sp. 113.

A haloalkane dehalogenase (EC 3.8.1.5) specific for (*S*)-3-chloro-1,2-propanediol in whole cells of *Pseudomonas* sp. was used by Diasco in Japan to accumulate optically pure 'untouched' substrate (*R*)-3-chloro-1,2-propanediol, which was converted nonenzymatically to highly optically active (*R*)-glycidol (99.3% e.e.) (Suzuki *et al.*, 1993).

2.4. Lyases

Lyases catalyze additions to double bonds or eliminations that generate double bonds and are specific for carbon–carbon, carbon–oxygen, carbon–nitrogen, carbon–sulfur, carbon–halide, and phosphorus–oxygen bonds. A strong enantiofacial discrimination during addition reactions is typical of lyases, and lyase-catalyzed eliminations are as a rule also highly enantioselective, which makes them useful for chiral synthesis (van der Werf *et al.*, 1994). Unfortunately, most lyases have a narrow substrate spectrum, which limits the applicability of these enzymes in biocatalytic processes (Faber, 1995).

2.4.1. Aldolases

Aldolases (EC 4.1.2.–) are carbon–carbon lyases common to *Pseudomonas* which play a central role in glycolysis and gluconeogenesis. These enzymes are also involved in degrading, for instance, aromatic hydrocarbons (Harayama and Rekik, 1990; Powlowski *et al.*, 1993), but we have not encountered a biotransformative reaction based on *Pseudomonas* aldolases.

2.4.2. Oxynitrilases

Addition of hydrogen cyanide to the *re* or *si* face of a carbonyl functionality is catalyzed by *R*- and *S*-oxynitrilases, respectively, which use either HCN or a cyanohydrin (transcyanation) as a donor of the cyano group. The reaction usually proceeds with high enantiofacial discrimination, and this provides an efficient route to optically active cyanohydrins, which are versatile building blocks for asymmetric synthesis of α-hydroxyacids, acyloins, α-hydroxyaldehydes, vicinal diols, and β-amino alcohols. Highly specific *R*-oxynitrilases from sweet almonds (EC 4.1.2.10) and *S*-oxynitrilases (EC 4.1.2.11) from *S. bicolor* or *X. americana* have been used by Solvay Duphar (van Scharrenburg *et al.*, 1993).

A process of Nitto based on oxynitrilase containing psuedomonads which additionally contain a nitrilase and thus produce α-hydroxycarboxylic acids from aldehydes and HCN has been discussed in section 2.3.4 (Fig. 15). Chemical hydrolysis can replace the nitrilase, which is illustrated by the acid-catalyzed formation of (*R*)-α-hydroxy-α-methyl carboxylic acids from (*R*)-α-hydroxy-α-methyl cyanohydrins produced from racemic aliphatic ketones and HCN by an (*R*)-oxynitrilase from *Rhodococcus* (Effenberger *et al.*, 1991).

2.4.3. Ammonia Lyases

Ammonia-lyases are useful for synthesizing α-amino acids from α,β-unsaturated acids. Histidine ammonia-lyase (EC 4.3.1.3.; histidase) from *P. fluorescens*, which catalyzes the formation of L-histidine from urocanic acid, has been studied in considerable detail (Furuta *et al.*, 1992; Hernandez and Phillips, 1994), but the *Pseudomonas* enzyme has not been developed into an application. In contrast, L-phenylalanine ammonia-lyase (EC 4.3.1.5) from the yeast *R. glutinis* has gained much commercial interest because the market for L-phenylalanine has increased due to the use of this amino acid for producing the artificial sweetener aspartame (L-aspartyl-L-phenylalanine methyl ester). The Hoechst process for L-phenylalanine production, discussed in section 2.2.3, uses a *P. fluorescens* containing aspartate-ammonia lyase (EC 4.3.1.1; aspartase) (Marquardt *et al.*, 1988; Takagi *et al.*, 1984). Wild-type strains of *P. trefolii* and *P. putida* have also been successfully employed to produce L-aspartate from fumarate and ammonia (Drauz and Waldmann, 1995).

A direct process for producing aspartame based on aspartate ammonia lyase, using fumarate, ammonia, and L-phenylalanine methyl ester as starting compounds, was realized by Toso in Japan (Tuneo *et al.*, 1989). This process integrates the L-aspartate synthesis and coupling to the L-phenylalanine methyl ester in one reaction vessel (Drauz and Waldmann, 1995).

2.4.4. Hydrolyases

Hydrolyases are characterized by water elimination or by the addition of H_2O to a double bond. The reactions usually proceed with high enantioselectivity, but even when there is no enantiofacial discrimination, the hydrolyases provide a 'green' and often qualitatively superior alternative to chemical catalysts.

A maleic acid hydrolyase (EC 4.2.1.31) from *P. pseudoalcaligenes* NCIMB9867, which synthesizes the chiral building block (*R*)-malic acid from maleic acid, has been investigated in an enzymatic production process (van der Werf *et al.*, 1994). By varying the counterions in the reaction medium, the reaction equilibrium can be pulled to the product side as the product precipitates (van der Werf *et al.*, 1995).

Carnitine hydrolyase (EC 4.2.1.89) containing whole cells of *Pseudomonas sp.* is used by Lonza to produce the drug L-carnitine (Hoeks, 1991; Kulla, 1991). The hydrolyase enzyme catalyzes the highly enantioselective addition of water to the double bond in crotonobetaine. A hydrolyase from *P. putida*, 2-keto-4-pentenoate hydratase (EC 4.2.1.80), catalyzes the addition of water to 2-keto-*cis*-4-alkenoic acids to produce 2-keto-4-hydroxycarboxylic acids (Drauz and Waldmann, 1995; Kunz *et al.*, 1981).

The most renowned hydrolyase is acrylonitrile hydrolase (EC

4.2.1.84) which is by the Nitto company used to produce acrylamide (Fig. 16) from acrylonitrile on a scale of 30,000 tons per year (Drauz and Waldmann, 1995). Although the product is a relatively simple achiral chemical that can easily be produced nonenzymatically, the biocatalytic route produces a product of higher quality and at lower energy cost (Faber, 1995; Yamada and Tani, 1983). For a considerable period, *P. chlororaphis* B23 was used as a catalyst for the reaction, but since 1991 *R. rhodochrous* J1 has been used on a production scale because it has superior reaction characteristics (Nagasawa *et al.*, 1993).

Pseudomonas sp. B21C9 from the Sumitomo company constitutes a similar biocatalyst, which also contains a nitrile hydratase with poor enantioselectivity but in combination with an S-specific amidase produces (*S*)-2-isopropyl-4'-chlorophenylacetic acid from racemic isopropyl-4'-chlorophenylacetonitrile (Masutomo *et al.*, 1995).

2.4.5. Decarboxylases

The Tanabe company has developed a two-biocatalyst process to produce L-alanine (Takamatsu *et al.*, 1986). An *E. coli* aspartase (section 2.4.3) catalyzes the conversion of fumarate and ammonium to L-aspartate, and subsequently L-aspartate-β-decarboxylase (EC 4.1.1.12) of *P. dacunhae* causes the decarboxylation of L-aspartate to L-alanine. By using only the latter biocatalyst, a process for simultaneously producing D-aspartate and L-alanine from racemic aspartate was obtained (Senuma *et al.*, 1989).

2.4.6. Halolyases

Halolyases, which catalyze the addition/elimination of halides are relatively uncommon in synthetic applications. Nevertheless, a synthetic application of a C–X lyase has been developed using *P. putida* CR1-1 cells containing a 3-chloro-D-alanine chloride-lyase (EC 4.5.1.2), which catalyzes the synthesis of D-cysteine and 3-chloro-L-alanine from racemic 3-chloro-alanine using sodium hydrosulfide as a nucleophile (Fig. 17).

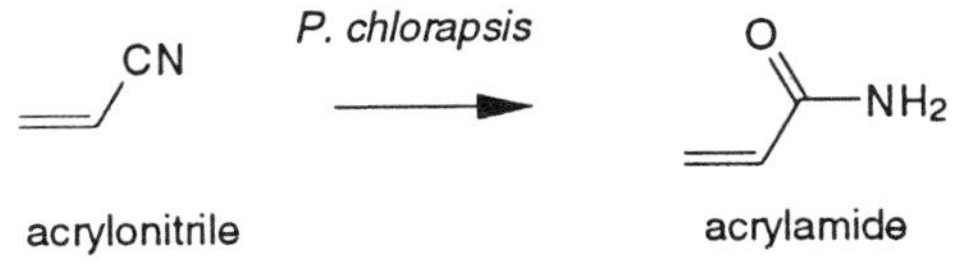

Figure 16. The production of acrylamide from acrylonitrile by Nitto using *P. chlorapsis* or *R. rhodochrous* J1 acrylonitrile hydrolyase (nitrile hydratase) (Yamada and Tani, 1983).

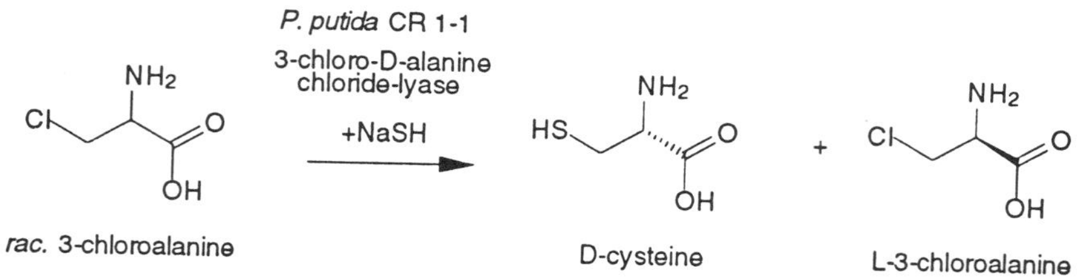

Figure 17. Synthesis of D-cysteine (Y = 94%) and 3-chloro-L-alanine (Y = 81%) using a halolyase from *P. putida* CR1-1 (Nagasawa *et al.*, 1983c).

The formation of L-cysteine was prevented by adding 5mM phenylhydrazine to the cells (Nagasawa *et al.*, 1983c). Other D-cysteine-related compounds are also accessible by applying this biocatalyst with nucleophiles other than NaSH (Nagasawa *et al.*, 1983a). This is illustrated by the synthesis of an important building block for the semisynthetic production of cephalosporin, (S)-carboxymethyl-D-cysteine, that can be performed using *P. putida* CR1-1 3-chloro-D-alanine chloride-lyase with ethylthioglycolate as a nucleophile (Nagasawa *et al.*, 1983b).

2.5. Isomerases and Racemases

2.5.1. Isomerases

2.5.1a. Carbohydrate Isomerases. The carbohydrate isomerase most prominently used in industry is glucose isomerase (EC 5.3.1.5), which is used to produce high fructose corn syrup from glucose on a multimillion tonne scale (Roberts *et al.*, 1995). The enzymes stem from a number of microorganisms and fungi (a.o. *Streptomyces*, *A. missouriensis*, *Microbacterium*, and *Irpex mollis*) and are also available in recombinant strains.

The utilization of carbohydrate isomerases from pseudomonads is limited, but some processes have been described. A process for producing of D-fructose from D-mannose, patented by Allied Signal (Allenza, 1991) is based on an immobilized mannose isomerase from *P. cepacia*. A thermostable variant of the same enzyme has been isolate and applied to synthesizing D-mannose from D-fructose (Takasaki *et al.*, 1993). Another process is based on D-tagatose 3-epimerase from *Pseudomonas* sp. ST-24, which has been used to convert D-fructose to D-psicose (Itoh *et al.*, 1995). In combination with glucose isomerase, a process was developed that allows the two-step conversion of D-glucose to D-psicose.

2.5.2. Racemases

2.5.2a. Amino Acid Racemases. Amino acid racemases (EC 5.1.1.10) catalyze the interconversion of D- and L-amino acids and play an important role in the industrial production of some amino acids because racemization of the undesired or uncoverted enantiomer makes a 100% yield of the desired enantiomer attainable (Drauz and Waldmann, 1995).

Amino acid racemases with a broad substrate specificity have been found in *P striata* (Roise *et al.*, 1984) and *P. putida* ATCC 17642 (Lim *et al.*, 1993). The L-specific amidase-based production route to natural and unnatural L-amino acids of DSM (see section 2.3.1 and Fig. 10) also produces the D-amino acid amide, which has to be recycled. This can be done chemically, by forming a Schiff's base and isolating the insoluble product, or enzymatically with an aspecific racemase from *Pseudomonas* ATCC17642 (Kamphuis *et al.*, 1992).

A racemase specific for alanine (EC 5.1.1.1) has been isolated from *P. putida*. This enzyme, or rather a thermostable variant of the enzyme from *B. stearothermophilus*, can be used to synthesize D-amino acids from their keto precursors in a multienzyme system containing alanine dehydrogenase (EC 1.4.1.1), D-amino acid aminotransferase (EC 2.6.1.21) and formate dehydrogenase (EC 1.2.1.2) (Drauz and Waldmann, 1995). Another multienzyme catalyst from *Pseudomonas sp.*, containing an amino acid racemase (EC 5.1.1.10) and an S-adenosyl-L-homocysteine hydrolase (EC 3.3.1.1) has been applied by Nippon Zeon to synthesize S-adenosyl-L-homocysteine from D-homocysteine and adenosine (Yamada *et al.*, 1986).

P. putida serine racemase was used on an industrial scale by Mitsui Toatsu chemicals in a multienzymatic production route to L-tryptophan from racemic serine and indole. The tryptophan synthase (EC 4.2.1.20) of *E. coli* catalyzes the reaction of l-serine into L-tryptophan and *P. putida* amino acid racemase simultaneously converts d- to L-serine in a 'one pot' system (Fig. 18) (Ishiwata *et al.*, 1990).

2.5.2b. Hydantoin Racemases. The resolution of racemic 5-substituted hydantoins to D- or L-carbamoylic acids by hydantoinases (EC 5.1.1.−; see section 2.3.1 and Fig. 11) provides a powerful route to D- and L-amino acids. Isomerization of the hydantoins after separation of the unreacted hydantoins from the carbamoylic acids can easily be effected at high pH, and subsequent reaction of the carbamoylic acids with HNO_2 releases optically pure amino acids (compiled in Drauz and Waldmann, 1995).

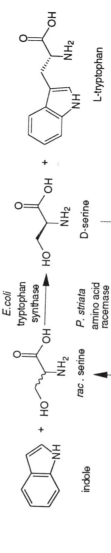

Figure 18. The Mitsui Toatsu process for producing L-tryptophan using serine racemase from *P. striata* (Ishiwata *et al.*, 1990).

A *Pseudomonas* biocatalyst (*Pseudomonas sp.* NS671) that contains an L-hydantoinase, a L-N-carbamoyl hydrolase, and a hydantoin racemase and thus completely eliminates the two chemical steps outlined above has been isolated by Nippon Soda (Ishikawa *et al.*, 1993). This strain converts racemic 5-(2-methyl-thioethyl)hydantoin to the desired product L-methionine at 93% yield. Other amino acids (L-isoleucine, L-leucine, L-phenylalanine, and L-valine) can also be produced from the corresponding 5-substituted hydantoins albeit at lower yields due to product consumption (Ishikawa *et al.*, 1993).

2.6. Ligases

Ligases catalyze the joining of two molecules at the expense of the hydrolysis of a high-energy phosphate bond and are of particular use in molecular biology. In biotransformation, however, these enzymes are applied to a very limited extent (Faber, 1995).

Phenyl acetyl CoA ligase (EC 6.2.1.21) from *P. putida* M has been used in conjunction with *Penicillum chrysogenum* acyl-CoA:6-APA acyltransferase (see section 2.2.1) to produce a variety of penicillin analogs (Martinez-Blanco *et al.*, 1990). Because of the wide substrate range of the enzymes involved (Fernandez-Valverde *et al.*, 1993), this system allows in vitro synthesis of a wide variety of β-lactam antibiotics, such as 3-furylmethylpenicillin and several ketoalkylpenicillins (Ferrero *et al.*, 1991a,b; Martinez-Blanco *et al.*, 1991).

3. THE IMPACT OF MOLECULAR BIOLOGY ON BIOCATALYST DEVELOPMENT

3.1. Constructing "Tailor-Made" Biocatalysts

Elucidating anabolic and catabolic pathways and characterizing the enzymes that catalyze individual reactions are an ever developing field. Newly discovered enzymatic activities are now easily characterized at the molecular and genetic level. As a result, a vast arsenal of cloned enzymes is available and many industrial enzymes are now supplied by genetically engineered enzyme producers. However, to date relatively few industrial processes that utilize recombinant whole-cell biocatalysts have been realized on an industrial scale.

Genetically engineered whole-cell biocatalysts can be beneficial when undesired side reactions of the host organism hamper a process and when space-time yields are insufficient. The desired enzymes can be cloned and expressed at appropriate levels in suitable hosts. Furthermore, regulation of enzyme systems can be controlled and multistep reactions can be assembled using genetic information from various sources. In short, genetic engineering enables the construction of whole-cell biocatalysts specifically tailored to a process.

3.1.1. Recombinant Biocatalysts

Biotransformations with recombinant strains are mostly performed with variants of well-established organisms, such as *Escherichia coli*, but pseudomonads are also very useful. In the first place, sophisticated genetic engineering tools for *Pseudomonas*, similar to those for *Escherichia coli*, are available (de Lorenzo and Timmis, 1994; Mermod *et al.*, 1986; West *et al.*, 1994), and genes from *Pseudomonas* are frequently more efficiently transcribed and translated in the original species than in heterologous hosts. Furthermore, pseudomonads, more than gram-positive or other gram-negative microorganisms, are suited for use in aqueous/solvent systems because of their inherent solvent resistance (Inoue and Horikoshi, 1991; Witholt *et al.*, 1990). Specific *Pseudomonas* isolates that resist high concentrations of toxic solvents, like styrene and toluene, have been isolated (Inoue and Horikoshi, 1989; Ramos *et al.*, 1995; Weber *et al.*, 1993). Such solvent-tolerant strains could serve as excellent hosts to harbor recombinant DNA encoding enzymes that are useful for converting water-insoluble substrates.

3.1.2. Protein Engineering

Knowledge of the structure and understanding the catalytic mechanism of industrial enzymes from *Pseudomonas* are central to improving them by protein engineering. The X-ray structures of several of the industrially relevant enzymes from *Pseudomonas* listed in Table I have been elucidated. Of the oxidoreductases, the X-ray structure of *p*-cresol methylhydroxylase from *P. putida* (EC 1.14.−.−) has been solved (Mathews *et al.*, 1991), and crystals of catechol-2,3-dioxygenase (EC 1.13.11.2) have recently been obtained (Earhart *et al.*, 1994). Within the hydrolase family, the structures of *P. fluorescens* (EC 3.1.1.2) esterase (Kim *et al.*, 1993) and of the haloacid dehalogenase (EC 3.8.1.2) (Asmara *et al.*, 1993) have been determined. Considerable information is available for the lipases (EC 3.1.1.3). X-ray structures of the lipases from *P. glumae* (Cleasby *et al.*, 1992), *P. fluorescens* (Larson *et al.*, 1991), *P. aeruginosa* (Misset *et al.*, 1994), *Pseudomonas sp.* (Kordel *et al.*, 1991), and *P. cepacia* (Kim *et al.*, 1992) are currently available. By comparing the substrate binding pocket of lipase (active site modeling), the 'proper' enzyme for a given substrate can be selected.

Using site-directed mutagenesis, the reaction mechanisms of some of these enzymes have been clarified (Asmara *et al.*, 1993; Noble *et al.*, 1994), and a number of these enzymes have been made better suited for industrial processes by protein engineering. Examples are the engineering of *P. glumae* lipase by Unilever to obtain a protease-resistant variant (Frenken *et al.*, 1993) and the modification of *Pseudomonas* sp. lipase by Genencor to make it better suited for synthesizing peracid bleaches (Estell, 1993).

In general then, protein engineering will enable the development of industrial enzymes that are tuned to and optimally suited for a specific biocatalytic conversion, a purpose that nature did not have in mind.

REFERENCES

Abalain, J. H., Distefano, S., Amet, Y., Quemener, E., Abalaincolloc, M. L., and Floch, H. H., 1993, Cloning, DNA sequencing, and expression of 3,17-β-hydroxysteroid dehydrogenase from *Pseudomonas testosteroni*, *J. Steroid Biochem. Mol. Biol.* **44**: 133–139.

Abbott, B. J., and Hou, C. T., 1973, Oxidation of 1-alkenes to 1,2-epoxyalkanes by *Pseudomonas oleovorans*, *Appl. Microbiol.* **26**: 86–91.

Adger, B., Bes, M. T., Grogan, G., McCague, R., Pedragosamoreau, S., Roberts, S. M., Villa, R., Wan, P. W. H., and Willetts, A. J., 1995, Application of enzymatic Baeyer–Villiger oxidations of 2-substituted cycloalkanones to the total synthesis of (R)- (+)-lipoic acid, *J. Chem. Soc. Chem. Commun.* **15**: 1563–1564.

Allen, C. C. R., Boyd, D. R., Dalton, H., Sharma, N. D., Haughey, S. A., McMordie, R. A. S., McMurray, B. T., Sheldrake, G. N., and Sproule, K., 1995, Sulfoxides of high enantiopurity from bacterial dioxygenase-catalyzed oxidation, *J. Chem. Soc. Chem. Commun.* **2**: 119–120.

Allenza, P., 1991, Conversion of mannose to fructose, U.S. Patent 5 049 494.

Allenza, P., Clifft, C. G., and Morrell, M. J., 1991, Some novel producers of cyclodextrin glycosyltransferases, U.S. Patent 5 008 195.

Allison, D. G., and Goldsbrough, M. J., 1994, Polysaccharide production in *Pseudomonas cepacia*, *J. Basic. Microbiol.* **34**: 3–10.

Alvarez-Jacobs, J., Court, D., and Guarneros, G., 1990, Lysine and methionine overproduction by an *Escherichia coli* strain transformed with *Pseudomonas acidovorans* DNA, *Biotechnol. Lett.* **12**: 425–430.

Amato, I., 1991, Bacterial indigo gives the blues to industrial chemists, *Science* **266**: 1213.

Amemura, A., and Futai, M., 1992, Polypeptide possessing isoamylase activity and its use in the hydrolysis of amylaceous substances, U.S. Patent 5 118 622.

Anderson, A. J., and Dawes, E. A. 1990, Occurrence, metabolism, metabolic role, and industrial uses of bacterial polyhydroxyalkanoates, *Microbiol. Rev.* **54**: 450–472.

Andreoni, V., Bernasconi, S., and Bestetti, G., 1995, Biotransformation of ferulic acid and related compounds by mutant strains of *Pseudomonas fluorescens*, *Appl. Microbiol. Biotechnol.* **42**: 830–835.

Anonymous, 1985, Biopol polymers made by fermentation, *Eur. Plast. News* **12**: 38.

Anonymous, 1986, ICI leads chemicals from biotech thrust, *Eur. Chem. News* **46**: 17.

Asano, Y., Yamamoto, Y., and Yamada, H., 1994, Catechol 2,3-dioxygenase-catalyzed synthesis of picolinic acids from catechols, *Bioscience Biotechnol. Biochem.* **58**: 2054–2056.

Asmara, W., Murdiyatmo, U., Baines, A. J., Bull, A. T., and Hardman, D. J., 1993, Protein engineering of the 2-haloacid halidohydrolase IVa from *Pseudomonas cepacia* MBA4, *Biochem. J.* **292**: 69–74.

Aust, S. D., Bourquin, A., Loper, J. C., Salanitro, J. P., Suk, W. A., and Tiedje, J., 1994, Biodegradation of hazardous wastes, *Environ. Health Perspect.* **102**: 245–252.

Baba, N., Mimura, M., Hiratake, J., Uchida, K., and Oda, J., 1988, Enzymic resolution of racemic hydroperoxides in organic solvent, *Agric. Biol. Chem.* **52**: 2685–2687.

Baird, J. K., Sandford, P. A., and Cottrell, I. W., 1983, Industrial applications of some new microbial polysaccharides, *Bio/Technol.* **1**: 778–783.

Ballard, D. G. H., Courtis, A., Shirley, I. A., and Taylor, S. C., 1983, A biotech route to polyphenylene, *J. Chem. Soc. Chem. Commun.* **634**: 954–955.

Baptist, J. N., Gholson, R. K., and Coon, M. J., 1963, Hydrocarbon oxidation by a bacterial enzyme system. I. Products of octane oxidation, *Biochim. Biophys. Acta* **69**: 40–47.

Bartilson, M., Nordlund, I., and Shingler, V., 1990, Location and organization of the dimethylphenol catabolic genes of *Pseudomonas* CF600, *Mol. Gen. Genet.* **220**: 294–300.

Beecher, J., Richardson, P., and Willetts, A., 1994, Baeyer–Villiger monooxygenase-dependent biotransformations: Stereospecific heteroatom oxidations by camphor-grown *Pseudomonas putida* to produce chiral sulfoxides, *Biotech. Lett.* **16**: 909–912.

Bestetti, G., and Galli, E., 1987, Characterization of a novel TOL-like plasmid from *Pseudomonas putida* involved in 1,2,4-trimethylbenzene degradation, *J. Bacteriol.* **169**: 1780–1783.

Bestetti, G., Galli, E., Leoni, B., Pelizzoni, F., and Sello, G., 1992, Regioselective hydroxylation of chlorobenzene and chlorophenols by a *Pseudomonas putida*, *Appl. Microbiol. Biotechnol.* **37**: 260–263.

Bevinakatti, H. S., and Banerji, A. A., 1992, Lipase catalysis in organic-solvents-application to the synthesis of (*R*)-atenolol and (*S*)-atenolol, *J. Org. Chem.* **57**: 6003–6005.

Bianchi, D., Bosetti, A., Cesti, P., Golini, P., and Spezia, S., 1993, Enzymic process for separating the optical isomers of racemic 1,2-diols using lipase, U.S. Patent 5 231 027.

Bosetti, A., Beilen, J. B. V., Preusting, H., Lageveen, R. G., and Witholt, B., 1992, Production of primary aliphatic alcohols with a recombinant *Pseudomonas* strain, encoding the alkane hydroxylase enzyme system, *Enzyme Microb. Technol.* **14**: 702–708.

Boyd, D., Sharma, N., Dorrity, M., Hand, M., Mcmordie, R., Malone, J., Porter, H., Dalton, H., Chima, J., and Sheldrake, G., 1993a, Structure and stereochemistry of *cis*-dihydrodiol and phenol metabolites of bicyclic azaarenes from *Pseudomonas putida* UV4, *J. Chem. Soc. Perkin. Trans. I*: **9**: 1065–1071.

Boyd, D., Sharma, N., Stevenson, P., Chima, J., Gray, D., and Dalton, H., 1991, Bacterial oxidation of benzocycloalkanes to yield monol, diol, and triol metabolites, *Tetrahedron Lett.* **32**: 3887–3890.

Boyd, D. R., Sharma, N. D., Boyle, R., Malone, J. F., Chima, J., and Dalton, H., 1993b, Structures and stereochemical assignments of some novel chiral synthons derived from the biotransformation of 2,3-dihydrobenzofuran and benzofuran by *Pseudomonas putida*, *Tetrahedron Asymm.* **4**: 1307–1324.

Boyd, D. R., Sharma, N. D., Hand, M. V., Groocock, M. R., Kerley, N. A., Dalton, H., Chima, J., and Sheldrake, G. N., 1993c, Stereodirecting substituent effects during enzyme-catalyzed synthesis of *cis*-dihydrodiol metabolites of 1,4-disubstituted benzene substrates, *J. Chem. Soc. Chem. Commun.* **11**: 974–976.

Bradshaw, C. W., Fu, H., Shen, G. J., and Wong, C. H., 1992a, A *Pseudomonas* sp. alcohol-dehydrogenase with broad substrate-specificity and unusual stereospecificity for organic synthesis, *J. Org. Chem.* **57**: 1526–1532.

Bradshaw, C. W., Lalonde, J. J., and Wong, C.-H., 1992b, Enzymatic synthesis of (*R*)- and (*S*)-1-deuterohexanol, *Appl. Biochem. Biotechnol.* **32**: 15–24.

Brand, J. M., Cruden, D. L., Zylstra, G. J., and Gibson, D. T., 1992, Stereospecific hydroxylation of indan by *Escherichia coli* containing the cloned toluene dioxygenase genes from *Pseudomonas putida* F1, *Appl. Env. Microbiol.* **58**: 3407–3409.

Britton, L. N., and Markovetz, A. J., 1977, A novel ketone monooxygenase from *Pseudomonas cepacia*. Purification and properties, *J. Biol. Chem.* **252**: 8561–8566.

Bushell, M. E., 1983, *Progress in Industrial Microbiology*, Vol. 18: Microbial polysaccharides, Elsevier, Amsterdam.

Chakrabarty, A. M., 1976, Plasmids in *Pseudomonas*, *Ann. Rev. Genet.* **10**: 23–29.

Chan, H. W., and Salazar, F. H., 1991, Cloning, expression and sequencing of an ester hydrolase gene in *Escherichia coli*, Patient EP 414 247.

Cleasby, A., Garman, E., Egmond, M. R., and Batenburg, M., 1992, Crystallization and preliminary X-ray study of a lipase from *Pseudomonas glumae*, *J. Mol. Biol.* **224**: 281–282.

Connors, M. A., and Barnsley, E. A., 1982, Naphthalene plasmids in *Pseudomonads*, *J. Bacteriol.* **149**: 1096–1101.

Cozzone, A. J., 1993, ATP-dependent protein kinases in bacteria, *J. Cell. Biochem.* **51**: 7–13.

Davies, J. I., and Evans, W. C., 1964, Oxidative metabolism of naphthalene by soil *Pseudomonads*, *Biochem. J.* **91**: 251–261.

de Lorenzo, V., and Timmis, K. N., 1994, Analysis and construction of stable phenotypes in gram-negative bacteria with Tn-5- and Tn*10*-derived minitransposons, in: *Methods in Enzymology*, 235, (V. L. Clark and P. M. Bavoil, eds.), Academic Press, pp. 386–405.

de Smet, M. J., 1983, A biotechnological approach to the synthesis of epoxides: Bioconversion of hydrocarbons by *Pseudomonas oleovorans* during growth in a multiphase system, *Biotechnol. Bioeng.* **25**: 1161–1162.

de Smet, M. J., Eggink, G., Witholt, B., Kingma, J., and Wynberg, H., 1983a, Characterization on intracellular inclusions formed by *Pseudomonas oleovorans* during growth on octane, *J. Bacteriol.* **154**: 870–878.

de Smet, M. J., Kingma, J., Wynberg, H., and Witholt, B., 1983b, *Pseudomonas oleovorans* as a tool in bioconversions of hydrocarbons: Growth morphology and conversion characteristics in different two-phase systems, *Enzyme Microb. Technol.* **5**: 352–360.

de Smet, M. J., Wijnberg, H., and Witholt, B., 1981, Synthesis of 1,2-epoxyoctane by *Pseudomonas oleovorans* during growth in a two phase system containing high concentrations of 1-octene, *Appl. Env. Microbiol.* **42**: 811–816.

Deamici, M., Demicheli, C., Molteni, G., Pitre, D., Carrea, G., Riva, S., Spezia, S., and Zetta, L., 1991, Chemoenzymatic synthesis of the 8 stereoisomeric muscarines, *J. Org. Chem.* **56**: 67–72.

den Dooren de Jong, L. E., 1926, PhD Thesis, Leiden University, The Netherlands.

Deretic, V., Martin, D. W., Schurr, M. J., Mudd, M. H., Hibler, N. S., *et al.*, 1993, Conversion to mucoidy in *Pseudomonas aeruginosa*, *Biotechnology* **11**: 1133–1136.

Drauz, K., and Waldmann, H., 1995, *Enzyme Catalysis in Organic Synthesis, A Comprehensive Handbook*, VCH Verlagsgesellschaft, Weinheim.

Earhart, C. A., Hall, M. D., Michaude Soret, I., Que, L., Jr., and Ohlendorf, D. H., 1994, Crystallization of catechol-1,2 dioxygenase from *Pseudomonas arvilla* C-1, *J. Mol. Biol.* **236**: 377–378.

Effenberger, F., Horsch, B., Weingart, F., Ziegler, T., and Kuhner, S., 1991, Enzyme-catalyzed synthesis of (R)-ketone-cyanohydrins and their hydrolysis to (R)-α-hydroxy-α-methyl-carboxylic acids, *Tetrahedron Lett.* **32:** 2605–2608.

Eggink, G., 1987, Thesis, University of Groningen, The Netherlands.

Eggink, G., Engel, H., Meijer, W. G., Otten, J., Kingma, J., and Witholt, B., 1988, Alkane utilization in *Pseudomonas oleovorans*: Structure and function of the regulatory locus *alkR*, *J. Biol. Chem.* **263:** 13400–13405.

Endo, T., and Tamura, K., 1991, Process for producing R- (−)-mandelic acid and derivatives thereof, Patent EP 449648.

Ensley, B. D., 1994, Biosynthesis of the textile dye indigo by a recombinant bacterium, *Chimia* **48:** 491–492.

Ensley, B. D., and Gibson, D. T., 1983, Naphthalene dioxygenase: Purification and properties of a terminal oxygenase component, *J. Bacteriol.* **155:** 505–511.

Ensley, B. D., Ratzkin, B. J., Osslund, T. D., Simon, M. J., Wackett, L. P., and Gibson, D. T., 1983, Expression of naphthalene dioxygenase genes in *Escherichia coli* results in biosynthesis of indigo, *Science* **222:** 167–169.

Estell, D. A., 1993, Engineering enzymes for improved performance in industrial applications, *J. Biotechnol.* **28:** 25–30.

Evans, C., Roberts, S., Shoberu, K., and Sutherland, A., 1992, Potential use of carbocyclic nucleosides for the treatment of AIDS. Chemoenzymatic syntheses of the enantiomers of carbovir, *J. Chem. Soc. Perkin Trans.* **1:** 589–592.

Evans, C. T., Peterson, W., Choma, C., and Misawa, M., 1987, Biotransformation of phenylpyruvic acid to L-phenylalanine using a strain of *Pseudomonas fluorescens* ATCC 11250 with high transaminase activity, *Appl. Microbiol. Biotechnol.* **26:** 305–312.

Fa, Y. H., Kusel, J. P., and Demain, A. L., 1984, Dependence of betaine stimulation of vitamin B12 overproduction on protein synthesis, *Appl. Environ. Microbiol.* **47:** 1067–10699.

Faber, K., 1995, *Biotransformations in Organic Chemistry*, Springer Verlag, Berlin.

Favre Bulle, O., Schouten, T., Kingma, J., and Witholt, B., 1991, Bioconversion of *n*-octane to octanoic acid by a recombinant *Escherichia coli* cultured in a two-liquid phase bioreactor, *Bio/Technology* **9:** 367–371.

Fernandez-Valverde, M., Reglero, A., Martinez-Blanco, H., and Luengo, J. M., 1993, Purification of *Pseudomonas putida* actyl coenzyme A ligase active with a range of aliphatic and aromatic substrates, *Appl. Env. Microbiol.* **59:** 1149–1154.

Ferrero, M. A., Reglero, A., Martinez-Blanco, H., Fernandez-Valverde, M., and Luengo, J. M., 1991a, In vitro enzymatic synthesis of new penicillins containing keto acids as side chains, *Antimicrob. Agents Chemother.* **35:** 1931–1932.

Ferrero, O., Reglero, A., Martin-Villacorta, J., Martinez-Blanco, H., and Luengo, J. M., 1991b, Synthesis of 3-furylmethylpenicillin using an enzymatic procedure, *FEMS Microbiol. Lett.* **83:** 1–6.

Fetzner, S., and Linigens, F., 1994, Bacterial dehalogenases: Biochemistry, genetics, and biotechnological applications, *Microbiol. Rev.* **58:** 641–685.

Florent, J., 1986, Vitamins, in: *Biotechnology*, Vol. 4, (H.–J. Rehm and G. Reed, eds.), VCH Verlaggesellschaft, Weinheim, pp. 117–158.

Forney, F. W., and Markovetz, A. J., 1969, An enzyme system for aliphatic methyl ketone oxidation, *Biochem. Biophys. Res. Commun.* **37:** 31–38.

Francalanci, F., Ricci, M., Cesti, P., and Venturello, C., 1993, Process for preparing L (−)-carnitine chloride, U.S. Patent 5 248 601.

Frenken, L. G. J., Egmond, M. R., Batenburg, A. M., and Verrips, C. T., 1993, *Pseudomonas glumae* lipase: Increased proteolytic stability by protein engineering, *Protein Eng.* **6:** 637–642.

Fukagawa, M., Ono, H., Ishitani, Y., Tsumura, M., Iwani, M., and Kojo, H., 1992, Glutaryl-7ACA acylase, Patent EP 482 844.

Fukuda, M., Nishi, T., Igarashi, M., Kondo, T., Takagi, M., and Yano, K., 1989, Degradation of ethylbenzene by *Pseudomonas putida* harboring OCT plasmid, *Agric. Biol. Chem.* **53:** 3293–3299.

Furakawa, K., Hayashida, S., and Taira, K., 1992, Biochemical and genetic basis for the degradation of polychlorinated biphenyls in soil bacteria, in: *Pseudomonas: Molecular Biology and Biotechnology*, (E. Galli, S. Silver, and B. Witholt, eds.), American Society for Microbiology, Washington D. C., pp. 259–267.

Furuhashi, K., 1992, Biological routes to optically active epoxides, in: *Chirality in Industry*, (A. N. Collins, G. N. Sheldrake, and J. Crosby, eds.), John Wiley & Sons, London, pp. 167–186.

Furuta, T., Takahashi, H., Shibasaki, H., and Kasuya, Y., 1992, Reversible stepwise mechanism involving a carbanion intermediate in the elimination of ammonia from L-histidine catalyzed by histidine ammonia-lyase, *J. Biol. Chem.* **267:** 12600–12605.

Gagnon, R., Grogan, G., Roberts, S. M., Villa, R., and Willetts, A. J., 1995, Enzymatic Baeyer–Villiger oxidations of some bicycle[2.2.1] heptan-2-ones using monooxygenases from *Pseudomonas putida* NCIMB 10007: Enantioselective preparation of a precursor of azadirachtin, *J. Chem. Soc. Perkin Trans.* **1:** 1505–1511.

Geary, P. J., and Haives, J. E., 1988, Microbial preparation of catechols, Patent EP 268 331.

Geary, P. J., and Pryce, R. J., 1990, *Pseudomonas putida* cells for microbial production of catechols, Br. Patent 2 222 176.

Gibson, D. T., 1991, The role of oxygenases in the microbial oxidation of aromatic compounds, *Abstr. Pap. Am. Chem. Soc.* **201:** 46.

Gibson, D. T., Resnick, S. M., Lee, K., Brand, J. M., Torok, D. S., Wackett, L. P., Schocken, M. J., and Haigler, B. E., 1995, Desaturation, dioxygenation, and monooxygenation reactions catalyzed by naphthalene dioxygenase from *Pseudomonas sp* strain-9816-4, *J. Bacteriol.* **177:** 2615–2621.

Gibson, D. T., and Subramanian, V., 1984, Microbial degradation of aromatic hydrocarbons, in: *Microbial Degradation of Organic Compounds* (D. T. Gibson, ed.), Marcel Dekker, New York, pp. 181–252.

Glazer, A. N., and Nikaido, H., 1995, *Microbial Biotechnology: Fundamentals of Applied Microbiology*, W. H. Freeman, New York.

Grogan, G., Roberts, S., Wan, P., and Willetts, A., 1993, Camphor-grown *Pseudomonas putida*, a multifunctional biocatalyst for undertaking Baeyer–Villiger monooxygenase-dependent biotransformations, *Biotechnol. Lett.* **15:** 913–918.

Grogan, G., Roberts, S., and Willetts, A., 1992, Biotransformation by microbial Baeyer–Villiger monooxygenases: stereoselective lactone formation *in vitro* by coupled enzyme systems, *Biotechnol. Lett.* **14:** 1125–1130.

Hagedorn, S. R., East, A. J., and Barer, S. J., 1989, Production of picolinic acid and pyridine products via *Pseudomonas*, U.S. Patent 4 859 592.

Haigler, B. E., and Spain, J. C., 1991, Biotransformation of nitrobenzene by bacteria containing toluene degradative pathways, *Appl. Environ. Microbiol.* **57:** 3156–3162.

Hailes, A. M., and Bruce, N. C., 1993, Biological synthesis of the analgesic hydromorphone, an intermediate in the metabolism of morphine, by *Pseudomonas putida* M10, *Appl. Environ. Microbiol.* **59:** 2166–2170.

Hansen, C., Fortnagel, P., and Wittich, R. M., 1992, Initial reactions in the mineralization of 2-sulfobenzoate by *Pseudomonas sp* RW611, *FEMS Microbiol. Lett.* **92:** 35–40.

Harayama, S., and Don, R. H., 1985, Catabolic plasmids: Their analysis and utilization in the manipulation of bacterial metabolic activities, *Genet. Eng.* **7:** 283–307.

Harayma, S., and Rekik, M., 1990, The *meta* cleavage operon of TOL degradative plasmid pWW0 comprises 13 genes, *Mol. Gen. Genet.* **221:** 113–120.

Harayama, S., Rekik, M., and Timmis, K. N., 1986, Genetic analysis of a relaxed substrate specificity aromatic ring dioxygenase, toluate-1,2-dioxygenase, encoded by TOL plasmid pWW0 of *Pseudomonas putida*, *Mol. Gen. Genet.* **202:** 226–234.

Hashimoto, Y., Endo, T., Tamura, K., and Hirata, Y., 1994, Process for producing optically active α-hydroxycarboxylic acid having a phenyl group, Patent EP 610 049.

Hernandez, D., and Phillips, A. T., 1994, Ser-143 is an essential active site residue in histidine ammonia-lyase of *Pseudomonas putida*, *Biochem. Biophys. Res. Commun.* **201:** 1433–1438.

Hirose, J., Kimura, N., Suyama, A., Kobayashi, A., Hayashida, S., and Furukawa, K., 1994, Functional and structural relationship of various extradiol aromatic ring-cleavage dioxygenases of *Pseudomonas* origin, *FEMS Microbiol. Lett.* **118:** 273–277.

Hoeks, F., 1991, Process for the microbiological discontinual preparation of L-carnitine, Patent EP 410 430.

Hoeks, F., and Venetz, D., 1992, Process for the production of 6-hydroxynicotinic acid, U.S. Patent 5 151 351.

Holloway, B., Escuadra, M., Morgan, A., Saffery, R., and Krishnapillai, V., 1992, New approaches to whole genome anlaysis of bacteria, *FEMS. Microbiol. Lett.* **100:** 101–105.

Hou, C. T., 1993, Screening of microbial esterases for asymmetric hydrolysis of 2-ethyl-hexyl butyrate, *J. Ind. Microbiol.* **11:** 73–81.

Hudlicky, T., Luna, H., Barbieri, G., and Kwart, L. D., 1988, Enantioselective synthesis through microbial oxidation of arenes. I. Efficient preparation of terpene and prostanoid synthons, *J. Am. Chem. Soc.* **110:** 4735–4741.

Hudlicky, T., Mandel, M., Rouden, J., Lee, R. S., Bachmann, B., Dudding, T., Yost, K. J., and Merola, J. S., 1994a, Microbial oxidation of aromatics in enantiocontrolled synthesis. 1. Expedient and general asymmetric synthesis of inositols and carbohydrates via an unusual oxidation of a polarized diene with potassium permanganate, *J. Chem. Soc. Perkin. Trans.* **1:** 1553–1567.

Hudlicky, T., Olivo, H. F., and McKibben, B., 1994b, Microbial oxidation of aromatics in enantiocontrolled synthesis. 3. Design of amino cyclitols (exo-nitrogenous) and total synthesis of (+)-lycoricidine via acylnitrosyl cycloaddition to polarized 1-halo-1,3-cyclohexadienes, *J. Am. Chem. Soc.* **116:** 5108–5115.

Hur, H. G., Sadowsky, M. J., and Wackett, L. P., 1994, Metabolism of chlorofluorocarbons and polybrominated compounds by *Pseudomonas putida* G786 (pHG-2) via an engineered metabolic pathway, *Appl. Environ. Microbiol.* **60:** 4148–4154.

Ichikawa, S., Yamamoto, K., and Matsuyama, K., 1988, Process for preparing a 7-aminocephalosporanic acid compound, U.S. Patent 4 774 179.

Idemitsu, 1986, Production of sulfoxides having a sulfide bond at alpha-position from formaldehyde mercaptal using bacterium or fungus, J. Patent 61 195 694.

Inagaki, M., Hiratake, J., Nishioka, T., and Oda, J., 1992, One-pot synthesis of optically active cyanohydrin acetates from aldehydes via lipase-catalyzed kinetic resolution coupled with in situ formation and racemization of cyanohydrins, *J. Orig. Chem.* **57:** 5643–5649.

Inoue, A., and Horikoshi, K., 1989, A *Pseudomonas* thrives in high concentrations of toluene, *Nature* **338:** 264–266.

Inoue, A., and Horikoshi, K., 1991, Estimation of solvent tolerance of bacteria by the solvent parameter Log P, *J. Fermentation Bioeng.* **71:** 194–197.

Irie, S., Doi, T., Yorifuyi, M., Takagi, M., and Yano, K., 1987a, Nucleotide sequencing and characterization of soluble benzene-oxidizing system from a strain of *Pseudomonas putida, J. Bacteriol.* **169:** 5174–5179.

Irie, S., Shirai, K., Doi, S., and Yorifuji, T., 1987b, Cloning of genes encoding oxidation of benzene in *Pseudomonas putida* and their expression in *Escherichia coli* and *P. putida, Agric. Biol. Chem.* **51:** 1489–1493.

Ishikawa, T., Watabe, K., Mukohara, Y., Kobayashi, S., and Nakamura, H., 1993, Microbial conversion of DL-5-substituted hydantoins to the corresponding L-amino-acids by *Pseudomonas* sp. strain NS671, *Bioscience Biotechnol. Biochem.* **57:** 982–986.

Ishiwata, K. I., Fukuhara, M., Shimada, M., Makiguchi, N., and Soda, K., 1990, Enzymic production of L-tryptophan from DL-serine and indole by a coupled reaction of tryptophan synthase and amino acid racemase, *Biotechnology Appl. Biochem.* **12:** 141–149.

Itoh, H., Sato, T., and Izumori, K., 1995, Preparation of D-psicose from D-fructose by immobilized D-tagatose 3-epimerase, *J. Fermentation Bioeng.* **80:** 101–103.

Jacobs, M. H., Van den Wijngaard, A. J., Pentenga, M., and Janssen, D. B., 1991, Characterization of the epoxide hydrolase from an epichlorohydrin-degrading *Pseudomonas* sp, *Eur. J. Biochem.* **202:** 1217–1222.

Jaeger, K. E., and Wohlfarth, S., 1993, Bacterial lipases: Biochemistry, molecular genetics, and applications in biotechnology, *Bioengineering* **9:** 39–46.

Janssen, D. B., Pries, F., and Vanderploeg, J. R., 1994, Genetics and biochemistry of dehalogenating enzymes, *Annu. Rev. Microbiol.* **48:** 163–191.

Jenkins, R. O., Stephens, G. M., and Dalton, H., 1987, Production of toluene *cis*-glycol by *Pseudomonas putida* in glucose batch-fed culture, *Biotechnol. Bioeng.* **29:** 873–883.

Johnson, B. F., and Mondello, F. J., 1991, Biological and chemical method for hydroxylating 4-substituted biphenyls and products obtained therefrom, U.S. Patent 4 981 793.

Johnston, J. B., and Renganathan, V., 1987, Production of substituted catechols from substituted benzenes by a *Pseudomonas* species, *Enzyme Microb. Technol.* **9:** 706–708.

Johnstone, S. L., Phillips, G. T., Robertson, B. W., Watts, P. D., Bertola, M. A., Koger, H. S., and Marx, A. F., 1987, Stereoselective synthesis of S-(−)-β-blockers via microbially produced epoxide intermediates, in: *Biocatalysis in Organic Media*, (C. Laane, J. Tramper, and M. D. Lilly, eds.), Elsevier, Amsterdam, pp. 387–392.

Kamphuis, J., Meijer, E. M., Boesten, W. H., Sonke, T., van den Tweel, W. J., and Schoemaker, H. E., 1992, New developments in the synthesis of natural and unnatural amino acids, *Ann. N.Y. Acad. Sci.* **672:** 510–527.

Kanerva, L. T., Rahiala, K., and Sundholm, O., 1994, Optically active cyanohydrins and enzyme catalysis, *Biocatalysis* **10:** 169–180.

Katopodis, A. G., Wimalasena, K., Lee, J., and May, S. W., 1984, Mechanistic studies on non-heme iron monooxygenase catalysis: Epoxidation, aldehyde formation, and demethylation by the ω-hydroxylation system of *Pseudomonas oleovorans., J. Am. Chem. Soc.* **106:** 7928–7935.

Kawamori, M., Hashimoto, Y., Katsumata, R., Okachi, R., and Takayama, K., 1983, Enzymatic production of amoxicillin by beta-lactamase-deficient mutants of *Pseudomonas melanogenum* KY 3987, *Agric. Biol. Chem.* **47:** 2503–2509.

Kelly, C. T., and Fogarty, W. M., 1983, Microbial alpha-glucosidases, *Process Biochem.* **18:** 6–12.

Kerkhoffs, P. L., and Boesten, W. H. J., 1989, Process for the preparation of L-α-amino acid and D-α-amino acid amide, U.S. Patent 4 880 737.

Kiener, A., 1992a, Enzymatic oxidation of methyl groups on aromatic heterocycles: a versatile method for the preparation of heteroaromatic carboxylic acids, *Angew. Chem. Int. Ed. Engl.* **31:** 774–775.

Kiener, A., 1992b, Microbial oxidation of alkyl groups in heterocycles, Patent EP466115.

Kiener, A., 1992c, Mikrobiologisches Verfahren zur terminalen Hydroxylierung von Ethylgruppen an aromatischen 5- oder 6-Ring-Heterocyclen, Patent EP 502524.

Kiener, A., 1995, Biosynthesis of functionalized aromatic N-heterocycles, *Chemtech* **Sept:** 31–35.

Kieslich, K., 1991, Biotransformations of industrial use, *Acta Biotechnol.* **11:** 559–570.

Kim, K., Hwang, K., Choi, K., Kang, J., Yoo, O., and Sush, S., 1993, Crystallization and preliminary X-ray crystallographic analysis of arylesterase from *Pseudomonas fluorescens, Protein Struct. Functional Genet.* **15:** 213–215.

Kim, K. K., Hwang, K. Y., Jeon, H. S., Kim, S., Sweet, R. M., Yang, C. H., and Suh, S. W., 1992, Crystallization and preliminary X-ray crystallographic analysis of lipase from *Pseudomonas cepacia, J. Mol. Biol.* **227:** 1258–1262.

Kloosterman, M., Elferink, V. H., van Lersel, J., Roskam, J. H., and Meijer, E. M., 1988, Lipases in the preparation of β-blockers, *Trends Biotechnol.* **6:** 251–256.

Kohler, H. P. E., Maarel, M. J. E. C. V. d., and Kohler–Staub, D., 1993, Selection of *Pseudomonas* species strain HBP1 Prp for metabolism of 2-propylphenol and elucidation of the degradative pathway, *Appl. Environ.* Microbiol. **59:** 860–866.

Kok, M., Eggink, G., Witholt, B., Owen, D. J., and Shapiro, J., 1984, Transposable elements as cloning vectors. Characterization of the *Pseudomonas oleovorans alk*-regulon., in: *Progress in Industrial Microbiology* (E. H. Houwink and R. R. Van der Meer, eds.), Elsevier Scientific, Amsterdam, pp. 373–380.

Kok, M., Oldenhuis, R., van der Linden, M. P. G., Meulenberg, C. H. C., Kingma, J., and Witholt, B., 1989a, The *Pseudomonas oleovorans alkBAC* operon encodes two structurally related rubredoxins and an aldehyde-dehydrogenase, *J. Biol. Chem.* **264:** 5442–5451.

Kok, M., Oldenhuis, R., van der Linden, M. P. G., Raatjes, P., Kingma, J., and van Lelyveld, P. H., 1989b, The *Pseudomonas oleovorans* alkane-hydroxylase gene: Sequence and expression, *J. Biol. Chem.* **264:** 5435–5441.

Kok, M., Shaw, J. P., and Harayama, S., 1992, Comparison of two hydrocarbon-monooxygenases of *Pseudomonas putida*, in: *Pseudomonas. Molecular Biology, and Biotechnology* (E. Galli, S. Silver, and B. Witholt, eds.), American Society for Microbiology, Washington D.C., pp. 214–222.

Konishi, M., Tomita, K., Oka, M., and Numata, K.-I., 1992, Peptide antibiotics. U.S. Patent 5 079 148.

Kordel, M., Hofmann, B., Schomburg, D., and Schmid, R. D., 1991, Extracellular lipase of *Pseudomonas sp.* strain ATCC 21808: Purification, characterization, crystallization, and preliminary X-ray diffraction data, *J. Bacteriol.* **173:** 4836–4841.

Krell, H. W., and Rasor, P., 1993, Microbial esterase for the enantioselective cleavage of 1-arylalkyl esters, Patent WO 9 324 648.

Kropp, K. G., Goncalves, J. A., Andersson, J. T., and Fedorak, P. M., 1994, Bacterial transformations of benzothiophene and methylbenzothiophenes, *Environ. Science Technol.* **28:** 1348–1356.

Kula, M. R., and Peters, J., 1993, New ketonic ester reductase, its preparation and use for enzymic redox reactions, Patent WO 9 318 138.

Kulla, G. H., 1991, Enzymatic hydroxylations in industrial application, *Chimia* **86:** 295–323.

Kunz, D. A., and Chapman, P. J., 1981, Catabolism of pseudocumene and 3-ethyltoluene by *Pseudomonas putida* (*arvilla*) mt-2: Evidence for new functions of the TOL (pWWO) plasmid, *J. Bacteriol.* **146:** 179–191.

Kunz, D. A., Ribbons, D. W., and Chapman, P. J., 1981, Metabolism of allyglycine and *cis*-crotylglycine by *Pseudomonas putida* (*arvilla*) mt-2 harboring a TOL plasmid, *J. Bacteriol.* **148:** 72–82.

Kusel, J. P., Fa, Y. H., and Demain, A. L., 1984, Betaine stimulation of vitamin B_{12} biosynthesis in *Pseudomonas denitrificans* may be mediated by an increase in activity of δ-aminolaevulinic acid synthetase, *J. Gen. Microbiol.* **130:** 835–841.

Kusunose, M., Ichihara, K., Kusunose, E., Nozaka, J., and Matsumoto, J., 1967, The possible role of flavin on the hydroxylation of hydrocarbon by bacterial enzyme system, *Agric. Biol. Chem.* **31:** 990–992.

Kutsuki, H., Sawa, I., Hasegawa, J., and Watanabe, K., 1986, Asymmetric hydrolysis of (DL)-1-acyloxy-2-halo-1-phenylethanes with lipases, *Agric. Biol. Chem.* **50:** 2369–2373.

Lageveen, R. G., Huisman, G. W., Preusting, H., Ketelaar, P., Eggink, G., and Witholt, B., 1988, Formation of polyesters by *Pseudomonas oleovorans*: Effect of substrates on formation and composition of poly (*R*)-3-hydroxyalkanoates and poly (*R*)-3-hydroxyalkenoates, *Appl. Environ. Microbiol.* **54:** 2924–2934.

Lapointe, G., Viau, S., Leblanc, D., Robert, N., and Morin, A., 1994, Cloning, sequencing, and expression in *Escherichia coli* of the D-hydantoinase gene from *Pseudomonas putida* and distribution of homologous genes in other microorganisms, *Appl. Environ. Microbiol.* **60:** 888–895.

Larson, S., Day, J., Greenwood, A., Oliver, J., Rubingh, D., and McPherson, A., 1991, Preliminary investigation of crystals of the neutral lipase from *Pseudomonas fluorescens*, *J. Mol. Biol.* **222:** 21–22.

Layh, N., Stolz, A., Forster, S., Effenberger, F., and Knackmuss, H. J., 1992, Enantioselective hydrolysis of *o*-acetylmandelonitrile to *o*-acetylmandelic acid by bacterial nitrilases, *Arch. Microbiol.* **158:** 405–411.

Lee, K., Brand, J. M., and Gibson, D. T., 1995, Stereospecific sulfoxidation by toluene and naphthalene dioxygenases, *Biochem. Biophys. Res. Commun.* **212:** 9–15.

Lee, M., and Chandler, A. C., 1941, A study of the nature, growth, and control of bacteria in cutting compounds, *J. Bacteriol.* **41:** 373–386.

Lenn, M. J., and Knowles, C. J., 1994, Production of optically active lactones using cyclo-alkanone oxygenases, *Enzyme Microb. Technol.* **16:** 964–969.

Lenz, R. W., Kim, Y. B., and Fuller, R. C., 1992, Production of unusual bacterial polyesters by *Pseudomonas oleovorans* through cometabolism, *FEMS Microbiol. Rev.* **103:** 207–214.

Lim, Y. H., Yokoigawa, K., Esaki, N., and Soda, K., 1993, A new amino acid racemase with threonine α-epimerase activity from *Pseudomonas putida*: Purification and characterization, *J. Bacteriol.* **175:** 4213–4217.

Mader, R. A., and Tautvydas, K. J., 1990, Biological production of novel cyclohexa-dienediols, Patent EP 400 779.

Marquardt, R., Then, J., Braeu, B., Praeve, P., and Woehner, G., 1988, Production of phenylalanine by recombinant bacteria, Patent EP 289 846.

Marques, A. M., Estanol, I., Alsina, J. M., Fuste, C., and Simon-Pujol, D., 1986, Production and rheological properties of the extracellular polysaccharide synthesized by Pseudomonas sp. strain EPS-5028, *Appl. Environ. Microbiol.* **52:** 1221–1223.

Martinez-Blanco, H., Reglero, A., Martin-Villacorta, J., and Luengo, J. M., 1990, Design of an enzymatic hybrid system: A useful strategy for the biosynthesis of benzylpenicillin in vitro, *FEMS Microbiol. Rev.* **72:** 113–116.

Martinez-Blanco, H., Reglero, A., and Luengo, J. M., 1991, *In vitro* synthesis of different naturally occurring, semisynthetic and synthetic penicillins using a new and effective enzymatic coupled system, *J. Antibiotics* **44:** 1252–1258.

Martins, L. O., and Sa-Correia, I., 1991, Alginate biosynthesis in mucoid recombinants of *Pseudomonas aeruginosa* overproducing GDP-mannose dehydrogenase, *Enzyme Microb. Technol.* **13:** 385–389.

Maruo, S., Yamamoto, H., Toda, M., Tachikake, N., Kojima, M., and Ezure, Y., 1993,

Enzymic synthesis of high purity maltotetraose using moranoline (1-deoxyno-jirimycin), *Biosci. Biotechnol. Biochem.* **57**: 499–501.

Masutomo, S., Inoue, A., Kumagai, K., Murai, R., and Mitsuda, S., 1995, Enantioselective hydrolysis of (*R,S*)-2-isopropyl-4'-chlorophenylacetonitrile by *Pseudomonas* sp B21C9, *Biosc. Biotechnol. Biochem.* **59**: 720–722.

Mathews, F. S., Chen, Z. W., Bellamy, H. D., and McIntire, W. S., 1991, Three-dimensional structure of *p*-cresol methylhydroxylase (flavocytochrome c) from *Pseudomonas putida* at 3.0-Å resolution, *Biochemistry* **30**: 238–247.

Matsubara, M., Masukawa, T., Adachi, N., Fukuta, M., and Kariya, M., 1991, Process for producing *cis*-4,5-dihydro-4,5-dihydroxyphthalic acid, U.S. Patent 5 037 748.

Maxwell, P. C., 1988, Process for the production of muconic acid, U.S. Patent 4 731 328.

May, S. W., and Abbott, B. J., 1972, Enzymatic epoxidation. I. Alkane epoxidation by the ω-hydroxylation system of *Pseudomonas oleovorans*, *Biochem. Biophys. Res. Commun.* **48**: 1230–1234.

May, S. W., and Katopodis, A. G., 1986, Oxygenation of alcohol and sulfide substrates by a prototypical non-heme iron monooxygenase: Catalysis and biotechnological potential, *Enzyme Microb. Technol.* **8**: 17–21.

May, S. W., and Swartz, R. D., 1974, Stereoselective epoxidation of octadiene catalyzed by an enzyme system of *Pseudomonas oleovorans*, *J. Am. Chem. Soc.* **96**: 4031–4032.

McIntire, W., Hopper, D. J., Craig, J. C., Everhart, E. T., Webster, R. V., Causer, M. J., and Singer, T. P., 1984, Stereochemistry of 1-(4'-hydroxyphenyl)ethanol produced by hydroxylation of 4-ethylphenol by *p*-cresol methylhydroxylase, *Biochem. J.* **224**: 617–621.

McIntire, W., Hopper, D. J., and Singer, T. P., 1985, *p*-Cresol methylhydroxylase. Assay and general properties, *Biochem. J.* **228**: 325–335.

McKenna, E. J., and Coon, M. J., 1970, Enzymatic ω-oxidation. IV. Purification and properties of the ω-hydroxylase of *Pseudomonas oleovorans*, *J. Biol. Chem.* **245**: 3882–3889.

McKibben, B. P., Barnosky, G. S., and Hudlicky, T., 1995, Unusual dehalogenation of a bridgehead halide—biocatalytic conversion of halocyclohexadiene-*cis*-diols to the *trans*-iosmers and synthesis of optically pure cyclohexadiene-*trans*-diols, *Synlett:* **8**: 806–808.

Mermod, N., Harayama, S., and Timmis, K. N., 1986, New route to bacterial production of indigo, *Bio/Technology* **4**: 321–324.

Mermod, N., Ramos, J. L., Lehrbach, P. R., and Timmis, K. N., 1986, Vector for regulated expression of cloned genes in a wide range of gram-negative bacteria, *J. Bacteriol.* **167**: 447–454.

Misset, O., Gerritse, G., Jaeger, K. E., Winkler, U., Colson, C., Schanck, K., Lesuisse, E., Dartois, V., Blaauw, M., Ransac, S., and Dijkstra, B. W., 1994, The structure-function relationship of the lipases from *Pseudomonas aeruginosa* and *Bacillus subtilis*, *Protein Eng.* **7**: 523–529.

Mitsuda, S., Matsuo, N., and Hirohara, H., 1991, Process for producing optically active benzyl alcohol compounds, U.S. Patent 4 985 365.

Morin, A., Hummel, W., Schuette, H., and Kula, M. R., 1986, Characterization of hydantoinase from *Pseudomonas fluorescens* strain DSM 84, *Biotechnol. Appl.* Biochem. **8**: 564–574.

Morin, A., Leblanc, D., Paleczek, A., and Hummel, W., 1990, Comparison of seven microbial D-hydantoinases, *J. Biotechnol.* **16**: 37–48.

Morris, V. J., 1992, Bacterial polysaccharides, *Agro Food Ind. Hi-Tech* **3**: 3–8.

Motosugi, K., Esaki, N., and Soda, K., 1984, Enzymatic preparation of D- and L-lactic acid from racemic 2-chloropropionic acid, *Biotechnol. Bioeng.* **26**: 805–806.

Murakami, N., 1993, Method for the preparation of an optically active 2-substituted carboxylic acid, U.S. Patent 5 238 828.

Murdock, D., Ensley, B. D., Serdar, C., and Thalen, M., 1993, Construction of metabolic operons catalyzing the *de novo* biosynthesis of indigo in *Escherichia coli*, *Bio/Technology* **11**: 381–385.

Nagasawa, T., Hosono, H., Ohkishi, H., Tani, Y., and Yamada, H., 1983a, Synthesis of D-cysteine-related amino acids by 3-chloro-D-alanine chloride-lyase of *Pseudomonas-putida* CR 1-1, *Biochem. Biophys. Res. Commun.* **111**: 809–816.

Nagasawa, T., Hosono, H., Ohkishi, H., and Yamada, H., 1983b, Synthesis of S-(carboxymethyl)-D-cysteine by 3-chloro-D-alanine chloride-lyase of *Pseudomonas putida* CR 1-1, *Appl. Biochem. Biotechnol.* **8**: 481–489.

Nagasawa, T., Hosono, H., Yamamo, H., Ohkishi, H., Tani, Y., and Tamada, H., 1983c, Synthesis of D-cysteine from a racemate of 3-chloroalanine by phenylhydrazine-treated cells of *Pseudomonas putida* CR 1-1, *Agric. Biol. Chem.* **47**: 861–868.

Nagasawa, T., Shimizu, H., and Yamada, H., 1993, The superiority of the third-generation catalyst, *Rhodococcus rhodochrous* J1 nitrile hydratase, for industrial production of acrylamide, *Appl. Microbiol. Biotechnol.* **40**: 189–195.

Nakayama, K., Honda, H., Ogawa, T., Ozawa, T., and Ohta, T., 1988, Method for producing carnitine, L-carnitineamide hydrolase and method for producing same, Br. Patent 2 195 630.

Nanba, H., Yamada, Y., Takano, M., Ikenaka, Y., and Takahashi, S., 1992, Process for producing D-α-amino acid, Patent WO 9 210 579.

Noble, M. E. M., Cleasby, A., Johnson, L. N., Egmond, M. R., and Frenken, L. G. J., 1994, Analysis of the structure of *Pseudomonas glumae* lipase, *Protein Eng.* **7**: 559–562.

Olivieria, R., Eletti Bianchi, G., Fascetti, E., and Centini, F., 1985, Process for the preparation of L-α-amino acids, Patent EP 152 977.

Onda, M., Motosugi, K., and Nakajima, H., 1990, A new approach for enzymatic synthesis of D-3-chlorolactic acid from racemic 2,3-dichloropropionic acid by halo-acid dehydrogenase, *Agr. Biol. Chem.* **54**: 3031–3033.

Owen, D. J., Eggink, G., Hauer, B., Kok, M., Yang, Y. L., and Shapiro, J. A., 1984, Physical structure, genetic content, and expression of the *alkBAC* operon, *Mol. Gen. Genet.* **197**: 373–383.

Patel, R. N., Banerjee, A., Ko, R. Y., Howell, J. M., Li, W. S., Comezoglu, F. T., Partyka, R. A., and Szarka, L., 1994, Enzymatic preparation of (3R-*cis*)-3-(acetyloxy)-4-phenyl-2-azetidinone-a taxol side-chain synthon, *Biotechnol. Appl. Biochem.* **20**: 23–33.

Patel, R. N., Liu, M., Banerjee, A., and Szarka, L. J., 1992, Stereoselective enzymatic hydrolysis of (*exo,exo*)-7-oxabicyclo[2.2.1]heptane-2,3-dimethanol diacetate ester in a biphasic system, *Appl. Microbiol. Biotechnol.* **37**: 180–183.

Pedragosa-Moreau, S., Archelas, A., and Furstoss, R., 1993, Microbiological transformations. 28. Enantiocomplementary epoxide hydrolyses as a preparative access to both enantiomers of styrene oxide, *J. Org. Chem.* **58**: 5533–5536.

Peterson, J. A., Kusunose, M., Kusunose, E., and Coon, M. J., 1967, Enzymatic ω-oxidation. II. Function of rubredoxin as the electron carrier in ω-hydroxylation, *J. Biol. Chem.* **242**: 4334–4340.

Philips, G. T., Robertson, B. W., Bertola, M. A., Koger, H. S., Marx, A. F., and Watts, P. D., 1986a, Arylglycidyl ethers and 3-substituted 1-alkylamino-2-propanols, Patent EP 193227.

Phillips, G. T., Robertson, B. W., Bertola, M. A., Koger, H. S., Marx, A. F., and Watts, P. D., 1986b, 4-(2-Methoxyethyl)phenylglycidyl ether and metropolol preparation by stereoselective microbiological epoxidation of corresponding phenylallyl ether, Patent EP 193 228.

Phillips, G. T., Bertola, M. A., Marx, A. F., and Koger, H. S., 1988, Process for the preparation of esters of 4-(2,3-epoxypropoxy)phenylacetic acid and 4-(2-hydroxy-3-isopropylaminopropoxy)-phenylacetic acid and/or atenolol in stereospecific form, Patent EP 256586.

Powlowski, J., Sahlman, L., and Shingler, V., 1993, Purification and properties of the physically associated meta-cleavage pathway enzymes 4-hydroxy-2-ketovalerate aldolase and aldehyde dehydrogenase (acylating) from *Pseudomonas* sp. strain CF600, *J. Bacteriol.* **175:** 377–385.

Ramos, J. L., Diaz, E., Dowling, D., Delorenzo, V., Molin, S., Ogara, F., Ramos, C., and Timmis, K. N., 1994, The behavior of bacteria designed for biodegradation, *Bio/Technology* **12:** 1349–1356.

Ramos, J. L., Duque, E., Huertas, M. J., and Haidour, A., 1995, Isolation and expansion of the catabolic potential of a *Pseudomonas putida* strain able to grow in the presence of high concentrations of aromatic hydrocarbons, *J. Bacteriol.* **177:** 3911–3916.

Reeve, C. D., Carver, M. A., and Hopper, D. J., 1989, The purification and characterization of 4-ethylphenol methylenehydroxylase, a flavocytochrome from *Pseudomonas putida* JD1, *Biochem. J.* **263:** 431–437.

Reeve, C. D., Carver, M. A., and Hopper, D. J., 1990, Stereochemical aspects of the oxidation of 4-ethylphenol by the bacterial enzyme 4-ethylphenol methylenehydroxylase, *Biochem. J.* **269:** 815–819.

Reichlin, F., and Kohler, H. P. E., 1994, *Pseudomonas* sp. strain HBP1 prp degrades 2-isopropylphenol (*ortho*-cumenol) via *meta* cleavage, *Appl. Environ. Microbiol.* **60:** 4587–4591.

Reineke, W., 1986, Construction of bacterial strains with novel degradative capabilities for chloroaromatics, *J. Basic Microbiol.* **26:** 551–567.

Resnick, S. M., Torok, D. S., and Gibson, D. T., 1993, Oxidation of carbazole to 3-hydroxycarbazole by naphthalene 1,2-dioxygenase and biphenyl 2,3-dioxygenase, *FEMS Microbiol. Lett.* **13:** 297–302.

Resnick, S. M., Torok, D. S., Lee, K., Brand, J. M., and Gibson, D. T., 1994, Regiospecific and stereoselective hydroxylation of 1-indanone and 2-indanone by naphthalene dioxygenase and toluene dioxygenase, *Appl. Environ. Microbiol* **60:** 3323–3328.

Rhee, H. I., Murata, K., and Mimura, A., 1987, Formation of the herbicide, δ-aminolevulinate, from L-alanine and 4,5-dioxovalerate by *Pseudomonas riboflavina*, *Agric. Biol. Chem.* **51:** 1701–1702.

Roberts, S. M., and Willetts, A. J., 1993, Development of the enzyme-catalyzed Baeyer–Villiger reaction as a useful technique in organic synthesis, *Chirality* **5:** 334–337.

Roberts, S. N., Turner, N. J., Willets, A. J., and Turner, M. K., 1995, *Introduction to Biocatalysis Using Enzymes and Microorganisms*, Cambridge University Press, Cambridge.

Robinson, D. S., 1964, Oxidation of selected alkanes and related compounds by a *Pseudomonas* strain, *Anton Leeuwenhoek* **30:** 303–316.

Roise, D., Soda, K., Yagi, T., and Walsh, C. T., 1984, Inactivation of the *Pseudomonas striata* broad specificity amino acid racemase by D and L isomers of β-substituted alanines: Kinetics, stoichiometry, active site peptide, and mechanistic studies, *Biochemistry* **23:** 5195–5201.

Romanov, V., and Hausinger, R. P., 1994, *Pseudomonas aeruginosa* 142 uses a three-component ortho-halobenzoate 1,2-dioxygenase for metabolism of 2,4-dichloro- and 2-chlorobenzoate, *J. Bacteriol.* **176:** 3368–3374.

Ruettinger, R. T., Griffith, G. R., and Coon, M. J., 1977, Characterization of the ω-hydroxylase of *Pseudomonas oleovorans* as a non-heme iron protein, *Arch. Biochem. Biophys.* **183:** 528–537.

Ruettinger, R. T., Olson, S. T., Boyer, R. F., and Coon, M. J., 1974, Identification of the ω-hydroxylase of *Pseudomonas oleovorans* as a non-heme iron protein requiring phospholipid for catalytic activity, *Biochem. Biophys. Res. Commun.* **57:** 1011–1017.

Rustemov, S. A., Golovleva, L. A., Alieva, R. M., and Baskunov, B. P., 1992, New pathways of styrene oxidation by *Pseudomonas putida* culture, *Mikrobiol.* **61:** 5–10.

Saftic, S., Fedorak, P. M., and Andersson, J. T., 1992, Diones, sulfoxides, and sulfones from the aerobic cometabolism of methylbenzothiophenes by *Pseudomonas* strain BT1, *Environ. Sic. Technol.* **26:** 1759–1764.

Sakashita, K., Nakamura, T., and Watanabe, I., 1993, Process for producing L-amino acids, U.S. Patent 5 215 897.

Sano, K., Yokozeki, K., Tamura, F., Yasuda, N., Noda, I., and Mitsugi, K., 1977, Microbial conversion of DL-2-amino-Δ^2-thiazoline-4-carboxylic acid to L-cysteine and L-cystine: Screening of microorganisms and identification of products, *Appl. Environ. Microbiol.* **34:** 806–810.

Schoemaker, H. E., Boesten, W. H. J., Kaptein, B., Hermes, H. F. M., Sonke, T., Broxterman, Q. B., Vandentweel, W. J. J., and Kamphuis, J., 1992, Chemoenzymatic synthesis of amino acids and derivatives, *Pure Appl. Chem.* **64:** 1171–1175.

Schofield, J. A., 1989, Biochemical process to produce catechol or 1,2-dihydroxy-cyclohexa-3,5-diene compounds, U.S. Patent 4 863 851.

Schurr, M. J., Martin, D. W., Mudd, M. H., and Deretic, V., 1994, Gene cluster controlling conversion to alginate-overproducing phenotype in *Pseudomonas aeruginosa*: Functional analysis in a heterologous host and role in the instability of mucoidy, *J. Bacteriol.* **176:** 3375–3382.

Schwartz, R. D., 1973, Octene epoxidation by a cold-stable alkane-oxidizing isolate of *Pseudomonas oleovorans*, *Appl. Microbiol.* **25:** 574–577.

Schwartz, R. D., and McCoy, C. J., 1973, *Pseudomonas oleovorans* hydroxylation-epoxidation system: Additional strain improvements, *Appl. Microbiol.* **26:** 217–218.

Senuma, M., Otsuki, O., Sakata, N., Furui, M., and Tosa, T., 1989, Industrial production of D-aspartic acid and L-alanine from DL-aspartic acid using a pressurized column reactor containing immobilized *Pseudomonas dacunhae* cells, *J. Fermentation Bioeng.* **67:** 233–237.

Serdar, C. M., Murdock, D. C., and Ensley, B. D., Jr., 1992, Enhancement of naphthalene dioxygenase activity during microbial indigo production, U.S. Patent 5 173 425.

Sheldrake, G. N., 1992, Biologically derived arene *cis*-dihydrodiols as synthetic building blocks, in: *Chirality in Industry* (A. N. Collins, G. N. Sheldrake, and J. Crosby, eds.), John Wiley & Sons, London, pp. 127–138.

Shimizu, H., Funita, C., and Watanabe, I., 1993, Process for preparing glycine from glycinonitrile, U.S. Patent 5 238 827.

Shimizu, S., Kim, J. M., Shinmen, Y., and Yamada, H., 1986, Evaluation of two alternative metabolic pathways for creatinine degradation in microorganisms, *Arch. Microbiol.* **145:** 322–328.

Shimizu, S., and Yamada, H., 1991, Microbial carbonyl reductases—their diversity and application to the synthesis of optically active alcohols, *J. Syn. Org. Chem. Japan* **49:** 52–70.

Shingler, V., Franklin, F. C. H., Tsuda, M., Holroyd, D., and Bagdasarian, M., 1989, Molecular analysis of plasmid-encoded phenol hydroxylase from *Pseudomonas* CF600, *J. Gen. Microbiol* **135:** 1083–1092.

Shipston, N. F., Lenn, M. J., and Knowles, C. J., 1992, Enantioselective whole cell and isolated enzyme catalyzed Baeyer–Villiger oxidation of bicyclo[3.2.0]hept-2-en-6-one, *J. Microbiol. Methods* **15:** 41–52.

Soberonchavez, G., and Palmeros, B., 1994, *Pseudomonas* lipases: Molecular genetics and potential industrial applications, *Crit. Rev. Microbiol* **20:** 95–105.

Stanier, R. Y., Palleroni, N. J., and Doudoroff, M., 1966, The aerobic *Pseudomonads*: A taxonomic study, *J. Gen. Microbiol.* **43:** 159–271.

Steinbuchel, A., and Valentin, H. E., 1995, Diversity of bacterial polyhydroxyalkanoic acids, *FEMS Microbiol. Lett.* **128:** 219–228.

Stephens, G. M., Sidebotham, J. M., Mann, N. Y., and Dalton, H., 1989, Cloning and expression in *Escherichia coli* of the toluene dioxygenase gene from *Pseudomonas putida* NCIB11767, *FEMS Microbiol. Lett.* **57:** 295–300.

Stieglitz, B., Linn, W. J., Jobst, W., Fried, K. M., and Fallon, R. J., 1992, Process for producing enantiomers of 2-aryl-alkanoic acids, Patent WO 9 201 062.

Sunjic, V., and Gelo, M., 1995, Chemoenzymic process for production of S-fen-propimorph, Patent EP 645 458.

Sutherland, I. W., 1990, The properties and potential of microbial exopolysaccharides, *Chimica Oggi* **8:** 9–14.

Sutherland, I. W., 1991, *Biotechnology of Microbial Exopolysaccharides*, Cambridge University Press, Cambridge.

Sutherland, I. W., Conti, E., Flaibani, A., and O'Regan, M., 1994, Alginate from *Pseudomonas fluorescens* and *P. putida*: Production and properties, *Microbiology* **140:** 1125–1132.

Suye, S.-I., Kawagoe, M., and Inuta, S., 1992, Enzymatic production of L-alanine from malic acid with malic enzyme and alanine dehydrogenase with coenzyme regeneration, *Can. J. Chem. Eng.* **70:** 306–312.

Suye, S. I., Yokoyama, S., and Obayashi, A., 1989, NADH Production from NAD+ using malic enzyme of *Pseudomonas diminuta* IFO- and 13182, *J. Fermentation Bioeng.* **68:** 301–304.

Suzuki, K., Gomi, T., Kaidoh, T., and Itagaki, E., 1991a, Hydroxylation of ortho-halo-genophenol and ortho-nitrophenol by salicylate hydroxylase. *J. Biochem.* **109:** 348–353.

Suzuki, M., Hayakawa, T., Shaw, J. P., Rekik, M., and Harayama, S., 1991b, Primary structure of xylene monooxygenase: Similarities to and differences from the alkane hydroxylation system, *J. Bacteriol.* **173:** 1690–1695.

Suzuki, T., Kasai, N., Yamamoto, R., and Minamiura, N., 1993, Production of highly optically active (*R*)-3-chloro-1,2-propanediol using a bacterium assimilating the (S)-isomer, *Appl. Microbiol.* Biotechnol. **40:** 273–278.

Swanson, P. E., 1992, Microbial transformation of benzocyclobutene to benzocyclo-butene-1-ol and benzocyclobutene-1-one, *Appl. Environ. Microbiol.* **58:** 3404–3406.

Syldatk, C., Cotoras, D., Dombach, G., Gross, C., Kallwass, H., and Wagner, F., 1987, Substrate- and stereospecificity, induction and metallo-dependence of a microbial hydantoinase, *Biotech. Lett.* **9:** 25–30.

Takagi, J. S., Fukunaga, R., Tokushige, M., and Katsuki, H., 1984, Purification, crystalliza-tion, and molecular properties of aspartase from *Pseudomonas fluorescens*, *J. Biochem.* **96:** 545–552.

Takahashi, S., 1983, Microbial synthesis of D-amino acids from DL-5-substituted hydan-toins, *Hakko Kogaku Kaishi* **61:** 139–151.

Takamatsu, S., Tosa, T., and Chibata, I., 1986, Industrial production of L-alanine from ammonium fumarate using immobilized microbial cells of two kinds, *J. Chem. Eng. Japan* **19:** 31–36.

Takasaki, Y., Hinoki, K., Kataoka, Y., Fukuyama, S., and Nishimura, N., 1993, Enzymic production of D-mannose from D-fructose by mannose isomerase, *J. Fermentation Bioeng.* **76:** 237–239.

Tan, H., Tang, H., Joannou, C., Abdelwahab, N., and Mason, J., 1993, The *Pseudomonas putida* ML2 plasmid-encoded genes for benzene dioxygenase are unusual in codon usage and low in G + C content, *Gene* **130:** 33–39.

Taylor, D. G., and Trudgill, P. W., 1986, Camphor revisited: Studies of 2,5-diketocamphane 1,2-monooxygenase from *Pseudomonas putida* ATCC 17453., *J. Bacteriol.* **165:** 489–497.

Theil, F., 1995, Lipase-supported synthesis of biologically active compounds, *Chem. Rev.* **95:** 2203–2227.

Timmis, K. N., 1995, Environmental biotechnology, *Bio/Technol.* **13:** 105.

Timmis, K. N., Steffan, R. J., and Unterman, R., 1994, Designing microorganisms for the treatment of toxic wastes, *Annu. Rev. Microbiol* **48:** 525–557.

Tsuchiya, Y., and Nishio, N., 1980, Vitamin B_{12} production from methanol by continuous culture of *Pseudomonas* AM-1, *J. Fermentation Technol.* **58:** 485–487.

Tuneo, H., Hisao, T., and Tatsuo, I., 1989, Process for the production of aspartyl phenylalaninyl alkylesters, Patent EP B1 0 102 529.

Ube, Y., 1982, Process for the production of L-methylphenylalanine, Patent DE 3 217 908.

van Beilen, J. B., Kingma, J., and Witholt, B., 1994, Substrate specificity of the alkane hydroxylase system of *Pseudomonas oleovorans* GPo1, *Enzyme Microb. Technol.* **16:** 904–911.

van den Wijngaard, A. J., Janssen, D. B., and Witholt, B., 1989, Degradation of epichlorohydrin and halohydrins by bacterial cultures isolated from freshwater sediment, *J. Gen. Microbiol* **135:** 2199–2208.

van der Linden, A. C., and Huybrechtse, R., 1967, Induction of alkane inducible and alpha-olefin-opoxidizing enzymes by a non-hydrocarbon in a *Pseudomonas*, *Antox Leeuwenhoek* **33:** 381–385.

van der Meer, J. R., 1994, Genetic adaptation of bacteria to chlorinated aromatic compounds, *FEMS Microbiol. Rev.* **15:** 239–249.

van der Werf, J., van den Tweel, W. J. J., Kamphuis, J., Hartmans, S., and de Bont, J. A. M., 1994, The potential of lyases for the industrial production of optically active compounds, *Trends Biotechnol.* **12:** 95–103.

van der Werf, M. J., Hartmans, S., and van den Tweel, W. J. J., 1995, Effect of maleate counterion on malease activity: Production of D-malate in a crystal-liquid two-phase system, *Enzyme Microb. Technol.* **17:** 430–436.

van Scharrenburg, G. J. M., Sloothaak, J. B., Kruse, C. G., Smitskampwilms, E., and Brussee, J., 1993, The potential of (*R*)-oxynitrilases and (*S*)-oxynitrilases for the enzymatic synthesis of optically-active cyanohydrins, *Ind. J. Chem. Section B* **32:** 16–19.

Verkhovskaya, M. A., and Yamskov, I. A., 1991, Enzymic methods for separation of racemic amino acids and their derivatives, *Uspekhi Khimii* **60:** 2250–2280.

Wandrey, C., and Bossow, B., 1986, Continuous cofactor regeneration—utilization of polymer bound NAD(H) for the production of optically active acids, *Biotechnol. Bioind.* **3:** 8–13.

Warhurst, A. M., and Fewson, C. A., 1994, Microbial metabolism and biotransformations of styrene, *J. Appl. Bacteriol.* **77:** 597–606.

Watanabe, M., Morinaga, Y., and Enei, H., 1987a, L-Serine production by a methorninieauxotrophic mutant of methylotrophic *Pseudomonas*, *J. Fermentation Technol.* **65:** 617–620.

Watanabe, M., Morinaga, Y., Takenouchi, T., and Enei, H., 1987b, Efficient conversion of glycine to L-serine by a glycine-resistant mutant of a methylotroph using cobalt as an inhibitor of L-serine degradation, *J. Fermentation Technol.* **65**: 563–567.

Weber, F. J., Ooijkaas, L. P., Schemen, R. M. W., Hartmans, S., and De Bont, J. A. M., 1993, Adaptation of *Pseudomonas putida* S12 to high concentrations of styrene and other organic solvents, *Appl. Environ. Microbiol.* **59**: 3502–3504.

West, S. E. H., Schweizer, H. P., Dall, C., Sample, A. K., and Runyenjanecky, L. J., 1994, Construction of improved *Escherichia-Pseudomonas* shuttle vectors derived from pUC18/19 and sequence of the region required for their replication in *Pseudomonas aeruginosa, Gene* **148**: 81–86.

Whited, G. M., and Gibson, D. T., 1991, Toluene-4-monooxygenase, a 3-component enzyme-system that catalyzes the oxidation of toluene to para-cresol in *Pseudomonas-mendocina* KR1, *J. Bacteriol.* **173**: 3010–3016.

Williams, M. G., Olson, P. E., Tautvydas, K. J., and Bitner, R. M., 1990, The application of toluene dioxygenase in the synthesis of acetylene-terminated resins, *Appl. Microbiol. Biotechnol.* **34**: 316–321.

Winter, R. B., Yen, K. M., and Ensley, B. D., 1989, Microbial degradation of trichloroethylene, Patent EP 336 718.

Witholt, B., and Lageveen, R. G., 1988, Production of compounds containing a terminal hydroxyl or epoxy group using microorganisms, e.g. *Pseudomonas oleovorans*, genetically engineered so that they are unable to convert the oxidation product further, Patent EP 277 674.

Witholt, B., de Smet, M. J., Kingma, J., van Beilen, J. B., Kok, M., Lageveen, R. G., and Eggink, G., 1990, Bioconversions of aliphatic compounds by *Pseudomonas oleovorans* in multiphase bioreactors: Background and economic potential, *Trends Biotechnol.* **8**: 46–52.

Wong, C. H., and Whitesides, G. M., 1994, *Enzymes in Synthetic Organic Chemistry*, Pergamon Press, Oxford.

Worsey, P. A., and Williams, M. J., 1975, Metabolism of toluene and xylenes by *Pseudomonas putida* (*arvilla*)mt-2: Evidence for a new function of the TOL plasmid, *J. Bacteriol.* **124**: 7–13.

Wright, M. A., Taylor, I. N., Lenn, M. J., Kelly, D. R., Mahdi, J. G., and Knowles, C. J., 1994, Baeyer–Villiger monooxygenases from microorganisms, *FEMS Microbiol Lett.* **116**: 67–72.

Wubbolts, M. G., and Timmis, K. N., 1990, Biotransformation of substituted benzoates to the corresponding cis-diols by an engineered strain of *Pseudomonas oleovorans* producing the TOL plasmid-specified enzyme toluate-1,2-dioxygenase, *Appl. Environ. Microbiol.* **56**: 569–571.

Wubbolts, M. G., Hoven, J., Melgert, B., and Witholt, B., 1994a, Efficient production of optically active styrene epoxides in two-liquid phase cultures, *Enzyme Microb. Technol.* **16**: 887–894.

Wubbolts, M. G., Reuvekamp, P., and Witholt, B., 1994b, TOL plasmid-specific xylene oxygenase is a wide substrate range monooxygenase capable of olefin epoxidation, *Enzyme Microb. Technol.* **16**: 608–615.

Wubbolts, M. G., Noordman, R., van Beilen, J. B., and Witholt, B., 1995, Enantioselective oxidation by non-heme iron monooxygenases from *Pseudomonas, Recl. Trav. Chim. Pays-Bas* **114**: 139–144.

Yamada, H., Shimizu, S., and Shiozaki, S., 1986, Process for producing S-adenosyl-L-homocysteine, U.S. Patent 4 605 625.

Yamada, H., and Tani, Y., 1983, Process for biologically producing amide, Patent EP 93 782.

Yamada, H., and Tani, Y., 1987, Process for biological preparation of amides, U.S. Patent 4 637 982.

Yamagami, T., Kobayashi, E., and Endo, T., 1993, Biological process for preparing optically active lactic acid, U.S. Patent 5 234 826.

Yamaoto, K., Otsubo, K., and Orshi, K., 1990, Process for producing optically active α-substituted organic acids, Patent EP 348 901.

Yen, K.-M., Blatt, L. M., and Karl, M. R., 1992, Bioconversions catalyzed by the toluene monooxygenase of *Pseudomonas mendocina* KR-1, Patent WO 9 206 208.

Yokozeki, K., Kamimura, A., Eguchi, C., and Kubota, K., 1988, Asymmetric synthesis of S-carboxymethyl-L-cysteine by a chemo-enzymic method, *Agric. Biol. Chem.* **52:** 2367–2368.

Yokozeki, K., Nakamori, S., Eguchi, C., Yamada, K., and Mitsugi, K., 1987a, Screening of microorganisms producing D-*p*-hydroxyphenylglycine from DL-5-(*p*-hydroxyphenyl)-hydantoin, *Agric. Biol. Chem.* **51:** 355–362.

Yokozeki, K., Nakamori, S., Yamanaka, S., Eguchi, C., Mitsugi, K., and Yoshinaga, F., 1987b, Optimal conditions for the enzymic production of D-amino acids from the corresponding 5-substituted hydantoins, *Agric. Biol. Chem.* **51:** 715–719.

Zeyer, J., Lehrbach, P. R., and Timmis, K. N., 1985, Use of cloned genes of *Pseudomonas* TOL plasmid to effect biotransformation of benzoates to *cis*-dihydrodiols and catechols by *Escherichia coli* cells, *Appl. Environ. Microbiol* **50:** 1409–1413.

Zimmermann, T., Kiener, A., and Harayama, S., 1992, Hydroxylation of methyl groups in aromatic heterocyclic compounds by microorganisms, Patent EP 477828.

Zimmerman, T., Kiener, A., and Harayama, S., 1993, Hydroxylation of methyl groups in aromatic heterocycles by microorganisms, U.S. Patent 5 217 884.

Index

Alginate, 85–97
Alkanes, 11
  degradation, 12
  oct plasmid, 11
Antibiotics
  antibiotic resistance, 2
    plasmids, 2–4
  imipenem resistance, 4
  outer membranes, 139–140
Amino acids, transport, 174–185

β-ketoadipate pathway, transport, 187–190
Biotransformations, 271–272
  biocatalysis versus chemical catslysis, 273
    enzyme classes and their practical application, 273–279
    sources of pseudomonas biocatalysts, 279
  constructing 'tailor-made' biocatalyst, 311–312
    protein engineering, 312–313
    recombinant biocatalysts, 312
  metabolites of pseudomonas in biotech, 272–273
  oxidoreductases, 279
    dehydrogenases, 279–280
    ester hydrolysis, 298–304
    dehalogenases, 304
    epoxide hydrolases, 301–302
    esterases, 298
    lipases, 299–301
    nitrile hydrolases, 303
    proteases, 299
    hydrolases, 294–311
    amidases, 294–295
    amide hydrolysis, 294–298
    aminoacylases, 297–298
    hydantoinases, 295–297
    lactamases, 298

Biotransformations (cont.)
  oxidoreductases (cont.)
    the impact of molecular biology on biocatalyst development, 311
    isomerases, 308
    carbohydrate isomerases, 308
    ligases, 311
    and racemases, 308
    lyases, 304–308
    aldolases, 305
    ammonia lyases, 305–306
    decarboxylases, 307
    dioxygenases, 287
    halolyases, 307–308
    hydrolyases, 306–307
    monooxygenases, 282
    oxynitrilases, 305
    oxygenases, 282
    racemases, 309–311
    amino acid racemases, 309
    hydantoin racemases, 309–311
    and isomerases, 308
    reduction of aldehydes and ketones, 280–281
    transferases, 291–294
    acyltransferases, 291,
    alkyltransferases, 291–292
    aminotransferases, 292
    glycosyltransferases, 292
    phosphotransferases, 292–294

Capsular polysaccharides, 85–87
Carbohydrate catabolism, 41
  catabolite repression control (CRC), 64–65
  central cycle
    Entner–Doudoroff aldolase, 59
    Entner–doudoroff dehydratase, 58
    fructose 1, 6-bisphosphatase, 57–58
    fructose 1, 6-bisphosphate aldolase, 57

Carbohydrate catabolism (*cont.*)
central cycle (*cont.*)
glucose 6-phosphate dehydrogenase,
58–59
lower Embden–Meyerhoff–Parnas
(EMP) pathway, 60
glyceraldehyde 3-phosphate
dehydrogenase, 60
phosphoglycerate kinase, 61
phosphoglycerate mutase (pgm), enolase
(eno), and pyruvate kinase (pyc), 61
pyruvate carboxylase, 61
phosphoglucoisomerase, 58
clustering of genes for glycolytic enzymes, 65–66
fructose utilization, 46–47
agm R, 55–56
glpR, 56
chromosomal mapping of glycerol metabolism genes, 56–57
g1pM protein, 53
glycerol kinase, 54–55
regulatory loci, 55
glucose utilization via phosphorylation,
41–45
phosphorylation, 44
transport, 41
glucose binding protein, 41
glucose transport regulator, 42
porin B, 41
glycerol and glycerol 3-phosphate utilization, 48
glycerol and glycerol-p uptake, 49
glycerol-p dehydrogenase, 52
glycerol transport, 49
uptake of glycerol-p, 52
mannitol utilization, 47
oxidative pathway of glucose utilization, 45
recycling of glyceraldehyde 3-phosphate, 61
regulation of central and lower emp
pathwasy, 62
hex regulon, 62
Chemotaxis, 260
Copper resistance, plasmids, 7–8

Enzymes, biotransformations, 271–301
Environment
lipids, 130–134
opr, (see outer membrane proteins),
139–160
Ethylene forming enzymes, 9

Fatty acids, 115–126
Flagellum, 245
antigenicity and immunogenicity, 262–264
antigenicity, 262–263
immunogenicity and passive protection, 263–264
biochemistry, 245–250
amino acid composition, 248–250
general morphology, 245–246
molecular weights of flagellins, 246–247
purification, 246
comparative annalysis of the fli C genes,
254–260
flagellin a-type genes, 257–260
flagellin b-type genes, 254–256
flagella and chemotaxis genes and regulation, 260–262
chemotaxis genes, 262
regulation of flagella biosynthesis and
assembly, 260–262
other pseudomonads, 265
posttranslational modifications of
flagellins, 250–254
glycosylation, 252–254
phosphorylation, 250–252
virulence, 264–265

Hg°-Hg^{2+} resistance, 6–7
plasmids, 6–7
Hydrocarbons
alkanes, 11–13
aromatic, TOL, NAH
benzene, 17
NAH and SAL, 16
TOL, 13–15

Inorganic ions, transport, 185–188
Iron metabolism and siderophores, 140,
201, 207
biosynthesis, genetic organizations, and
regulation of siderophores, 217–
220
biosynthetic pathway of pyoverdine,
202, 217, 219, 229
genetics and regulation, 220–222
intracellular release of iron from
ferrisiderophores, 215
iron uptake from other sources, 215–216
translocation through outer and inner
membranes
internalization of iron, 214–215, 227

Iron, metabolism and siderophores (*cont.*)
iron, siderophores, and biotechnology, 203, 227
iron, siderophores and human pathogenicity, 227
iron, siderophores, and plant-related biocontrol, 227
siderotyping and searching for new siderophores back to the bench, 229
siderophores at the bench, 202–209
detection of siderophores, 203–204
optimizationof siderophore production, 202–203
purification of sidephores, 204–209
siderophore-mediated iron uptake systems in nonfluorescent pseudomonas and related strains, 223
Burkholderia (formerly *Pseudomonas*) *cepacia* and related strains, 224–226
other strains, 226
Pseudomonas stutzeri, 223–224
siderophore-mediaated iron uptake systems in fluorescent pseudomonas, 209
fluorescent pseudomonas siderophores, 209
pyoverdines, 204, 207, 209–213, 220
pyochelin and salicylate 140, 204, 213–214

Lipids, 111
alteration of lipids in response to environmental conditions, 130
antibiotic resistance, 133–134
growth temperature, 130–131
desiccation, 131–132
oxygen tension, 131
nutrient deprivation, 132
solvent tolerance, 133
lipids of the genus pseudomonas, 111
exolipids, 128
ornithine amide lipids, 113–115
rhamnolipid 128–130, 134
fatty acids, 115
hydroxy fatty acids, 116–126
non-hydroxy fatty acids, 115–116
membrane lipids, 111
ornithine amide lipids 113–115
phospholipids, 111–113, 133
storage lipids, 126
polyhydroxyalkoanates (PHA), 126–128
Lipopolysaccharides, 73–85

Metabolism, carbohydrate, 41–62
Embden–Meyerhof, 57–61
fructose, 46–47
glucose, 41
glycerol, 48–52
mannose, 52
transport, 171
Metals resistance, 6
copper, 7–8
mercury resistance, 6
Myo-inositol, transport, 190–191

Outer Membrane Proteins, 139
role in antibiotic susceptibilityn 139–140
algE, 146
E2, 155
IROMPS; FpvA, FptA, PfeA 140–145, 155, 157, 214, 216, 226, 228
oprB, 41, 50, 151–152
oprC, 145
oprd, 45, 140 , 145, 152–153, 155
oprE, 145, 153–155
oprF, 145, 155–159,161, 216
oprG, 159
oprH, 159–160
oprI, 156, 160–161
oprJ, 146
oprK, 146–147, 148
oprL, 156, 160
oprM, 146–147, 148
oprN, 146
oprO, 149
oprP, 148, 149

Plasmids, 1
aminoglysides, 3
biosynthesis, 9
coronatine production, 9
ethylene-forming enzyme, 9
(IAA) production, 9
malonate assimilation, 10
P. syringae, 9, 88
syrinolide production, 10
cloning vectors, 17
incompatibility group P1, 19
incompatibility group P4/Q, 16
incompatibility group W, 19
PGV1124, 20
RSF1010, 4, 18, 20
conclusions and perspectives, 20
P. putida DOT-T1, 20

Plasmids (*cont.*)
  conclusions and perspectives (*cont.*)
    P. putida idaho, 20
  degradative, 10
    CAM, 11
      p.putide PpG1, 11
    NAH (NAH7) and SAL (SAL1), 16
    PHMT112, 17
    OCT, 11
    TOL (pWWO), 13
  DNA preparation, 2
  p1 plasmids, 3
    Agrobacterium tumefaciens, 6
    psa, 6
    rp1, rp2, rk2, 3
      pmg1, 4
      and rlb679, 4
  replication and conjugation, 1
  r-plasmids (metal ions, pesticides, and uv
      light), 6
    organophosphorus compounds
      (pCMS1), 8
    pMG2, 8
    pmr1, pmerph, 6–7
    ppt23d, ppsr1, 7
R-plasmids resistance, 2
Phospholipids, 111–113, 126–128
Polysaccharides, 74–78
  capsular polysaccharides and slime, 85–
      97
    alginate-biosynthesis, 86, 90–92, 93, 272
    alginate function, 90, 93, 95–97
    alginate-intro, 88
    alginate-structure and physical proper-
      ties, 88–90
    immunology of alginate, 95–96
    introduction, 85–86
    O-capsule as an antigen, 87–88
    O-capsule function, 87
    O-capsular polysaccharide structure,
      86–87
    regulation of alginate synthesis, 92–95
  lipopolysaccharide, 73–78, 83, 111
    biological activity of
      lipopolysaccharide, 84
    biosynthesis and genetic regulation
      lps, 80–82
    heterogeneity, 78–80
    immunology of lps, 84–85
    physical properties, 83–84
    structure, 74–78

Post-transcational, flagellin modification,
    250–254
Pseudomonas strains
  acidovorans, 182, 184, 294
  aeruginosa, 1–3, 8, 35, 42–43, 47–50, 53–
      54, 63–65, 73,85, 87,90, 92–93,
      95–96, 127,131, 133, 139–152,
      155–157, 159, 161, 170–176, 181–
      187, 202–204, 207, 210, 215, 219,
      228, 249, 251–255, 264, 265, 272,
      283–285, 299, 312
    AK1401, 79
    agmR, 44, 55
    ATTC 15692, 204, 210, 213–214, 216,
      220, 229
    ATTC 27853, 214, 229
    glpD, 49, 52–54
    glpF, 49, 50, 52
    glpFK, 52, 54
    flpK, 49–50, 52, 54–55
    glpM, 53, 54
    glpR, 49, 52
    glpR2, 49, 55
    glpR8, 55
    GN17203, 4
    PAO, 179–181, 217, 219, 220, 222
    PAO1, 49–50, 53–54, 62, 75, 254
    PAO104, 52
    PAO1162, 17
    PAO151, 53
    PAO206, 54
    PAO311, 50
    PML, 179–181
    NM48, 50
  amyloderamosa, 304
  aureofaciens, 131
  cepacia, 213, 224, 227, 272, 285, 299,
      308, 312
  chlororaphis, 203, 212, 307
    ATTC 9446, 212
    B23, 307
  daccunhae, 307
  denitrificans, 131, 186, 283
  diminuta, 8, 111, 113, 132, 226, 280
  Doudoroffi, 173
  fluorescens, 1, 10, 17, 88, 113, 130, 132,
      156–157, 184, 186–187, 201–214,
      207, 210, 212, 227, 229, 251, 272,
      295, 298–299, 304–306, 312
    ATTC 13525, 212, 216, 229, 265
    CHAO, 204, 207, 213–214

Pseudomonas strains (*cont.*)
  fluorescens (*cont.*)
    DSM 84, 295, 298
    UK1, 184
  fluorescensw, 115
  fragi, 115, 226, 298
    IFO 3458, 298
    5R, 17
  glumae, 312–313
  melanogenum, 298
    KY3987, 298
  mendocina, 88, 226
  mildenbergii, 210
  oleovorans, 126, 127, 283
  ovalis, 210, 303
  putida, 1–2, 13, 16, 18–19, 21, 41, 43, 48, 88, 126–127, 133, 144, 151, 182–183, 185, 187–190, 201, 209–210, 212, 227, 229, 272, 280, 284–285, 290–291, 295, 304, 306–309, 311–312
    ATTC 12633, 295
    ATTC 17642, 309
    ATTC 17453, 285
    ATTC 31916, 290
    CR1-1, 307–308
    DOT-T1, 20
    IH-2000, 20
    KT2440, 17
    PpG1, 11
    PR52000, 265
    ML2, 17
    MT15, 15
    MT53, 15
    NCIMB 11767, 285
    NCIMB 10007, 286
    S12, 133
    TMB, 15
    WCS358, 214, 220
  pseudoalcaligenes, 306
    NCIMB9867, 306
  pseudomallei, 265
  putrefaciens, 7
  rubescens, 116
  saccharophila, 126
  striata, 295, 309

Pseudomonas strains (*cont.*)
  stutzeri, 145, 186, 207, 223–224, 282
    ATTC 11607, 224
    ATTC 14405, 224
    ATTC 17588, 223–224
  syringae, 1, 7–10, 73–74, 77–79, 89, 218–219, 251, 265
  testosteroni, 191–192
  vesicularis, 111, 113, 130
Pyochelin (and salacylate), 213–214
Pyoverdine, 209–213

Siderophores, 140, 201–243
Steroids, transport, 191–192

Transport Systems, 169
  amino acids, 174–185
    aromatic amino acids, 184
    basic amino acids, 182–183
    branched-chain amino acids, 174–185
      LIV-1,174, 176–179
      LIV-2, 174, 179–180
      LIV-3, 174, 180–181
    methionine, 184–185, 248, 249
    proline, 183–184, 248, 249
  inorganic ions, 185–188
    anions, 185–187
      nitrate, 186–187
      phosphate, 185–186
      sulfate, 187
    cations, 185
  other compounds, 187–192
    catabolized by β-ketoadipate pathway, 187–190
      benzoate, 187–188
      4-hydroxybenzoate, 188–189
      β-ketoadipate, 189–190
    myo-inositol, 190–191
    steroids, 191–192
  sugars, 171–174
    fructose and mannitol, 173–174
    glucose and gluconate, 41 –42, 171–173
    glycerol, 49, 52

UV resistance, plasmids, 8